Disaster Risk Reduction

Methods, Approaches and Practices

Series Editor

Rajib Shaw, Keio University, Shonan Fujisawa Campus, Fujisawa, Japan

Disaster risk reduction is a process that leads to the safety of communities and nations. After the 2005 World Conference on Disaster Reduction, held in Kobe, Japan, the Hyogo Framework for Action (HFA) was adopted as a framework for risk reduction. The academic research and higher education in disaster risk reduction has made, and continues to make, a gradual shift from pure basic research to applied, implementation-oriented research. More emphasis is being given to multi-stakeholder collaboration and multi-disciplinary research. Emerging university networks in Asia, Europe, Africa, and the Americas have urged process-oriented research in the disaster risk reduction field. With this in mind, this new series will promote the output of action research on disaster risk reduction, which will be useful for a wide range of stakeholders including academicians, professionals, practitioners, and students and researchers in related fields. The series will focus on emerging needs in the risk reduction field, starting from climate change adaptation, urban ecosystem, coastal risk reduction, education for sustainable development, community-based practices, risk communication, and human security, among other areas. Through academic review, this series will encourage young researchers and practitioners to analyze field practices and link them to theory and policies with logic, data, and evidence. In this way, the series will emphasize evidence-based risk reduction methods, approaches, and practices.

All submitted proposals are carefully reviewed by the series editor.

Publication Ethics and Publication Malpractice Statement: https://www.springernature.com/gp/policies/book-publishing-policies

Ethical, Conflict-of-Interest, and Editorial Policy Statement: https://www.springer.com/us/authors-editors/journal-author/journal-author-helpdesk/before-you-start/before-you-start/1330#c14214

Mohammad Golam Kibria · Gulsan Ara Parvin ·
Md. Anwarul Abedin · Rajib Shaw
Editors

Urban Blue-Green Infrastructure Approach for Food Security and Climate Disaster Resilience

Springer

Editors
Mohammad Golam Kibria
Department of Soil Science
Bangladesh Agricultural University
Mymensingh, Bangladesh

Md. Anwarul Abedin
Department of Soil Science
Bangladesh Agricultural University
Mymensingh, Bangladesh

Gulsan Ara Parvin
College of Policy Science, Ritsumeikan University
Osaka Ibaraki Campus, Osaka, Japan

Rajib Shaw
Graduate School of Media and Governance
Keio University
Fujisawa, Kanagawa, Japan

ISSN 2196-4106 ISSN 2196-4114 (electronic)
Disaster Risk Reduction
ISBN 978-981-96-8745-9 ISBN 978-981-96-8746-6 (eBook)
https://doi.org/10.1007/978-981-96-8746-6

This Springer imprint is published by the registered company Springer Nature Singapore Pte Ltd.
The registered company address is: 152 Beach Road, #21-01/04 Gateway East, Singapore 189721, Singapore

Preface

Blue-Green Infrastructure (BGI) is a cornerstone for building resilient cities that can adapt to the challenges posed by climate change. By integrating natural systems into urban landscapes, blue-green infrastructure helps cities withstand and recover from extreme weather events. Urban BGI is considered as a strategy that can bring multiple benefits and help to build resilient urban food systems at the city region level. Cities have an important role to play in climate change mitigation and adaptation, disaster risk management and in enhancing the climate resilience of their vulnerable residents. Climate change together with a decrease in blue-green coverage in major cities poses serious threats to food security and negatively affecting the urban climate disaster resilience. Growing built-up surfaces associated with urbanization also reduce water infiltration and increase water runoff during rainstorms. With increasingly intense rainfall, flooding is common in cities that lack adequate drainage systems. Therefore, it important to develop climate resilient urban BGI approaches to ensure food security for the urban community people.

Growing international policy attention is focused on the role of cities in fostering more sustainable and resilient food systems capable of withstanding and recovering from crises, including natural disasters. The urban Blue-Green Infrastructure (BGI) approach presents a promising strategy for integrating urban food security with climate adaptation and disaster management efforts. In densely populated cities with limited space, rooftop gardens offer a viable solution to enhance thermal comfort in top-floor apartments while contributing to food production. A scenario suggests that utilizing half of a city's available rooftop space for urban agriculture could supply approximately 4% of the food needs for 10,000 people. When combined with hydroponic greenhouses, this potential rises to 60%. However, a comprehensive understanding of the complex interactions between urbanization, food security, and climate change is crucial for developing effective urban BGI strategies. This book marks the first step in exploring the prospects of urban BGI in ensuring food security and climate resilience.

Within the above-mentioned context, this book is a modest approach to provide an overview of the trends, challenges and potential strategies to ensure food security as well as disaster resilience, urban BGI approaches to address this issue across the globe, and finally the planning, resource management and financing the BGI approaches in developed and developing countries. The book is organized into four parts, which comprises a total of 15 chapters, each of which will emphasize the importance of urban BGI to ensure food security and climate resilience. Part I (2 chapters) highlights the complex interaction of urbanization and environmental degradations, and how it affects food security and climate disaster resilience. Part II (5 chapters) discusses about urban agriculture in context of South Asia, its impact on disaster risk reduction, BGI approaches to mitigate landslide risk and global actions for climate resilient BGI practices. Part III (7 chapters) deals with BGI in the informal urban areas of Asia, land use management for sustainable BGI practices to ensure food security, a case study on the assessment of urban resilience for climate disaster, urban waste management for BGI, smart agricultural approach for smart cities across the globe, cost and benefit for green infrastructure to mitigate urban heat and financing urban BGI approaches. Part IV comprises only one concluding chapter which highlights the lessons, recommendations and future action policies to promote urban BGI in different developed and developing countries across the globe.

This book is written for the students, researchers, practitioners, and policy makers in the fields of urban planning, disaster risk reduction and environmental studies. We sincerely hope and believe that this book will connect the urban blue green infrastructural approach to the enhanced climate resilience for future food security and policy related research. We will be happy if the readers find the book useful and relevant to their work. Finally, we are thankful to several of our mentors, and we would like to dedicate this book to our parents who helped us to shaping our life.

Mymensingh, Bangladesh — Mohammad Golam Kibria
Osaka, Japan — Gulsan Ara Parvin
Mymensingh, Bangladesh — Md. Anwarul Abedin
Fujisawa, Japan — Rajib Shaw

About This Book

Urbanization can have both positive and negative impacts on agriculture and aquaculture. It often leads to changes in land use, increased competition for resources, and shifts in market dynamics. Due to the increasing urbanization rate across the globe, the cultivable agricultural land is decreasing and threatening the food supply for the growing global population as well as increasing the disaster occurrence in urban areas. The city dwellers are mainly getting vulnerable to ensure food security and climate disaster resilience due to the rapid urbanization across the globe. However, the urbanization is not the same for the developed and developing countries, which plays a significant role in food security and disaster resilience for the urban areas in these two contrasting groups. Food security and climate disaster resilience are critical issues facing urban areas, and adopting an urban blue-green infrastructure approach can help address these challenges. This approach combines elements of both natural and built environments to create more sustainable and resilient cities. Adopting an urban **B**lue-**G**reen **I**nfrastructure (BGI) approach can enhance food security and climate disaster resilience by integrating nature-based solutions into urban planning and design.

To fill in the prodigious gap in knowledgebase, this book will include in-depth reviews as well as new data-based case studies, involving multidisciplinary research. This book will deliberate on the trends, challenges, and potential strategies to ensure food security as well as disaster resilience, urban BGI approaches to address this issue across the globe, and finally the planning (both engineering and social aspects), resource management, and financing the BGI approaches in developed and developing countries. Overall, this book will be a valuable resource material for students, researchers, academia, policymakers, and development practitioners.

Contents

Editors and Contributors

About the Editors

Dr. Mohammad Golam Kibria is an Associate Professor in the Department of Soil Science in Bangladesh Agricultural University, Bangladesh. He has a Ph.D. in soil fertility and plant nutrition from the University of Western Australia, Perth, Australia. After completing his Ph.D., his research focuses on urban agriculture, disaster risk reduction, climate change adaptation, integrated social aspects, and soil health assessment under changing global environment. Dr. Kibria has also worked for IUCN, CDRI, FAO, and APN in different research projects focusing on climate disaster resilience. Presently, he is working in three different collaborative projects funded by GATES Foundation, USAID, and APN, which focused on rural women entrepreneurship, urban horticulture and climate induced loss assessment in rice production. Now, he has active research collaboration with UC Davis, USA; Salford University, UK; Ritsumeikan University, Japan; AFU, Nepal; and many more. He has authored around 40 research articles, reviews, and book chapters in different national and international peer-reviewed journals and books. Among these publications, he has published eight book chapters and five research articles focusing on disaster risk reduction, climate change adaptations, sustainable infrastructure, and urban governance in some renowned journals and publishers such as international journal of disaster risk reduction (IJDRR), sustainable and resilient infrastructure, sustainability, Springer, Oxford university press, Elsevier, and SAGE publishers.

Gulsan Ara Parvin is a researcher and Instructor of Collage of Policy Science, Ritsumeikan University, Japan. She served different graduate schools and research centers of Kyoto University, Japan for about 12 years. She was an Associate Professor of Urban and Regional Planning Department of Bangladesh University of Engineering and Technology (BUET). She has conducted her Ph.D. from Urban Engineering Department of the University of Tokyo. Her expertise is in Disaster Risk Reduction and Community Resilience, social capital, women empowerment and microfinance in Bangladesh. She has published about 50 research papers in different journals and books. She has research and teaching experiences in Bangladesh, Japan, Malaysia, Thailand, and USA.

Md. Anwarul Abedin is a Professor and Leader of Laboratory of Environment and Sustainable Development in the Department of Soil Science, Bangladesh Agricultural University, Bangladesh. He has got his Ph.D. from Kyoto University, Japan and also did his JSPS-postdoc in Kyoto University and Commonwealth postdoc in Northumbria University, UK. Dr. Abedin was awarded the esteemed "Pan Asia Risk Reduction Fellowship" offered by START, USA and carried out collaborative research among the BAU, START, and Kyoto University, Japan. He also worked as JSPS Invitational Fellow at College of Policy Science, Ritsumeikan University, Osaka-Ibaraki Campus, Japan. He has 20 years teaching and research experiences in the field of agriculture, food safety and security and environmental issues, viz., Water resource management, solid waste management, environmental impact assessment, community-based adaptations and resilience mapping, climate change and disaster risk reduction, land use change, community participation. He has worked as consultant with FAO, JICA, IUCN, Oxfam and GCRF. Dr. Abedin is collaborating in a number of research projects funded by DFID, NERC, British Council, USAID, JICA-JST, CARRIA, ESRC-UKRI, World Bank, Kurita Foundation, International Science Council, and China Government. Currently, he has research collaboration with UCDavis, USA; Salford University, UK; Sichuan University, China; Ritsumeikan University, Japan; IIT, Roorkee, India; AFU, Nepal; and many more. He has served as Editorial Board, reviewers in good number of national and international journal. He has published more than 100 research articles, books, and book chapters in reputed journals and publishers.

Rajib Shaw a Japanese national of Indian origin, Rajib Shaw is the professor in Graduate School of Media and Governance in Keio University's Shonan Fujisawa Campus (SFC). He is also co-founder of a Delhi (India) based social entrepreneur startup Resilience Innovation Knowledge Academy (RIKA) and RIKA Institute, a not-for-profit entity of RIKA. Prof. Shaw is the recipient of prestigious "Pravasi Bharatiya Samman Award (PBSA)" in 2021 for his contribution in education sector. PBSA is the highest honor conferred on overseas Indians and persons of Indian origin from the President of India. Prof. Shaw is also the recipient of United Nations Sasakawa Award for Disaster Risk Reduction in 2022 for his lifetime contributions in the field of disaster risk reduction. Sasakawa Award is the highest UN award for disaster risk reduction and was established by 1986 by the founding Chairman of the Nippon Foundation, Mr. Ryoichi Sasakawa. He is also the Senior Fellow of Institute of Global Environmental Strategies (IGES) Japan, and the Chairperson of SEEDS Asia and CWS Japan, two Japanese NGOs, and a Distinguished Professor of Institute of Disaster Management and Reconstruction (IDMR) of Sichuan University, China. Earlier, he was the Executive Director of the Integrated Research on Disaster Risk (IRDR), a decade-long research program co-sponsored by the International Council for Science (ICSU), the International Social Science Council (ISSC), and the United Nations International Strategy for Disaster Reduction (UNISDR).

Previously, he served as a Professor in the Graduate School of Global Environmental Studies of Kyoto University. His expertise includes community-based disaster risk management, climate change adaptation, urban risk management, and disaster and environmental education. Prof. Shaw was the Chair of the United Nations Science Technology Advisory Group (STAG) for disaster risk reduction; and is the Co-chair of the Asia Pacific Science Technology Advisory Group (AP-STAG). He was also the CLA (Coordinating Lead Author) for the IPCC 6th Assessment Report on Climate Change (Working Group II, Asia Chapter). He is the editor of a book series on disaster risk reduction, published by Springer, Editor-in-chief of Progress in Disaster Science, a journal published by Elsevier. Prof. Shaw has published more than 61 books and over 400 academic papers and book chapters.

Contributors

Md. Anwarul Abedin Department of Soil Science, Bangladesh Agricultural University, Mymensingh, Bangladesh;
College of Policy Science, Osaka Ibaraki Campus, Ritsumeikan University, Osaka, Japan

Reazul Ahsan Department of City and Metropolitan Planning, University of Utah, Salt Lake City, USA

Irtija Alam Department of Urban and Regional Planning (URP), Khulna University of Engineering & Technology (KUET), Khulna, Bangladesh

Yunus Ali Pulpadan Department of Earth and Environmental Sciences, Indian Institute of Science and Education Research Mohali, Mohali, India

Sharmin Ara Department of Urban and Regional Planning (URP), Chittagong University of Engineering and Technology (CUET), Chittagong, Bangladesh;
Department of Geography, Texas A&M University, Texas, USA

Rumana Asad Architecture Discipline, Khulna University, Khulna, Bangladesh

Shazmin Shareena Ab. Azis Faculty of Built Environment & Surveying, Universiti Teknologi Malaysia, Skudai, Johor, Malaysia

Shastitharran Baskaran Faculty of Built Environment & Surveying, Universiti Teknologi Malaysia, Skudai, Johor, Malaysia

Ayush Baskota Department of Civil and Infrastructure Engineering (CIE), Asian Institute of Technology (AIT), Pathum Thani, Thailand

Mosa Lajina Begum Department of Soil Science, Bangladesh Agricultural University, Mymensingh, Bangladesh

Jotirmoy Chakrobortty Department of Soil Science, Khulna Agricultural University, Khulna, Bangladesh

Md. Esraz-Ul-Zannat Griffith School of Engineering & Built Environment, and Cities Research Institute, Griffith University - Gold Coast Campus, Southport, Australia

Soumaya Ezazaa Urbanist and Architect, Carta—reichen et robert associes architectes urbanistes, Rabat, Morocco

Sumaiya Farzana Graduate Training Institutue, Bangladesh Agricultural University, Mymensingh, Bangladesh

Liuyue He Ocean College, Zhejiang University, Zhoushan, Zhejiang, China

Md. Hosenuzzaman Department of Soil Science, Bangladesh Agricultural University, Mymensingh, Bangladesh;
UWA School of Agriculture and Environment, The University of Western Australia, Perth, Australia

Shahin Imran Departemnt of Agronomy, Khulna Agricultural University, Khulna, Bangladesh

Israt Jahan Department of Soil Science, Bangladesh Agricultural University, Mymensingh, Bangladesh

Sadiya Arefin Juthee Department of Crop Botany, Bangladesh Agricultural University, Mymensingh, Bangladesh

Mohammad Golam Kibria Department of Soil Science, Bangladesh Agricultural University, Mymensingh, Bangladesh

Yeonji Lee Department of Civil and Infrastructure Engineering (CIE), Asian Institute of Technology (AIT), Pathum Thani, Thailand

Kebei Liu Institute for Disaster Management and Reconstruction, Sichuan University, Chengdu, China

Kenichiro Onitsuka Graduate School of Global Environmental Studies (GSGES), Kyoto University, Kyoto, Japan

Indrajit Pal Department of Civil and Infrastructure Engineering (CIE), Asian Institute of Technology (AIT), Pathum Thani, Thailand

Gulsan Ara Parvin College of Policy Science, Ritsumeikan University, Osaka-Ibaraki Campus, Osaka, Japan

Newton Chandra Paul Departemnt of Agronomy, Khulna Agricultural University, Khulna, Bangladesh

Mohammad Saidur Rhaman Department of Seed Science and Technology, Bangladesh Agricultural University, Mymensingh, Bangladesh

Subir Sen Department of Humanities and Social Sciences, Indian Institute of Technology Roorkee, Roorkee, Uttarakhand, India

Shashi Kant Sharma Department of Humanities and Social Sciences, Indian Institute of Technology Roorkee, Roorkee, Uttarakhand, India

Shubham Sharma Department of Humanities and Social Sciences, Indian Institute of Technology Roorkee, Roorkee, Uttarakhand, India

Rajib Shaw Graduate School of Media and Governance, Keio University, Minato, Japan

Srikrishnan Siva Subramanian Centre of Excellence in Disaster Mitigation and Management, Indian Institute of Technology Roorkee, Roorkee, India

Seak Sopath Department of Civil and Infrastructure Engineering (CIE), Asian Institute of Technology (AIT), Pathum Thani, Thailand

Mousumi Jahan Sumi Department of Crop Botany, Khulna Agricultural University, Khulna, Bangladesh

Nina Takashino College of Policy Science, Osaka Ibaraki Campus, Ritsumeikan University, Osaka, Japan

Md Zayeed Hasan Talukdar Former Consultant, International Finance Corporation (IFC), Dhaka, Bangladesh

Md Moyeed Hasan Talukder Department of Horticulture and Landscape Architechture, Purdue University, West Lafayette, USA

Akshat Vashistha Centre of Excellence in Disaster Mitigation and Management, Indian Institute of Technology Roorkee, Roorkee, India

Jiacheng Xing Institute for Disaster Management and Reconstruction, Sichuan University, Chengdu, China

Jingyuan Xue Institute for Disaster Management and Reconstruction, Sichuan University, Chengdu, China

Yusen Ye Institute for Disaster Management and Reconstruction, Sichuan University, Chengdu, China

Part I
Urbanization, Food Security and Climate Disaster Resilience

Chapter 1
Triple Nexus of Urbanization, Food Security, and Climate Disaster Resilience: Present Trend, Necessity, and Challenges

Gulsan Ara Parvin and Sharmin Ara

Abstract The whole world is combating to minimize the challenges associated with rapid urbanization, climate change induced impacts and disasters and food insecurity. Urbanization, food security, and climate change are the three key concerning issues of all nations, irrespective of developed and developing. However, the trend and challenges associated with urbanization, food security, and climate-induced disasters are not same in all parts of the world. It is estimated that by 2050 about 2.5 billion people will be added in urban areas around the world and projected that 90% of this increase will take place in Asia and Africa, where poverty, food insecurity, and climate-induced disasters are relatively higher than other parts of the world. How urbanization, food security, and climate disaster resilience are shaping each other in different parts of the world is important to take need oriented integrated policy and action planning. Therefore, this chapter highlights the trend and challenges associated with the triple nexus of Urbanization, Food Security, and Climate Disaster Resilience around the world.

Keywords Urbanization · Food security · Climate change · Opportunities · Challenges

---

G. A. Parvin (✉)
College of Policy Science, Ritsumeikan University, Osaka-Ibaraki Campus, Osaka, Japan
e-mail: niruurp@yahoo.com

S. Ara
Department of Urban and Regional Planning (URP), Chittagong University of Engineering and Technology (CUET), Chittagong, Bangladesh

M. G. Kibria et al. (eds.), *Urban Blue-Green Infrastructure Approach for Food Security and Climate Disaster Resilience*, Disaster Risk Reduction,
https://doi.org/10.1007/978-981-96-8746-6_1

1.1 Introduction

Urbanization, the process of population concentration in cities and towns, has seen unprecedented growth worldwide in recent decades. In 1950, about 30% of people worldwide lived in urban areas; by 2018, that number had risen to 55%, and by 2050, it is expected to reach 68% (UNDESA 2018). While it brings economic growth and social opportunities, it also has a significant impact on the environment (Li et al. 2021) in different scales from local to global (Nagendra et al. 2018). As urban areas continue to grow, issues such as infrastructure development, resource management, social inequality, and environmental sustainability become increasingly critical to address (Cheshmehzangi et al. 2021; Wei and Ewing 2018).

The whole world is combating to minimize the challenges associated with rapid urbanization, climate change induced impacts and disasters and food insecurity. Urbanization, food security, and climate change are the three key concerning issues of all nations, irrespective of developed and developing. However, the trend and challenges associated with urbanization, food security, and climate-induced disasters are not same in all parts of the world. It is estimated that by 2050 about 2.5 billion populations will be added in cities around the world and projected that 90% of this increase will take place in Asia and Africa (UNDESA 2018), where poverty, food insecurity, and climate-induced disasters are relatively higher than other parts of the world. How urbanization, food security, and climate disaster resilience are shaping each other in different parts of the world is important to analyze for formulating need oriented and integrated policy and action planning. Therefore, this chapter highlights the trend and challenges associated with the triple nexus of Urbanization, Food Security, and Climate Disaster Resilience around the world.

1.2 Urbanization and Food Security: Trend and Challenges

1.2.1 Urbanization and Food Security Trends and Relationship

The rapid and widespread global trend of urbanization exerts a profoundly intricate and multifaceted influence on food security, encompassing diverse factors such as the conversion of agricultural land to urban development, high reliance on imported food sources, and loss of biodiversity (Ekpenyong 2015; FAO et al. 2023). Ruel et al. (2017) claimed that food insecurity is prevalent in urban area, particularly rapid rises in overweight and obesity. In addition, it amplified the vulnerability to food insecurity, particularly those marginalized within informal settlements or grappling with urban poverty (de Bruin et al. 2021; Kousar et al. 2021). There are three core aspects of food security: accessibility, affordability, and availability of food (Lang and Barling 2012; Opitz et al. 2016), these cannot be ensured in urban areas. Rapid urbanization has been changing the food system globally, particularly in Sub-Saharan Africa and

South Asia (Chin et al. 2024; Djurfeldt 2015; Ruel et al. 2017) as it is expected that the majority of the urban expansion will be in Asia and Africa (Engdaw and Kebede 2019; UN-Habitat 2016). Urban food security is a growing concern as the number of big cities is increasing. The 2023 State of Food Security and Nutrition in the World report explores links between urbanization and changing food systems (Ruel et al. 2017). Urbanization affects the entire food system (Kang et al. 2023) and rural livelihood in many ways. For instance, a substantial amount of people are living in sub-urban and peri-urban areas which previously worked as connecting nodes for rural–urban areas. As a result, a large population in the peri-urban areas is losing their agricultural land due to urban expansion. This loss of agricultural land is a significant threat to food security in two ways, (1) reduced land availability, and (2) strained natural resources. Regarding loss of land, as cities expand, fertile land on the outskirts gets swallowed for housing, industry, and infrastructure (Engdaw and Kebede 2019; Kang et al. 2023). This shrinks the area available for growing food, leading to a potential decrease in overall food production. In addition, farmers in peri-urban areas might be forced to sell their land for development projects, reducing the agricultural workforce and disrupting established food production systems (Abdulai et al. 2023). Moreover, there are direct and indirect land use land cover changes that lead to alterations in food demand. This changes has also an impact on migration, the labor market, and trading (Abdulai et al. 2023). As a result, households adapted various non-farm activities for survival and there are changes in income, employment, lifestyle, and food consumption behavior, which in turn makes a complex market structure (FAO et al. 2023; Kastner et al. 2012). This in turn increases the high reliance on imported food. Regarding strained resources, urban areas on average consume a lot of water. With the expansion of cities, uses for water resources (both households and agriculture) intensify, and this can leave less water for farming in peri-urban areas, affecting crop yields. Additionally, urban sprawl can lead to increased pollution from industries and households. This air and water pollution can harm nearby agricultural lands, impacting the quality and safety of food produced.

1.2.2 Scenario of Urbanization Impacts on Food Security in the Global South and the Global North

There are also global variations in food security. The challenges and opportunities of urban food security differ between developed and developing countries. The Global North countries may have more resources to invest in technology and infrastructure solutions, while the Global South countries might prioritize social safety nets and local production initiatives. There is also regional differences in terms of climate, geography, and cultural preferences. Cities in arid regions might prioritize water-efficient technologies, while coastal cities may focus on sustainable seafood production. The UN recognizes food security as a significant human security along with economic, health, and environmental security (UNDESA 2021). In 2022, nearly

29.6% of the global population were moderately or severely food insecure, among them about 11.3% (900 million individuals) who were highly food insecure (FAO et al. 2023). It is also projected that more than 600 million people worldwide are projected to face hunger in 2030 (UN 2022). The Sustainable Development Goal (SDG) 2 on "Zero Hunger" prioritizes food security issues within the broader SDGs 2030 Agenda. One of the targets of SDG 2 goals is to ensure sustainable food production systems and implement resilient agricultural practices that increase productivity, and production which in turn help maintain ecosystems, and strengthen adaptation to climate change, and extreme weather events. Hunger and food insecurity have been increasing since 2015 due to climate change, pandemic, cost of living, declining food production, and conflict.

1.2.3 Challenges and Opportunities of Food Security in Urban Areas

As urbanization is increasing, and it is projected by UN Department of Economic and Social Affairs (UN DESA) that seven in ten people will believing in urban area by 2050, there are both challenges and opportunities in urban area regarding food security. The major challenges and opportunities are discussed below and demonstrated in Fig. 1.1.

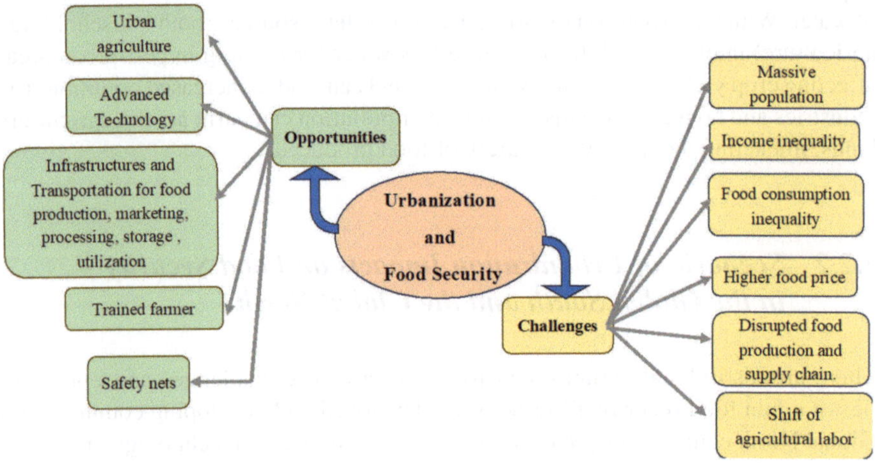

Fig. 1.1 Challenges and opportunities of urban area regarding food security

1.2.3.1 Challenges

- **Rapid Urbanization**: More people in cities means a strain on resources and infrastructure to deliver affordable, nutritious food.
- **Income Inequality**: Low-income residents often struggle to afford healthy food, even if it's available.
- **Inequalities in Food Consumption**: Urban areas face profound inequalities based on socioeconomic groups, ethnicity, migratory status, and location of residence (slums or formal settlements).
- **Access Disparities**: An abundance of food doesn't guarantee equal access to nutritious, diverse, and affordable diets. Urban poor are particularly vulnerable to food insecurity. Also, insufficient availability of vegetables and fruits to meet the daily standard requirements for a healthy living.
- **Disrupted Supply Chains**: Climate change, extreme weather events, and political instability can disrupt food supplies to cities.
- **Rise in Overweight and Obesity**: Urban areas witness alarming increases in overweight and obesity, alongside persistent under-nutrition, and micronutrient deficiencies.
- **Higher Food Prices**: With less land and potentially lower production, food prices can rise due to scarcity. This can disproportionately affect low-income urban dwellers who rely on affordable food from nearby peri-urban farms.

1.2.3.2 Opportunities

- **Urban Agriculture**: Growing food within cities through rooftop gardens, vertical farms, and community plots can increase access to fresh produce.
- **Advanced Information Technology and Innovation**: Several web-based applications and online platforms can connect consumers with local farmers and producers, reducing food waste and increasing access.
- **Off-farm Activity Generation**: Urbanization results to generate various off-farm activity in urban area. Specially woman and youth can engage themselves in various types of non-farm, unskilled income generating activities (Dadi et al. 2022). In addition, the proximity to urban areas can lead to greater exposure to new technologies and innovative business practices, which can be adopted in off-farm activities.
- **Trained Farmer**: Urbanization is commonly accompanied by better access to education and skills training, which equips individuals with the knowledge to start and manage farm and off-farm activities effectively. Moreover, off-farm employment can improve farmland production efficiency. For instance, when a significant portion of a household's time is dedicated to off-farm employment, it can lead to the adoption of labor-saving technologies in agriculture, thus increasing farmland productivity (Hou et al. 2023).
- **Social Safety Nets**: Strengthening social programs can ensure everyone has access to a basic level of food security.

Governments should engage with food system stakeholders at all levels to address the challenges created by urbanization. New measures of food security consider safe, nutritious, and sufficient food access, going beyond mere calorie availability. Resilient and adaptable urban food security depends on a number of factors including urban expansion trend, consumption behavior, connection between the rural and urban food supplies, and limitations of smallholder farmers (Wenban-Smith et al. 2016). In general, urban food security takes into account the whole food system—from production to consumption. Innovation, technology, and a focus on sustainability are key to ensuring food security for a growing urban population worldwide.

1.3 Climate Change and Food Security: Trend and Challenges

Food security means access by all people at all times to enough food for an active, healthy life (USDA 2023). Food security is combinedly determined by four pillars (Committee on World Food Security 2012) food availability, food access, food utilization and food supply. Since the history of humanity food security has been a concernig issues. This concerning issue has been aggravated by climate change, which threatens lives and livelihood of billions of people around the world. Climate change induced natural disasters, environmental degradation and extreme weather patterns disrupt harvests, deplete fisheries, erode livelihoods and spur infectious diseases and all these are menacing food security. According to FAO (2016) all dimensions of food security are negatively affected by climate change and increasing climate variability. Further, climate change affects the natural resources (e.g., water, land) used in agricultural production (Cadro et al. 2019; Čustović et al. 2012; Simunic et al. 2019). Agricultural productivity, food production, food system, natural resources, rural livelihoods associated with food production are combating with numerous challenges due to climate crisis. Figure 1.2 reveals the monthly global temperature anomalies (∘C) from 1980 to 2020. This figure is a clear indication of the trend of rapid global climate change.

According to the research findings of Dasgupta and Robinson (2022) temperature anomaly not only increases the probability of food insecurity but the magnitude of this impact is increasing over time. This study warns that the temperature anomaly may continue to increase due to future climate change, likely further intensifying the stress on food security. Study conducted by Hasegawa et al., argues that an additional 20–36 and 11–33% population may face hunger by 2050 under a once-per-100-yr extreme climate event under high and low emission scenarios, respectively. This extreme climate events like heatwave, flood, drought, wildfire, and rainfall all are in the trend of increase due to climate change (Clarke et al. 2022), which will severely affect all the four pillars (food availability, food access, food utilization, and food system stability) of food security (Bilali et al. 2020; Lin et al. 2022).

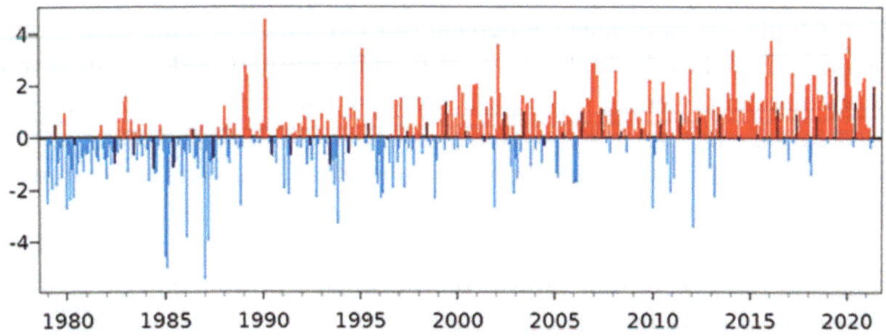

Fig. 1.2 Monthly global temperature anomalies (∘C). *Source* Dasgupta and Robinson (2022) (permission is received)

The intensity and extent of climate change and its impacts are not same in all-over the world. Further, along with climate change, other socioeconomic, topographical, environmental, and political issues have great influence to determine food security status of a community. Therefore, in food insecurity status wide variation is observed around the world. In 2021, more than half (425 million) of people in the world affected by hunger lived in Asia, and more than one-third (278 million) dwelled in Africa. Findings of another recent study (Pickson et al. 2023) reveal that the impacts of climatic variables on food production differ significantly from country to country. Figure 1.3 depicts the regional variation of the food insecurity.

According to FAO report in 2021, about 2.3 billion people in the world experience moderate or severe food insecurity. Amon these food in secured people half (1.15 billion) are in Asia; more than one-third (795 million) are in Africa. In Africa more than half of the population faces moderate to severe food insecurity. After Africa, about 12 percent (268 million) food insecured people live in Latin America and the Caribbean. Northern America and Europe together have the lowest number of foods

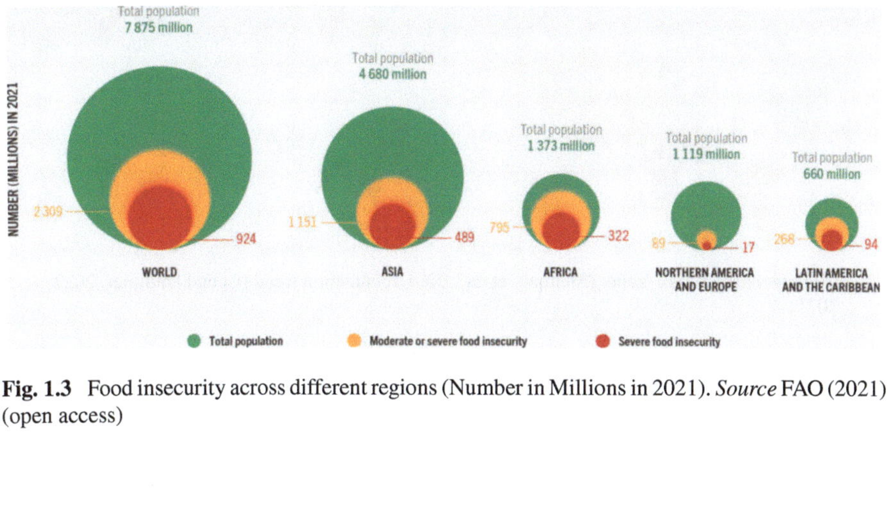

Fig. 1.3 Food insecurity across different regions (Number in Millions in 2021). *Source* FAO (2021) (open access)

insecured people which is nearly 4 percent (89 million). Compare to any other part of the world Africa has the higher level of food insecurity (moderate or severe).

In 2019 number of people suffering acute food insecurity was 135 milling, which increased to 345 milling in 82 countries by June 2022. By reducing agricultural productivity, climate change and climate change induced disasters and extreme events are affecting agricultural productivity, rural livelilihood and food security, especially in developing countries (Falco et al. 2019). Coupled with climate change, Covid-19 pandemic and war in different parts of the world are imposing severe challenges to ensure food security of the growing population of the world. Since last three to four years Covid-19 pandemic and conflicts in different parts of the world have aggravated the food insecurity situation.

1.4 Urbanization and Climate Disasters

Concept of climate disaster, delving into the various types such as extreme weather events, sea-level rise, and ecological breakdowns. Discuss the far-reaching impacts, including the loss of life, economic damage, and displacement of communities. These disasters such as cyclone, floods, drought exacerbate social inequalities and strain resources, particularly in densely populated urban areas. Study has argued migration as an adaptive strategy to climate change impacts and climate change induced disasters (Bettini 2014). Especially, Africa and Asia are vulnerable to climate change and facing rapid urbanization. Figure 1.4 (comparing image a and b) denotes that the countries that are most at risks in climate change are having and will have higher urbanization trend.

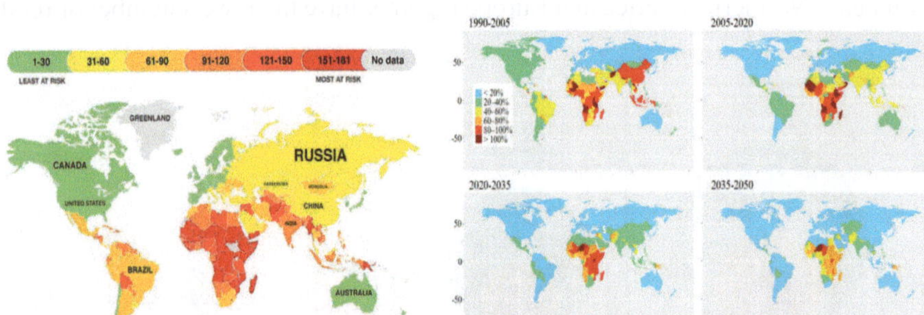

a. Climate change risk zone (Shafqat, et al., 2021)
b. Urbanization trend (Li and Dallimer, 2022)

Fig. 1.4 Overlapping of climate change risks and urbanization trends. *Note* Figures are from Open access sources, where permissions are given with proper citation

In Africa rural out-migration of young adults is the most sensitive to shifts in weather patterns, with lower rainfall, lower variability in rainfall, and higher temperatures increasing subsequent rural out-migration (Weinreb et al. 2020). A different study has been examined. It is estimated that an increase in the national urban share of 0.45% with a reduction in national rainfall of 1% (Barrios et al. 2006). In Asia, Bangladesh is one of the most climate change vulnerable countries, where climate induced disasters create hundreds of thousands of climate migrants every year (McDonnell 2019). Sudden-onset climatic disasters such as cyclones, floods, river erosion, and landslides are cusing displacement in allover the country. Slow-onset disasters like sea-level rise and salinization also create new displacements, especially in coastal areas of Bangladesh. It is projected that more than 35 million people in the coastal areas of Bangladesh will be at risk of human migration by 2050. As many as 19 coastal districts and those along the major river banks (Padma, Meghna and Jamuna) are among the highest levels of disaster risk. Cities, especially the major cities in Bangladesh, are obviously vulnerable to climate-induced migration. They suffer a lot environmentally, socioeconomically, and politically for climate change or disaster-induced rural to urban migration (Rana and Ilina 2021). Climate change is acting as push factors of migration and enhancing rapid and unplanned urbanization, disaster risks in urban areas and environmental degradation and pollution (Fig. 1.5). On the other hand urbanization is aggravating climate change by numerous ways.

Urbanization and climate disasters are intricately linked, creating a vicious cycle that requires a multifaceted and holistic approach. As cities grow, concentration of populations and infrastructures become major contributors to climate change through greenhouse gas emissions. This, in turn, accelerates the intensity and frequency of climate disasters like floods, heatwaves, cyclones, tornados, and droughts. These extreme events can disproportionately impact densely populated

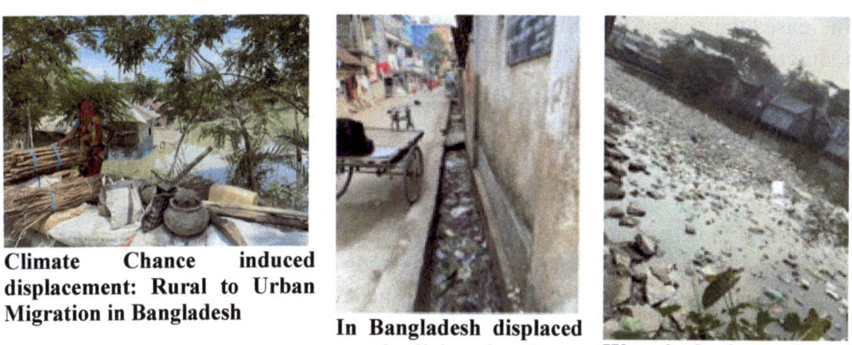

Climate Chance induced displacement: Rural to Urban Migration in Bangladesh

In Bangladesh displaced people living in slums. Drain in a slum causing pollution, water logging and enhancing urban flooding

Waterbody in a slum in Bangladesh-causing pollution, and waterlogging and urban flooding

Fig. 1.5 Climate change induced displacement, Slums in cities: enhancing unplanned urbanization, disaster risks, and pollution. *Source* Authors

urban areas, straining resources, damaging infrastructure, and displacing residents. The rapid rise of urban populations poses a significant challenge in the face of a changing climate. Urbanization exacerbates climate disasters in many ways:

I. Increase of Greenhouse Gas Emissions: Cities are major contributors to greenhouse gas emissions, the primary driver of climate change. This is due to energy consumption and transportation. Urban areas are hubs of energy use for buildings, transportation, and industries. Fossil fuel-powered electricity generation significantly contributes to CO_2 emissions (UN-Habitat 2022). Densely populated cities rely heavily on private vehicles, leading to high emissions from combustion engines. Public transportation, while beneficial, can also contribute depending on its reliance on fossil fuels.

II. Urban Heat Island Effect: Urban landscapes significantly alter the local climate (Chapman et al. 2017; Grimmond et al. 2015). Buildings, roads, and paved surfaces absorb and retain heat more effectively than natural landscapes, leading to the urban heat island (UHI) effect. This effect can raise urban temperatures by several degrees compared to surrounding rural areas (Li et al. 2022). The UHI effect not only increases discomfort to the residents but also exacerbates heatwaves, placing additional stress on urban infrastructure and public health.

III. Strained Infrastructure: Rapid urbanization often outpaces infrastructure development, leaving cities vulnerable to extreme weather events. Drainage systems designed for lower populations can become overwhelmed during heavy downpours, leading to flash floods. Similarly, limited water storage capacity can exacerbate droughts and water shortages.

IV. Loss of Natural lands: Rapid urbanization costs on natural ecosystems like forests and wetlands. Due to these land cover change, the ecosystem value has been decreasing significantly. These ecosystems play a crucial role in mitigating floods and droughts by absorbing rainwater and regulating water flow. Their loss increases the risk of flooding and intensifies drought impacts within urban areas (Mu et al. 2024).

Climate disasters affect urban areas in different ways. It turns cities strengths into vulnerabilities. Extreme climatic events create excessive pressure on cities in addition to dense populations and concentrated infrastructure. Sea level rise threatens coastal cities and inundates the low-lying land and displaces population. For instance, floods overwhelm drainage systems designed for lower capacities, causing flash floods. Floods and stagnant water create breeding grounds for mosquitoes and other disease vectors. Heatwaves, intensified by the urban heat island effect, strain power grids and endanger public health. These disasters not only cause immediate devastation but disrupt essential services, displacing residents and hindering economic activity for long. Climate change induced hazards also can affect agricultural areas, impacting food security in cities. Additionally, climate disasters increase the vulnerability of low income and informal settlements in the global south and global north (Cooper et al. 2008). Due to pre-existing condition, lack of infrastructure and services, poorer group are relatively in higher risk of extreme climatic events (Haque 2021).

1.5 Triple Nexus—Urbanization, Food Security, and Climate Disaster Resilience

Several studies including recent report of Food and Agricultural Organization have tried to establish the link between climate changes and food insecurity in different parts of the world. These food insecurity, climate change impacts, and climate induced natural disasters are triggering livelihood losses and displacement of people that are one of the principal causes of urbanization. Different researches examined the interconnections between climate change and urbanization and climate change and food security in separately. However, examination of the triple nexus of urbanization, food security, and climate change are not often noticed. Nature and pattern of interconnections of these three crucial issues and associated causes and effects are not well plotted yet. In order to have effective policy and action planning it is necessary to deal these three critical issue, climate change, urbanization, and food security in a holistic and integrated approach. Figure 1.6 depicts the triple nexus of Urbanization, Food Insecurity, and Climate disaster.

Migration has become a key component of livelihood strategies and climate change adaptation strategies of increasing numbers of rural households in many developing countries (Choithani 2017; Wiederkehr et al. 2018). Since agriculture is the main source of livelihood in developing countries and it is affected by climate change, in order to avoid climate stress, climate shoks, migration is one of the main strategy of adaptation (Cattaneo et al. 2019). Study conducted by Choithani (2017) in India claims that from the perspective of migration-food security relationship internal migration is far more significant than international migration. This internal migration strategy of climate change adaoptation is one of the main cause of urbanization in developing countries, especially in Africa and Asia. Migration has both positive and negative impact on food security. In the place of origin, rural to urban migration causes shortage of farmers and agricultural labors that has direct negative impact on food production (Mutisya et al. 2016). In the place of destination (city/

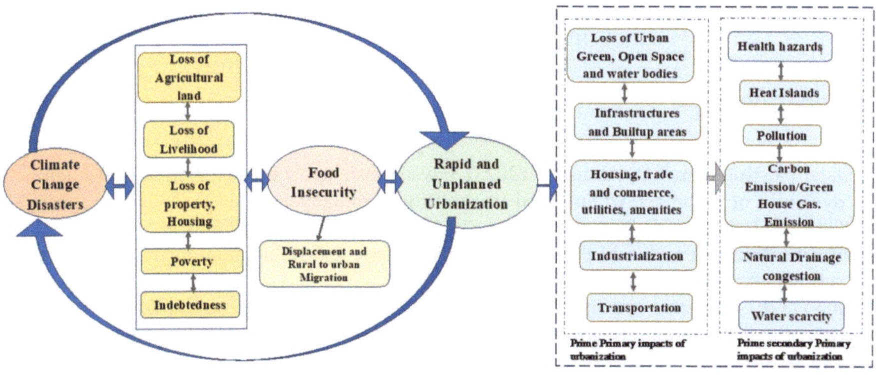

Fig. 1.6 Nexus of urbanization, climate change disasters, and food security

town) it is found that migrant households are more likely to be food insecure than non-migrant households. Sometimes, migrants in the city sacrifice their own food security in order to meet the food needs of their rural households. Further, some migrants have less income stability compared to no-migrant households. Migrant households are also lack of social services, such as basic income grants that can help poor households purchase food (Chikanda et al. 2017). Furthermore, to accommodate the augmented population and to provide services to the increased population, urban lands are rapidly encroaching agricultural areas within and around the cities (Parvin et al. 2022), which negatively affects food security. Szabo (2016) also claims that with the expansion of urban areas, food supplies are affected by the use of agricultural land in the urban peripheries for non-agricultural purposes and the diversion of water from agriculture to domestic and industrial uses due to its shortage. On the other side, climate disasters accelerate climate migration, displacement, and urbanizations that hinder sustainable urban development in the affected area.

Breaking this cycle requires a holistic and sustainable approach that tackles both urbanization patterns and climate mitigation strategies. It is needed to invest in green cities with sustainable practices alongside global efforts to curb emissions and prepare for the inevitable impacts of a changing climate. Nature-based solutions, like urban green spaces, wetlands restoration, and floodplains to regulate water flow are needed. Sustainable urban planning, greener infrastructure, energy-efficient buildings, climate-smart infrastructures such as building seawalls, flood barriers, and heat-resistant housing are few example of green initiatives to combat with adverse impact of urbanizations. Internal and International cooperation, sharing knowledge, resources, and technology to address the global challenges are also necessary. Climate actions and policies, urban agendas and policy action, food and wellbeing initiatives should goes hand in hands. However, most often there are absence of coordination among food and agriculture related organizations, urban governance and climate change actions and strategies both at global and local level. Furthermore, pandemic, conflicts, and wars in different parts of the world have jeopardize progress and initiatives taken at national and international level to handle rapid urbanization, climate change, and food insecurity.

References

Abdulai IA, Kuusaana ED, Ahmed A (2023) Toward sustainable urbanization: understanding the dynamics of urban sprawl and food (in)security of indigenous households in peri-urban Wa, Ghana. J Urban Aff 1–26. https://doi.org/10.1080/07352166.2023.2224023

Barrios S, Bertinelli L, Strobl E (2006) Climatic change and rural–urban migration: the case of sub-Saharan Africa. J Urban Econ 60(3):357–371

Bettini G (2014) Climate migration as an adaption strategy: de-securitizing climateinduced migration or making the unruly governable? Crit Stud Secur 2(2):180–195. https://doi.org/10.1080/21624887.2014.909225

Bilali EH, Bassole IHN, Dambo L, Berjan S (2020) Climate change and food security. Agric For/Poljoprivreda i Sumarstvo 66(3)

Cadro S, Uzunović M, Cherni-Cadro S, Zurovec J (2019) Changes in the water balance of Bosnia and Herzegovina as a result of climate change. Agric For 65(3):19–33

Cattaneo C, Beine M, Fröhlich CJ, Kniveton D, Martinez-Zarzoso I, Mastrorillo M et al (2019) Human migration in the era of climate change. Rev Environ Econ Policy

Chapman S, Watson JEM, Salazar A, Thatcher M, McAlpine CA (2017) The impact of urbanization and climate change on urban temperatures: a systematic review. Landscape Ecol 32:1921–1935

Cheshmehzangi A, Butters C, Xie L, Dawodu A (2021) Green infrastructures for urban sustainability: issues, implications, and solutions for underdeveloped areas. Urban For Urban Green 59:127028

Chin CF, Goh EV, Clarke AC, Ang MLE, Show PL, Supramaniam CV, Ali A, Azam-Ali S, Beyer R, Culaba A (2024) Transforming food systems in maritime Southeast Asia and Pacific Small Island developing states To Support Food Security And Sustainable Healthy diets. Front Sustain Food Syst 8:1304317

Chikanda A, Crush J, Frayne B (2017) Migration and urbanization: Consequences for food security. Food Nutr Secur S Afr Cities 48–65

Choithani C (2017) Understanding the linkages between migration and household food security in India. Geogr Res 55(2):192–205

Committee on World Food Security (2012) Coming to terms with terminology: food security, nutrition security, food security and nutrition. Food and nutrition security, Rome

Clarke B, Otto F, Stuart-Smith R, Harrington L (2022) Extreme weather impacts of climate change: an attribution perspective. Environ Res Clim 1(1):012001

Cooper PJM, Dimes J, Rao KPC, Shapiro B, Shiferaw B, Twomlow S (2008) Coping better with current climatic variability in the rain-fed farming systems of sub-Saharan Africa: an essential first step in adapting to future climate change? Agr Ecosyst Environ 126(1–2):24–35. https://doi.org/10.1016/j.agee.2008.01.007

Čustović H, Đikić M, Ljuša M, Žurovec O (2012) Effect of climate changes on agriculture of the Western Balkan countries and adaptation policies. Agric Forest 58(2):127–141

Dadi W, Mulegeta M, Simie N (2022) Urbanization and its effects on income diversification of farming households in Adama district, Ethiopia. Cogent Econ Finance 10(1):2149447. https://doi.org/10.1080/23322039.2022.2149447

Dasgupta S, Robinson EJ (2022) Attributing changes in food insecurity to a changing climate. Sci Rep 12(1):4709

de Bruin S, Dengerink J, van Vliet J (2021) Urbanisation as driver of food system transformation and opportunities for rural livelihoods. Food Secur 13(4):781–798. https://doi.org/10.1007/s12571-021-01182-8

Definitions of Food Security. USDA Economic Research Service. Updated October 25, 2023. Accessed 15 May 2024. https://www.ers.usda.gov/topics/food-nutrition-assistance/food-security-inthe-us/definitions-of-food-security.aspx

Djurfeldt AA (2015) Urbanization and linkages to smallholder farming in sub-Saharan Africa: implications for food security. Glob Food Sec 4:1–7

Ekpenyong AS (2015) Urbanization: its implication for sustainable food security, health and nutritional nexus in developing economies-a case study of Nigeria. J Stud Soc Sci 11(1)

Engdaw B, Kebede M (2019) The impact of urbanization on food (in) security in Amhara Regional State Metropolitan cities: Monitoring income generating capacity limitation as a mediating variable. Int J Adv Sci Res Manage 4(9):42–57

Falco C, Galeotti M, Olper A (2019) Climate change and migration: is agriculture the main channel? Glob Environ Chang 59:101995

FAO, IFAD, UNICEF, WFP, & WHO (2023) The State of Food Security and Nutrition in the World 2023. Urbanization, agrifood systems transformation and healthy diets across the rural–urban continuum

FAO (2021). https://openknowledge.fao.org/server/api/core/bitstreams/6ca1510c-9341-4d6a-b285-5f5e8743cc46/content/sofi-2022/food-security-nutrition-indicators.html

FAO (2016) The state of food and agriculture: climate change. Agriculture and Food Security, Rome

Grimmond CSB, Ward HC, Kotthaus S (2015) How is urbanization altering local and regional climate?
Haque AN (2021) Climate risk responses and the urban poor in the global South: the case of Dhaka's flood risk in the low-income settlements. Int J Disaster Risk Reduct 64:102534. https://doi.org/10.1016/j.ijdrr.2021.102534
Hasegawa T, Sakurai G, Fujimori S, Takahashi K, Hijioka Y, Masui T (2021) Extreme climate events increase risk of global food insecurity and adaptation needs. Nature Food 2(8):587–595
Hou Y, Ji X, Chen J, Zhang H (2023) The impact of off-farm employment on farmland production efficiency: an empirical study based in Jiangsu Province, China. Processes 11(1). https://mdpi-res.com/d_attachment/processes/processes-11-00219/article_deploy/processes-11-00219.pdf?version=1673335782
Kang J, Duan X, Yun R (2023) The impact of urbanization on food security: a case study of Jiangsu Province. Land 12(9). https://mdpi-res.com/d_attachment/land/land-12-01681/article_deploy/land-12-01681-v2.pdf?version=1695797310
Kastner T, Rivas MJI, Koch W, Nonhebel S (2012) Global changes in diets and the consequences for land requirements for food. Proc Natl Acad Sci 109(18):6868–6872. https://doi.org/10.1073/pnas.1117054109
Kousar S, Ahmed F, Pervaiz A, Bojnec Š (2021) Food insecurity, population growth, urbanization and water availability: the role of government stability. Sustainability 13(22). https://mdpi-res.com/d_attachment/sustainability/sustainability-13-12336/article_deploy/sustainability-13-12336.pdf?version=1636426004
Lang T, Barling D (2012) Food security and food sustainability: reformulating the debate. Geograph J 178(4):313–326. https://doi.org/10.1111/j.1475-4959.2012.00480.x
Li G, Fang C, Qi W (2021) Different effects of human settlements changes on landscape fragmentation in China: Evidence from grid cell. Ecol Ind 129:107927
Lin HI, Yu YY, Wen FI, Liu PT (2022) Status of food security in East and Southeast Asia and challenges of climate change. Climate 10(3):40
Li X, Stringer LC, Dallimer M (2022) The impacts of urbanisation and climate change on the urban thermal environment in Africa. Climate 10(11). https://mdpi-res.com/d_attachment/climate/climate-10-00164/article_deploy/climate-10-00164-v3.pdf?version=1667558401
McDonnell T (2019) Climate change creates a new migration crisis for Bangladesh. Natl Geogr 24. https://www.nationalgeographic.com/environment/article/climate-change-drives-migration-crisis-in-bangladesh-from-dhaka-sundabans. Accessed 29 May 2024
Mutisya M, Ngware MW, Kabiru CW, Kandala NB (2016) The effect of education on household food security in two informal urban settlements in Kenya: a longitudinal analysis. Food Secur 8:743–756
Mu Y, Li Y, Yan R, Luo P, Liu Z, Sun Y, Wang S, Zhu W, Zha X (2024) Analysis of the ongoing effects of disasters in urbanization process and climate change: China's floods and droughts. Sustainability 16(1). https://mdpi-res.com/d_attachment/sustainability/sustainability-16-00014/article_deploy/sustainability-16-00014.pdf?version=1702972369
Nagendra H, Bai X, Brondizio ES, Lwasa S (2018) The urban south and the predicament of global sustainability. Nat Sustain 1(7):341–349
Opitz I, Berges R, Piorr A, Krikser T (2016) Contributing to food security in urban areas: differences between urban agriculture and peri-urban agriculture in the Global North. Agric Hum Values 33(2):341–358. https://doi.org/10.1007/s10460-015-9610-2
Parvin F, Hassan MA, Tareq SM (2022) Risk assessment of microplastic pollution in urban lakes and peripheral Rivers of Dhaka Bangladesh. J Hazard Mater Adv 8:100187. https://doi.org/10.1016/j.hazadv.2022.100187
Rana MMP, Ilina IN (2021) Climate change and migration impacts on cities: lessons from Bangladesh. Environ Chall 5:100242
Ruel MT, Garrett J, Yosef S, Olivier M (2017) Urbanization, food security and nutrition. In de Pee S, Taren D, Bloem MW (eds) Nutrition and health in a developing world. Springer International Publishing, pp 705–735. https://doi.org/10.1007/978-3-319-43739-2_32

Shafqat W, Naqvi SA, Maqbool R, Haider MS, Jaskani MJ, Khan IA (2021) Climate change and citrus. Citrus-Res Dev Biotechnol 147
Simunic I, Likso T, Miseckaite O, Orlovic-Leko P, Ciglenecki I, Spalevic V (2019) Climate changes and soil water regime. J Agric For 65(3). https://doi.org/10.17707/AgricultForest, https://doi.org/10.17707/AgricultForest.65.3.01
Szabo S (2016) Urbanisation and food insecurity risks: assessing the role of human development. Oxf Dev Stud 44(1):28–48. https://doi.org/10.1080/13600818.2015.1067292
UN-Habitat (2016) World cities report 2016: urbanization and development-emerging futures, UN
UN-Habitat (2022) World cities report 2022: envisaging the future of cities. https://unhabitat.org/sites/default/files/2022/06/wcr_2022.pdf
UN. (2022). Inforgraphic: no hunger. Retrieved 30 Apr 2024, from https://www.un.org/sustainabledevelopment/wp-content/uploads/2023/08/2309739_E_SDG_2023_infographics-2-2.pdf
UNDESA (2018) World urbanization prospects: the 2018 revision (ST/ESA/SER.A/420). https://www.un.org/en/desa
UNDESA (2021) Decade of action: population, food security, nutrition and sustainable development (Policy Brief No 102)
Weinreb A, Stecklov G, Arslan A (2020) Effects of changes in rainfall and temperature on age-and sex-specific patterns of rural-urban migration in sub-Saharan Africa. Popul Environ 42:219–254
Wiederkehr C, Beckmann M, Hermans K (2018) Environmental change, adaptation strategies and the relevance of migration in Sub-Saharan drylands. Environ Res Lett 13(11):113003
Wei YD, Ewing R (2018) Urban expansion, sprawl and inequality. Landscape Urban Plan 177:259–265. https://doi.org/10.1016/j.landurbplan.2018.05.021
Wenban-Smith H, Faße A, Grote U (2016) Food security in Tanzania: the challenge of rapid urbanisation. Food Secur 8(5):973–984. https://doi.org/10.1007/s12571-016-0612-8

Chapter 2
Urbanization and Environmental Degradation in Khulna, Bangladesh: An Analysis of Land Use, Land Cover, and Microclimate Dynamics

Reazul Ahsan, Sharmin Ara, Rumana Asad, and Soumaya Ezazaa

Abstract Urbanization is expanding rapidly globally, particularly in developing countries within the Global South. This study explores the impacts of urbanization on land use, land cover (LULC), and microclimates in Khulna, Bangladesh, a fast-growing industrial city. By utilizing satellite imagery from 2000, 2010, and 2020, the research examines temporal changes in LULC and corresponding shifts in land surface temperature (LST) that affect the local microclimate. The analysis reveals substantial vegetation and forest cover reductions, accompanied by a notable expansion of built-up areas. These land use changes are directly associated with a significant increase in average LST over the two decades, indicating a clear link between urban expansion and rising surface temperatures. The study underscores the importance of sustainable urban planning to mitigate these negative environmental consequences, particularly in rapidly urbanizing regions like Khulna. As the city grows, strategies such as green infrastructure development and preservation of natural landscapes are critical to curbing the heat island effect and protecting local ecosystems. This case is a vital example for other cities in developing nations, where balancing urban growth with environmental sustainability is increasingly critical to ensuring long-term resilience against climate-related challenges.

R. Ahsan (✉)
Department of City and Metropolitan Planning, University of Utah, Salt Lake City, USA
e-mail: reazul.ahsan@utah.edu

S. Ara
Department of Geography, Texas A&M University, Texas, USA

Department of Urban and Regional Planning, CUET, Chittagong, Bangladesh

R. Asad
Architecture Discipline, Khulna University, Khulna, Bangladesh

S. Ezazaa
Urbanist and Architect, Carta—reichen et robert associes architectes urbanistes, Rabat, Morocco

M. G. Kibria et al. (eds.), *Urban Blue-Green Infrastructure Approach for Food Security and Climate Disaster Resilience*, Disaster Risk Reduction,
https://doi.org/10.1007/978-981-96-8746-6_2

Keywords Urbanization · Land Use Land Cover (LULC) · Land Surface Temperature (LST) · Blue-Green Infrastructure (BGI) · Khulna · Bangladesh

2.1 Introduction

Recent centuries have witnessed significant growth in global urbanization, with the percentage of individuals residing in cities projected to rise from 55 to 68% of the global population by 2050 (Esraz-Ul-Zannat et al. 2020). Regional urbanization expansion has been more pronounced in countries of the Global South compared to the Global North. An analysis of the relationship between urbanization and its social and environmental consequences, focusing on current and future global issues, reveals a troubling growth in built-up regions accompanied by a depletion of permeable soil and blue and green infrastructure. The consequences include increasing temperatures, intense flooding, precipitation events induced by climate change, and deterioration of bio-diversity (Haque et al. 2022). These modifications have an adverse impact on human health and quality of life, as well as the urbanized environment and natural meteorological patterns, which are significant global concerns. Bangladesh, a densely populated and fast urbanizing country, is not immune to these problems, most of which derive from insufficient administration of green and blue infrastructures in urban areas (Abdullah et al. 2022).

For urban dwellers, green and blue spaces offer several environmental benefits, including alleviating air pollution (Matos et al. 2019) and regulating local climate (Adnan et al. 2022); aiding in water storage and carbon sequestration (Wang et al. 2021), reducing noise (Koprowska et al. 2018), and managing the release of urban water flows. Furthermore, BGI can significantly contribute to enhance the visual aesthetics of any city (Sabbion 2018). Existence of water bodies leads to a reduction in temperature. Specifically, the existence of UBS within a distance of 30 m from the waterbody can reduce the ambient air temperature by 1–3 °C (Kleerekoper et al. 2012). almost the past five decades, almost 4000 ponds in Bangladesh's largest cities have been emptied due to unregulated landfilling driving fast urban growth (Akhtar et al. 2017).

Khulna, the third-largest industrial city in Bangladesh, is currently undergoing remarkable expansion in its unplanned urban zones. Approximately 91% of Khulna's urban environment has developed naturally and is adaptable to accommodate social and morphological changes (Sowgat and Roy 2020). Over time, a substantial percentage of new migrants settle in the outskirts of the city, therefore contributing to a massive realignment of the environment. Poor building standards and deteriorating thermal and ventilation performance, as well as the challenge of accommodating the growing number of migrants, can be attributed to insufficient construction regulations and lenient planning controls (Esraz-Ul-Zannat and Islam 2016). In addition to experiencing severe heat waves and climate change-related catastrophes, Khulna has already begun to lose its green open space and water bodies. Within the Khulna metropolitan city, around 1000 ponds of different dimensions have been flooded to increase the urban development space (Mahmud et al. 2020). The city saw a loss of 6.34% of open space between 1961 and 2012 in response to the pressures of rapid urbanization (Sowgat and Roy 2020). During the period from 1989 to 2014, demonstrated that the surface temperature within the city area fluctuated between 29.64° and 40.71°, whereas outside the city area it varied between 28.77° and 40.16° (Khan and Uddin 2015).

The primary objective of this study is to examine the effects of urbanization on the microclimate and urban landscape of Khulna, a tropical industrial city. The present study will initially investigate the analysis of land use. Land cover change (LULC) over a 20-year period from 2000 to 2020 using satellite imagery from 2000, 2010, and 2020 to examine the temporal changes in LULC and corresponding shifts in land surface temperature (LST), that affect the local microclimate and accelerate urban climate vulnerability through heat island effect. The research findings and the proposed recommendations aim to establish future development patterns that guarantee long-term sustainability.

2.2 Supporting Literature

An increasing number of cities worldwide are facing a substantial problem posed by climate change. Furthermore, this phenomenon is now widely acknowledged as a major catalyst for population movement, especially in susceptible areas where environmental hazards pose a severe threat to livelihoods and communities. Climate-induced displacements prompt individuals to seek shelter in metropolitan regions as a means of coping with environmental hazards (McLeman and Smit 2006). Urban regions, due to their enhanced infrastructure and abundant resources, attract climate-displaced persons who are in search of safety and economic prospects in the face of environmental instability (Warner et al. 2009). This coping mechanism encompasses complex dynamics. Although urban areas have the potential to provide enhanced access to services and employment opportunities, their rapid and unexpected urbanization can worsen vulnerabilities, resulting in socio-economic inequalities, insufficient infrastructure, and environmental deterioration (Neil Adger et al. 2005).

The projected impact of climate change on local livelihoods is anticipated to result in internal migration for almost 140 million individuals in Sub-Saharan Africa, South Asia, and Latin America by the year 2050. Moreover, almost 50% of the population residing along the southwest coast of Bangladesh would be forced to leave their homes only due to severe climatic events. They will already be living on the outskirts of the nearest city, namely in the peripheral areas of Khulan city (Ahsan et al. 2024a, b).

A significant percentage of these climate migrants will undergo relocation to urban areas. Internal migration from rural to urban regions is a crucial determinant of urbanization. Changes in production, particularly in the industrial sector, are the primary drivers of this migration (Montgomery 2008). Metropolitan areas offer more advantageous job opportunities, which appeal to young persons. Nevertheless, the lack of opportunities and adverse environmental conditions in rural regions drive people to settle in urban areas in search of better living conditions (Zhao 1999). Over the past few decades, this trend has led to significant growth of urban areas in countries including Bangladesh, China, India, and Indonesia. Based on estimates, over 70% of the global population is projected to live in urban areas by now. This development presents both opportunities and challenges, including significant environmental risks.

The process of urbanization amplifies climate risk and vulnerability through two primary mechanisms. From the outset, the increase in population size and the clustering of individuals in urban regions lead to an increased susceptibility to natural disasters. Urban areas often have more severe impacts from storms and flooding because of their higher population density and greater availability of important resources (Donner and Rodríguez 2008; Gu 2019). Moreover, urbanization has a substantial impact on the climatic and environmental system through the modification of population dynamics and lifestyle patterns. Enhanced consumption, waste production, and land-use changes resulting from mass migration can place strain on the local ecosystems of metropolitan areas where people relocate (Garling 1998). However, the existing research on the relationship between the rise in migrant population and its impact on the environment, specifically in regard to local air pollution and greenhouse gas emissions, is nonetheless unclear (Liang et al. 2020; Squalli 2021). The degradation of the environment, particularly resulting from the effects of migrant inflow and changes in land use patterns, can create difficulties in preserving environmental balance and greatly modify the microclimates by disrupting the balance between green and blue infrastructure. Within the context of green infrastructure, the decrease in green areas disturbs the equilibrium. The phenomenon known as the urban heat island is often observed as a result of diminished green areas and heightened impermeable surfaces (Akbari et al. 2001). A decrease in green infrastructure reduces the cooling influence of vegetation and worsens heat retention, therefore affecting local temperature patterns and air quality (Seto et al. 2012). Moreover, the encroachment of wetlands is the simultaneous cause of the damage on the blue infrastructure. Blue infrastructure, which includes water bodies and vegetated ground, is essential for controlling microclimates by facilitating cooling through evapotranspiration and minimizing stormwater runoff (Lin et al. 2017). The detrimental effects of climate displacement on the green and blue infrastructure have a substantial impact on the environmental balance and the dynamics of the microclimate.

Climate displacement issues leading to an imbalance of green and blue infrastructure are experienced by many cities worldwide grappling with climate change difficulties. The dual phenomenon of growing urbanization and the consequences of climate change pose significant challenges for Jakarta, the capital city of Indonesia. The city has undergone substantial urban growth and a surge in population as a result of climate-induced displacement, leading to the intensification of the urban heat island phenomenon. An uncontrolled expansion of urban areas and contemporary concrete planning are significant contributors to the intensification of urban heat island effects (Ahsan et al. 2024a, b). In 2005, Hurricane Katrina caused substantial displacement and modification of urban patterns in New Orleans, Louisiana,

United States. The reconstruction efforts mostly concentrated on levee systems and urban infrastructure, which had a significant impact on the dynamics of microclimate and the development of resilience techniques (Moser and Ekstrom 2010). Another instance is Dhaka, the capital of Bangladesh, a city that is extremely susceptible to climate change as a result of the rising sea levels and the monsoon season. Accelerated urbanization is propelled by climate displacement, resulting in a rise in urban heat islands. The expansion of human activities into wetlands and the decrease in natural areas have worsened the severity of temperature extremes and air pollution (Ahsan 2019; Tabassum et al. 2024).

2.3 Research Methods

2.3.1 Land-Use and Land-Cover (LULC) Change Analysis

In this study, we wanted to analyze land cover changes' spatial and temporal dynamics using The GLAD Global Land Cover and Land Use Change dataset. This dataset quantifies changes in forest extent and height, cropland, built-up lands, surface water, and perennial snow and ice extent from 2000 to 2020 at 30-m spatial resolution globally. For this study years 2000, 2010, and 2020 datasets (available in TIFF format) have been downloaded and processed for further analysis in June 2024. The given LULC dataset has 256 classes which are reclassified into six (6) major classes in Bangladesh. The six major LULC classes are Vegetation, Forest/Bushes, Water bodies, Cropland, Built-up, and Low-lying land. The classification has been validated using the ground truth points from Google Earth (78% accuracy). After reclassification, a change detection analysis has been conducted to identify the transitions in land cover between 2000, 2010, and 2020. All the processes have been done in ArcGIS Pro. Using global scale data, the approach provided insights into land cover dynamics and highlighted significant trends and patterns in land use transformation in Khulna, Bangladesh.

2.3.2 Land Surface Temperature (LST) Calculation and Change Analysis

This study also aims to assess spatial and temporal variations in Land surface temperatures (LST) for the years 2000–2020. Multi-temporal and multi-sensor Landsat satellites, e.g., Landsat 5 TM for 2005 and 2010 and Landsat 8 OLI/TIRS for 2020 images, were used to fulfill the objectives of this study (see Table 2.1). The projection system was the Universal Transverse Mercator (UTM, Zone 46 North), while the World Geodetic System (WGS)—1984 applied as a datum. The data was accessed via Google Earth Engine's data catalogue from January to April of the three selected

Table 2.1 Source of satellite data set for measuring the changes in LST in three different years (2000, 2010 and 2020)

Year	Satellite	Sensor	Thermal band	Temporal coverage
2000	Landsat 5	TM	120 m (Band 6)	January–April
2010				
2020	Landsat 8	OLI/TIRS	100 m (Band 10 and Band 11)	January–April

years because of the dry season (less cloud cover). Four images of the four consecutive months were used to calculate the average LST value of each year. The thermal infrared band of satellite images was used to demonstrate the dynamics of LST in three different years (Min et al. 2019). The details of the data are given below.

The land surface temperature was derived from the thermal band represented by Band 6. Nevertheless, Landsat 8 features two thermal TIRS bands, namely Band 10 and 11, which have a spatial resolution of 100 m. These bands have wavelengths ranging from 10.60 to 11.19 μm and 11.50 to 12.51 μm, respectively (Roy et al. 2014). The present work obtained Land Surface Temperature (LST) values from both TIRS bands (10 and 11) and subsequently computed the average LST value for these two bands (Ara et al. 2021).

2.4 Findings

2.4.1 Changes in LULC Over Three Decades

Vegetation class decreased drastically from 2000 to 2010 and slightly decreased from 2010 to 2020. About 5% of the vegetation land decreased between 2000 and 2010 and 1% between 2010 and 2020. In addition to vegetation cover, the forest cover also decreased over the study period from about 102–69 km^2. The change matrix indicates that these land cover types are predominantly converted to built-up land. On the other hand, the built-up area increased from 2000 to 2020 by 8% of the total land cover classes. From the classified maps, it is clear that rapid growth happened between 2010 and 2020. Although this is not that large, water bodies and cropland have gained areas over the period. The following table shows the changes in LULC over the last three decades in the study area. The table also presents the increase or decrease of land use and cover over time (Table 2.2).

Furthermore, the following maps illustrate land use and land cover change in the study area (Fig. 2.1).

Over the past thirty years, developed regions have expanded rapidly, accompanied by a reduction in vegetation, forest/bushland resources, and low-lying land. This suggests that the fast process of urbanization and the patterns of settlement alter the land uses in the local area, which directly affects the local microclimate. The graphic

Table 2.2 Changes in land use and land cover in three different years (2000, 2010 and 2020)

LULC type	Area in Km^2					Increase/decrease
	2000	2010	2020	2000–2010	2010–2020	
Vegetation	232.25	158.00	152.35	−74.25	−5.65	Decrease
Forest/bush	101.68	90.38	69.30	−11.30	−21.07	Decrease
Water	47.88	62.00	74.80	14.13	12.80	Increase
Cropland	665.62	688.73	680.21	23.10	−8.52	Decrease
Built up	220.14	264.11	339.02	43.96	74.91	Increase
Low lying land	184.88	189.25	136.78	4.36	−52.46	Decrease

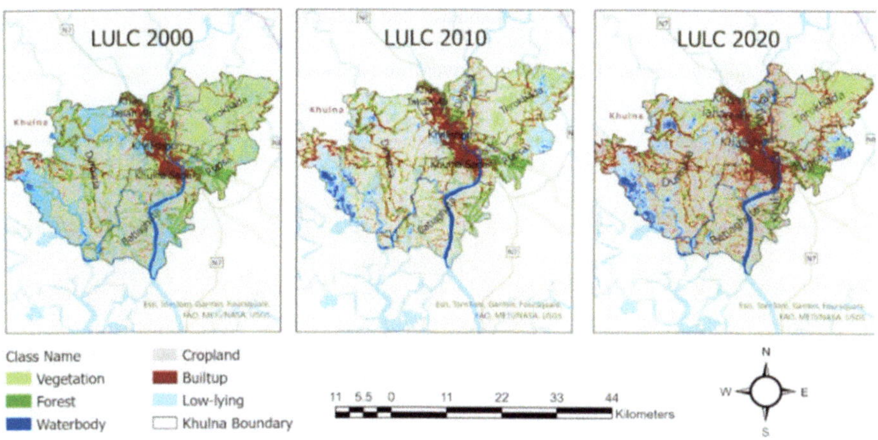

Fig. 2.1 Changes in LULC from 2000 to 2020 in Khulna and its surrounding area

below illustrates the fluctuations in land use and land cover observed over a period of three decades (Fig. 2.2).

2.4.2 Changes in LST

The mean Land Surface Temperature (LST) was computed for four consecutive months (January, February, March, and April) over the study period spanning from 2000 to 2020. The mean Land Surface Temperature (LST) values were 29.05, 31.12, and 33.95 °C for the years 2000, 2010, and 2020, respectively. An upward trend in the mean temperature was seen from the start to the year 2020. As mentioned in the preceding section, the conversion of vegetation and forest cover into built-up areas led to a rising temperature trend from 2000 to 2020 (Fig. 2.3).

Furthermore, a spatio-temporal association study between LULC and LST of three distinct years has been undertaken to understand how the LST evolves in different

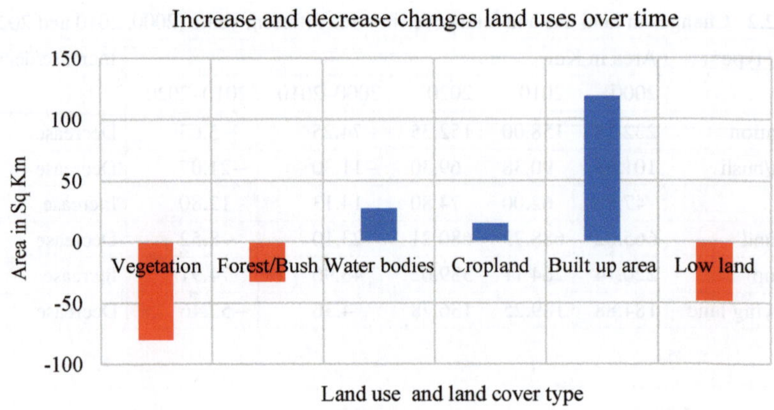

Fig. 2.2 The fluctuations in LULC observed over three decades

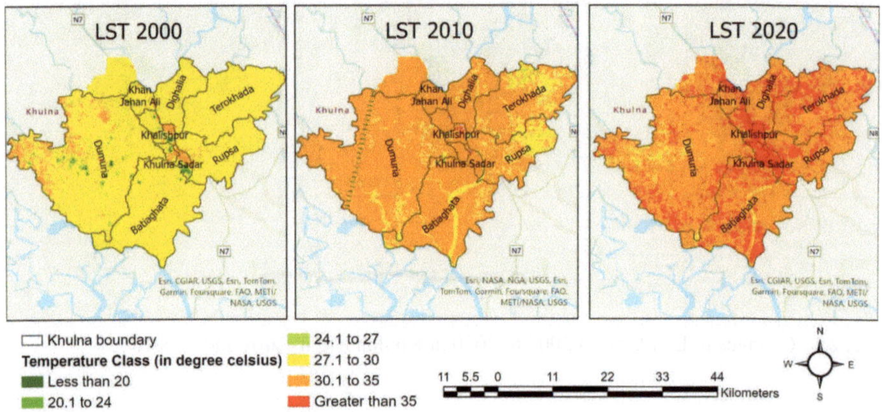

Fig. 2.3 Changes in LST from 2000 to 2020 in Khulna and its surrounding area due to LULC changes over a while

LULC categories. As the size of the city rises, the built-up area increases with time, resulting in higher average temperatures in urban areas. Moreover, deforestation is also visible in this region to accommodate the expanding population and urbanization, which is also a major cause for higher mean LST.

The following two graphs demonstrate the association between LULC and the change in LST. This unequivocally demonstrates the correlation between temperature fluctuations and the expansion of heavily populated regions. Greater urban heat island (UHI) impact is observed in areas with higher development density and lower vegetation cover. Compared to adjacent land uses and changes throughout time, urbanized zones suffer temperature rises ranging from 2 to 7 °C (or more). Such variation in average temperature alters the local microclimate and increases

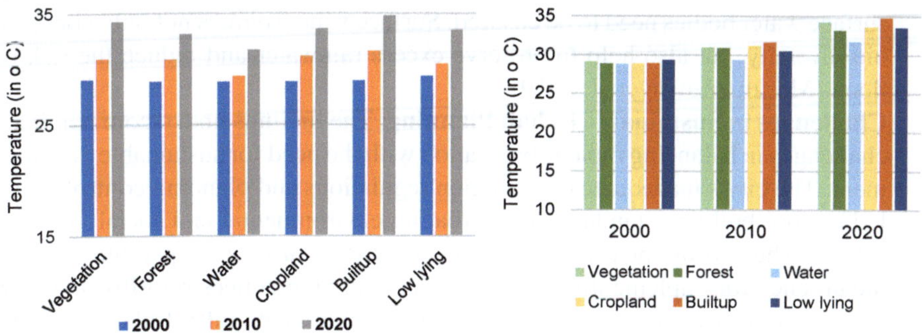

Fig. 2.4 Changes in LST due to the changes in different LULC classes over three decades

the susceptibility of the urban environment to climate change (Yang et al. 2023) (Fig. 2.4).

The figure illustrates that the presence of green vegetation, low-lying land, and water bodies has a cooling impact. In contrast, developed urban areas exhibit elevated temperatures attributed to the construction materials.

The aforementioned comparative study illustrates that the Land Surface Temperature (LST) has increased by around 5 °C in the selected sectors over the last 2020 years due to alterations in Land Use and Land Cover (LULC). Mean temperature fluctuations affect the local microclimate by modifying precipitation amounts, raising humidity, inducing droughts, restricting fresh water supply, and affecting agricultural production.

Urbanization Trends and Impact: The study highlights the rapid urbanization in Khulna, Bangladesh, driven by population growth and migration. Over the past two decades, this expansion has significantly changed land use and land cover (LULC). Specifically, there has been a marked reduction in vegetation and forest areas, with substantial portions being converted to built-up land. The built-up area increased by 8% from 2000 to 2020, illustrating the extent of urban sprawl.

1. **Environmental Consequences**: The loss of green infrastructure, particularly vegetation and forests, has contributed to environmental degradation. This includes the exacerbation of urban heat island effects, as evidenced by the increasing trend in land surface temperatures (LST). The study recorded a rise in mean LST from 29.05 °C in 2000 to 33.95 °C in 2020, correlating with the decline in natural land cover and the expansion of impervious surfaces. Tree plantation is necessary in urban areas. Otherwise, the city will transform toward a UHI very soon (Rana et al. 2022).
2. **Impact on Blue Infrastructure**: While the study notes an increase in water bodies over the period, this gain is overshadowed by the loss of other critical land covers like low-lying areas. The transformation of these areas into built-up land disrupts the natural hydrological cycles, potentially increasing the risk of flooding and further contributing to the urban heat island effect. Protection of

surface water bodies need to be ensured. Surface water sources not only conserve bio-diversity, but also help to preserve excess rainwater and reduce the risk of flood (Marufuzzaman et al. 2019).

3. **Challenges in Sustainable Urban Planning**: The findings underscore Khulna's challenges in balancing rapid urbanization with the need for sustainable development. The city's inadequate construction regulations and planning controls have led to poor building standards, exacerbating environmental and social vulnerabilities. The loss of open spaces and the increase in surface temperatures reflect the urgency for implementing better urban planning practices prioritizing green and blue infrastructure. A city-wide integrated blue-green infrastructure need to be developed for open space design, and ensure better storm water management practices. Designated spaces need to be conserved as open space to control the future encroachment of land for further development, as well as to conserve the native vegetation and bio-diversity.

2.5 Urban Policies for Addressing Land Use and Microclimate Dynamics in Khulna, Bangladesh

As previously stated, rapid urbanization in Khulna, fueled by rural-to-urban migration and economic expansion, has dramatically altered land use patterns and contributed to negative microclimatic changes. The Ministry of Local Government, Rural Development, and Cooperatives' Local Government Division (Rahman et al. 2021) proposed a set of national urban policies to address these difficulties. These policies promote balanced urbanization, reduce microclimatic consequences, and assure long-term urban growth. The authors have identified and stated the following major policy and implementation strategies for this research:

1. **Decentralized Urbanization**

Policy: Encourage decentralized urbanization by utilizing a hierarchy of urban settlements, from mega-cities to small towns, to create a balanced urban system and spatial pattern. Investments in urban services in rural areas and strengthening rural–urban linkages are essential to achieving this balance.

Implementation Strategies:

- Invest in infrastructure development and basic services in small towns and rural areas to reduce migration pressure on Khulna and other major cities.
- Strengthen transportation and communication networks to enhance rural–urban connectivity and enable equitable resource distribution.
- Develop regional growth centers to act as economic hubs, attracting investment, and employment opportunities.

2. **Urbanization and Migration**

Policy: Recognize the role of rural-to-urban migration in urbanization while managing its consequences. Properly guide migration to avoid overpopulation and overburdening of resources in cities like Khulna.

Implementation Strategies:

- Establish migration support centers in rural areas to inform and guide migrants about employment opportunities and housing in less populated urban areas.
- Implement policies for equitable resource allocation across cities to distribute migration impacts more evenly.
- Introduce incentives for industries and businesses to establish operations in smaller towns.

3. **Infrastructure Development and Urban Planning**

Policy: Establish urban planning departments in every municipality to prepare structure plans, master plans, and detailed area plans. Without local expertise, municipalities can seek support from central government agencies like UDD (Urban Development Directorate) or LGED (Local Government Engineering Department).

Implementation Strategies:

- Provide technical and financial support to municipalities for setting up urban planning units.
- Develop comprehensive urban plans with clear zoning, infrastructure layouts, and development guidelines.
- Train municipal officials and planners to ensure effective implementation of urban plans.

4. **Localized Planning for Urban Development**

Policy: Prepare detailed plans for specific sections of urban areas, focusing on rapid development and special projects. Plans should include road layouts, infrastructure networks, land use zones, and community facilities.

Implementation Strategies:

- Identify priority areas within Khulna for localized development, such as flood-prone or environmentally sensitive zones.
- Involve local communities in the planning process to ensure that development meets their needs and priorities.
- Use Geographic Information Systems (GIS) and other modern planning tools for precise and sustainable planning.

5. **Special Zones for the Urban Poor**

Policy: Allocate specific zones for low-income settlements, micro-enterprises, farmer's markets, and hawkers' stalls. Formalize and improve informal settlements to ensure better living conditions for the urban poor.

Implementation Strategies:

- Identify and designate land for special zones within Khulna's urban plans, prioritizing accessibility to jobs and services.
- Provide basic infrastructure and utilities in low-income zones to improve living conditions and promote social equity.
- Collaborate with non-governmental organizations (NGOs) and private entities to implement housing and livelihood programs for informal settlers.

2.6 Summary of Policy Implementation in Action

1. **Institutional Strengthening**: Establish well-equipped urban planning units in Khulna's municipal authorities to guide and enforce urban development policies.
2. **Stakeholder Engagement**: Engage local communities, NGOs, and private entities to ensure inclusive and participatory implementation.
3. **Resource Allocation**: Allocate adequate financial and technical resources to implement decentralized urbanization, infrastructure development, and special zone initiatives.
4. **Monitoring and Evaluation**: Use modern tools like GIS and remote sensing to monitor land use changes and assess the impacts of policies on urban microclimates.
5. **Capacity Building**: Train local government officials and planners to enhance technical expertise and ensure effective policy execution.

Khulna can accomplish sustainable urban expansion, reduce Blue-Green land degradation, and increase resilience to microclimate changes by putting these policies into action through coordinated efforts.

2.7 Conclusion

The local microclimate, denoting the geographical, atmospheric conditions inside a particular region, is significantly shaped by alterations in land use (how humans employ land) and land cover (the tangible elements present on the Earth's surface, such as flora, water, and structures). Modifying land use and land cover may substantially impact temperature, humidity, wind patterns, precipitation, and the general environmental quality. The present study reveals that the microclimate at the periphery of Khulna City has undergone significant changes due to fast urbanization and the intensifying climate change phenomenon. An important aspect contributing to urbanization in the fringe is internal migration, driven by economic considerations or coastal climatic uncertainties. The facts notwithstanding, it plays a crucial role in altering the urban atmosphere. The important elements below address changes in microclimate in Khulna's urban edge regions due to LULC changes, which impact the LST.

Khulna's findings are consistent with broader global patterns observed in rapidly urbanizing Global South regions such as Jakarta and Dhaka. These cities face similar difficulties, such as urban growth, environmental deterioration, and rising temperatures caused by climate change-induced migration. In Khulna, the lack of an organized land distribution system has resulted in the fragmentation of big agricultural holdings into small residential plots, emphasizing the critical need for smart urban growth management.

The study emphasizes the necessity of comprehensive strategies for integrated urban development and land use planning in reducing uncontrolled urban land use changes. Existing and proposed policies prioritize sustainable urbanization by addressing resource allocation, capacity building, and microclimate change mitigation techniques. These include encouraging high-density, multi-story residential projects, and wide-open spaces for community facilities.

Land use and land cover changes significantly impact microclimates, affecting both human life and ecosystems through difficulties such as heat stress, poor air quality, higher energy demands, and agricultural disturbances. To address these concerns, the study recommends sustainable land management techniques, urban planning incorporating green areas, and proactive actions to mitigate the negative effects of land use changes on microclimate. Cities like Khulna can handle the difficulties of climate change and urbanization more effectively by connecting urban growth with environmental resilience.

Land use and cover changes significantly impact microclimates, which affect human life and ecosystems. These changes contribute to various issues, including heat stress, poor air quality, increasing energy demands, and interruptions in agriculture. Addressing these concerns necessitates sustainable land management methods, urban planning with green areas, and efforts to offset the detrimental consequences of land use change on the microclimate.

References

Abdullah S, Adnan MSG, Barua D, Murshed MM, Kabir Z, Chowdhury MBH, Hassan QK, Dewan A (2022) Urban green and blue space changes: a spatiotemporal evaluation of impacts on ecosystem service value in Bangladesh. Eco Inform 70:101730

Adnan MSG, Dewan A, Botje D, Shahid S, Hassan QK (2022) Vulnerability of Australia to heatwaves: a systematic review on influencing factors, impacts, and mitigation options. Environ Res 213:113703

Ahsan R (2019) Climate-induced migration: impacts on social structures and justice in Bangladesh. South Asia Res 39(2):184–201. https://doi.org/10.1177/0262728019842968

Ahsan R, Lee D, Been C (2024) Urban forest, an ecosystem service-based solution for pollution and urban congestion. In: Bowo A, Dankers J, Timmer P, Jame E, Ahsan R, Dumrak J, Jer MM, Prajnawrdh TA (eds) Tourism heritage empowerment—visual colonial architecture & documentation, vol 1. Udayana University Press, pp 346–354

Ahsan R, Nokrek P, Asad A (2024) Climate change and long-term impact on women's reproductive health an unfolding and untold social crisis in the south coast region of Bangladesh. In Srikanth

R, Thompson L (eds) Climate justice and public health realities, responses, and reimagining for a better future, 1st ed, vol 1. University of Massachusetts Press, pp 302–321

Akbari H, Pomerantz M, Taha H (2001) Cool surfaces and shade trees to reduce energy use and improve air quality in urban areas. Sol Energy 70(3):295–310

Akhtar S, Saleem W, Nadeem VM, Shahid I, Ikram A (2017) Assessment of willingness to pay for improved air quality using contingent valuation method. Glob J Environ Sci Manage 3(3):279–286. https://doi.org/10.22034/gjesm.2017.03.03.005

Ara S, Alif MAUJ, Islam KMA (2021) Impact of tourism on LULC and LST in a coastal island of Bangladesh: a geospatial approach on St. Martin's Island of Bay of Bengal. J Ind Soc Remote Sens 49(10):2329–2345. https://doi.org/10.1007/s12524-021-01389-4

Donner W, Rodríguez H (2008) Population composition, migration and inequality: the influence of demographic changes on disaster risk and vulnerability. Soc Forces 87(2):1089–1114. https://doi.org/10.1353/sof.0.0141

Esraz-Ul-Zannat M, Abedin MA, Pal I, Zaman M (2020) Building resilience fighting back vulnerability in the coastal City of Khulna, Bangladesh: a perspective of climate-resilient city approach. Int Energy J 20(3A)

Esraz-Ul-Zannat M, Islam I (2016) A study on land use policies of khulna structure plan 2000–2020 in the light of climate change induced flood scenario. J Bangladesh Inst Plan 8(1):23–33

Garling S (1998) Immigration policy and the environment: the Washington D.C. metropolitan Area. Popul Environ 20(1):23–54. https://doi.org/10.1023/A:1023399801488

Gu D (2019) Exposure and vulnerability to natural disasters for world's cities. Department of Economic and Social Affairs United Nations

Haque MNSaroar M, Fattah MA, Morshed SR (2022) Environmental benefits of blue ecosystem services and residents' willingness to pay in Khulna city, Bangladesh. Heliyon 8(5):e09535. https://doi.org/10.1016/j.heliyon.2022.e09535

Khan S, Uddin MS (2015) Ecological stress of cities and its impact on urban growth management: a study on Khulna city. J Bangladesh Inst Plan 8(1):187–194

Kleerekoper L, van Esch M, Salcedo TB (2012) How to make a city climate-proof, addressing the urban heat island effect. Resour Conserv Recycl 64:30–38

Koprowska K, Łaszkiewicz E, Kronenberg J, Marcińczak S (2018) Subjective perception of noise exposure in relation to urban green space availability. Urban Forest Urban Green 31:93–102

Liang SA-O, Yang X, Qi JA-O, Wang Y, Xie W, Muttarak R, Guan D (2020) CO(2) emissions embodied in international migration from 1995 to 2015 1520–5851 (Electronic)

Lin P, Lau SSY, Qin H, Gou Z (2017) Effects of urban planning indicators on urban heat island: a case study of pocket parks in high-rise high-density environment. Landsc Urban Plan 168:48–60. https://doi.org/10.1016/j.landurbplan.2017.09.024

Mahmud A, Sikder S, Joardar JC (2020) Assessment of groundwater quality in Khulna city of Bangladesh in terms of water quality index for drinking purpose. Appl Water Sci 10(11):226. https://doi.org/10.1007/s13201-020-01314-z

Marufuzzaman M, Khanam MM, Hasan MK (2019) Exploring the impact of urban growth on surface waterbody area loss in Khulna city using GIS techniques. J Remote Sens GIS 8(2):262

Matos P, Vieira J, Rocha B, Branquinho C, Pinho P (2019) Modeling the provision of air-quality regulation ecosystem service provided by urban green spaces using lichens as ecological indicators. Sci Total Environ 665:521–530

McLeman R, Smit B (2006) Migration as an adaptation to climate change. Clim Change 76(1):31–53. https://doi.org/10.1007/s10584-005-9000-7

Min M, Lin C, Duan X, Jin Z, Zhang L (2019) Spatial distribution and driving force analysis of urban heat island effect based on raster data: a case study of the Nanjing metropolitan area, China. Sustain Cities Soc 50:101637

Montgomery MR (2008) The urban transformation of the developing world. Science 319(5864):761–764. https://doi.org/10.1126/science.1153012

Moser SC, Ekstrom JA (2010) A framework to diagnose barriers to climate change adaptation. Proc Natl Acad Sci 107(51):22026–22031

Neil Adger W, Arnell NW, Tompkins EL (2005) Successful adaptation to climate change across scales. Glob Environ Chang 15(2):77–86
Rahman A, Nazem NI, Committee on Urban Local Governments (2021) National Urban Sector Policy, 2021 (Draft). Government of the people's Republic of Bangladesh: Ministry of Local Government, Dhaka. Retrieved from https://urbanpolicyplatform.org/wp-content/uploads/2021/07/BANGLADESH-URB-POL-2011-NAT.pdf
Rana MS, Sarkar S, Sadat MN, Tabassum T, Fahim AU (2022) Predicting potential change in land cover and its effect on the temperature of the land surface: a case study of Khulna city corporation, Bangladesh. Environ Urban Asia 13(2):247–264
Roy DP, Wulder MA, Loveland TR, Ce W, Allen RG, Anderson MC, Helder D, Irons JR, Johnson DM, Kennedy R, Scambos TA, Schaaf CB, Schott JR, Sheng Y, Vermote EF, Belward AS, Bindschadler R, Cohen WB, Gao F et al (2014) Landsat-8: science and product vision for terrestrial global change research. Remote Sens Environ 145:154–172
Sabbion P (2018) Chapter 4.3—green streets social and aesthetic aspects. In: Pérez G, Perini K (eds) Nature based strategies for urban and building sustainability. Butterworth-Heinemann, pp 283–290. https://doi.org/10.1016/B978-0-12-812150-4.00026-4
Seto KC, Güneralp B, Hutyra LR (2012) Global forecasts of urban expansion to 2030 and direct impacts on biodiversity and carbon pools. Proc Natl Acad Sci 109(40):16083–16088. https://doi.org/10.1073/pnas.1211658109
Sowgat T, Roy S (2020) Khulna: the diversity and disparity of neighbourhoods from organic growth. Centre for Sustainable, Healthy and Learning Cities and Neighbourhoods (SHLC). https://www.centreforsustainablecities.ac.uk/wp-content/uploads/2020/10/SHLC_Research_Summary_09_KHULNA.pdf
Squalli J (2021) Disentangling the relationship between immigration and environmental emissions. Popul Environ 43(1):1–21. https://doi.org/10.1007/s11111-020-00369-z
Tabassum A, Park K, Seo JM, Han J-Y, Baik J-J (2024) Characteristics of the urban heat island in Dhaka, Bangladesh, and its interaction with heat waves. Asia-Pac J Atmos Sci. https://doi.org/10.1007/s13143-024-00362-8
Wang Y, Chang Q, Li X (2021) Promoting sustainable carbon sequestration of plants in urban greenspace by planting design: a case study in parks of Beijing. Urban Forest Urban Green 64:127291. https://doi.org/10.1016/j.ufug.2021.127291
Warner K, Ehrhart C, de Sherbinin A, Adamo S, Chai-Onn T (2009) In search of shelter: mapping the effects of climate change on human migration and displacement. CARE International, CIESIN Columbia University, UNHCR, UNU-EHS, The World Bank
Yang S, Wang L, Stathopoulos T, Marey AM (2023) Urban microclimate and its impact on built environment—a review. Build Environ 238:110334
Zhao Y (1999) Leaving the countryside: rural-to-urban migration decisions in China. Am Econ Rev 89(2):281–286. https://doi.org/10.1257/aer.89.2.281

Part II
Urban Blue-Green Infrastructure for Food Security and Climate Disaster Resilience

Chapter 3
Urban Agriculture: Current Developments and Importance in Food Security and Climate Resilience

Mohammad Saidur Rhaman, Shahin Imran, Mousumi Jahan Sumi, Newton Chandra Paul, and Mohammad Golam Kibria

Abstract Urban agriculture is an essential part of addressing the urgent demand for food security in the face of rapid urbanization. As cities burgeon and populations swell, traditional food systems strain to meet the demands of urban dwellers. Urban agriculture is a practical solution that involves using empty places within cities to grow fresh fruits, vegetables, and raise livestock. By integrating agriculture into the urban landscape, cities can reduce their reliance on distant food sources vulnerable to disruptions in supply chains. Furthermore, urban agriculture promotes resilience by decentralizing food production; ensuring communities have access to nutritious food even in times of crisis. In places with no good food options, urban farming saves the day by offering healthy, locally grown food instead of processed stuff. Moreover, it fosters community empowerment by encouraging locals to actively engage in the cultivation and distribution of food, which in turn strengthens social connections and encourages self-reliance. As the world's population becomes more concentrated in cities, urban agriculture plays an increasingly vital role in ensuring food security. It provides a sustainable and resilient solution to the difficulties of feeding growing urban populations. This chapter provides an in-depth study of the

Mohammad Saidur Rhaman and Shahin Imran have Contributed Equally.

M. S. Rhaman (✉)
Department of Seed Science and Technology, Bangladesh Agricultural University, Mymensingh, Bangladesh
e-mail: saidursst@bau.edu.bd

S. Imran · N. C. Paul
Departemnt of Agronomy, Khulna Agricultural University, Khulna, Bangladesh

M. J. Sumi
Department of Crop Botany, Khulna Agricultural University, Khulna, Bangladesh

M. G. Kibria
Department of Soil Science, Bangladesh Agricultural University, Mymensingh, Bangladesh

M. G. Kibria et al. (eds.), *Urban Blue-Green Infrastructure Approach for Food Security and Climate Disaster Resilience*, Disaster Risk Reduction,
https://doi.org/10.1007/978-981-96-8746-6_3

current state of urban agriculture practices, the obstacles they face, and their significance in promoting food security and improving resilience to climatic disasters in urban areas worldwide.

Keywords Urban agriculture · Food security · Food production · Urban population · Urban resilience

3.1 Introduction

By 2050, it is anticipated that roughly 70% of the population would live in cities which will change both the environment and human conditions (Caprotti 2018; Mamun et al. 2022). Cities worldwide face challenges due to urban growth and uncertain population expansion. In addition, climate change is causing extreme weather, affecting energy, water, and food security. These issues demand urgent attention for sustainable urban development (Erickson et al. 2013; Sarker et al. 2019). Besides, the increase in prices of essential food items has a significant impact on low-income urban households. This highlights the pressing need for more effective methods of providing financial assistance to individuals in developing nations (Dessus et al. 2008). Urban food security concerns are made worse by factors such as people moving from rural to urban areas, the environmental impact of large-scale farming, and a lack of access to safe and nutritious food (Stewart et al. 2013). As a result, urban communities globally are placing greater importance on implementing policies that safeguard agricultural land and support urban food production projects. These efforts aim to strengthen resilience, decrease climate-related risks to food supplies and boost regional economy (Buxton and Carey 2014; White and Hamm 2015).

Urban agriculture (UA) refers to the cultivation of crops and rearing of livestock in urban environments, such as cities and towns. This includes practices like using domestic waste to generate fresh and affordable food, as well as improving green spaces in metropolitan regions (Akin 2011; Zezza and Tasciotti 2010). The UA has the dual purpose of providing food for urban inhabitants and improving metropolitan environments through the expansion of green spaces (Sarker et al. 2019). It reduces the urban heat inland effect, improving local air and storm water quality, increasing pollinator populations, and providing climate mitigation services like carbon sequestration (Lovell 2010; Kulak et al. 2013; Harrison and Winfree 2015; Siegner et al. 2018). Moreover, it offers a new way for landscape architects and land use planners to work on urban development projects. In addition, community farms, rooftop gardens, allotment gardens, edible landscaping, urban woods, and other productive features in urban environments are made easier to establish in urban farming (Siegner et al. 2018). However, it has a number of advantages, including the provision of fresh produce, fruits, and other foods in close proximity to customers. This helps to solve the problem of food accessibility, which is especially important in places that lack grocery shops or markets, or "food deserts" (Whelan et al. 2002). Furthermore, it improves food security by providing producers with wholesome and culturally appropriate food directly (Gibbon 2009). Additionally, it creates revenue for locals and promotes economic vibrancy through local market sales (Deelstra et al. 2001; Brown et al. 2003; Lovell 2010).

The UA is becoming more recognized as a key to food security and poverty reduction, although it is still mostly unregulated and rarely considered in agricultural policy or urban planning. Considering this, this chapter discusses urban agriculture's current state and potential and how modern urban concerns are affecting food security. It is aimed at critically analyzing how urban agriculture may reduce food insecurity and promote sustainable urban growth. The chapter offers creative strategies and collaborative research to build resilient, food-secure cities.

3.2 Overview of the Present Condition of Urban Agricultural Approach

The UA is gaining popularity worldwide due to a growing awareness of local food production, food security, and sustainability. Innovative green urban architecture combines rooftop gardens, greenhouses, indoor farms, and other building-integrated forms, such as "ZFarming," to scale up food production (Specht et al. 2014). Utilizing cutting-edge technologies like aeroponics, hydroponics, aquaponics, and vertical farming, rooftop gardening revolutionizes urban agriculture by enabling cultivation in non-traditional spaces (Fig. 3.1). Rooftop gardening provides a sustainable solution for meeting urban food demands. In addition, techniques like organic farming, container gardens, and green roofs are used for this purpose (Quddus 2022). Again, soil-based beds and simplified hydroponic systems require minimal inputs, offering cost-effective and low-maintenance food production options. With their lightweight design, simplified hydroponic (using only water without soil) systems are especially well-suited for rooftop installations. These approaches hold promise for addressing poverty and food insecurity among low-income families (Rodríguez-Delfín et al. 2017). Besides, another method for urban farming is to set up community-managed plots where people can grow fruits, vegetables, and herbs. It has been reported that self-organized and group-organized gardens provide urban dwellers with other benefits in addition to locally grown food (Guitart et al. 2012; Lohrberg et al. 2015). Furthermore, the utilization of hydroponic or aeroponic systems to cultivate crops vertically indoors offers several benefits, including reduced pesticide usage, water conservation, and sanctuary for plants during severe weather conditions (Oh and Lu 2023). Eldridge et al. (2020) and Shamshiri et al. (2018a) suggest that this method allows for local production, reducing food supply chains and ensuring fresh, nutritious food all year.

In addition, with the advantages of time an enhancement in the strategy of urban agriculture has occurred with the progression of time. The most advanced approach to urban agriculture involves the integration of ICT (Information and Communication

Fig. 3.1 Most frequently used urban agriculture approaches in recent decades. Image credit: growing microgreens on table background; beekeeper working at bee farm; vertical aeroponic towers; community garden; hydroponic system; vertical farming with LED technology; rooftop gardening; aquaponics; and greenhouse archive at Freepik (PearPran, Premium license, https://www.freepik.com)

Technology) and LEDs (Light Emitting Diodes). This integration capitalizes on the declining expenses linked to information processing and the accessibility of cost-effective renewable energy sources. In addition, internet-connected plant factories and LED-lit greenhouses are essential for promoting effective closed plant production systems (CPPS) in the metropolitan areas (Kozai 2016).

Besides, aquaponics combines fish farming with vegetable, flower, or herb cultivation. Widely acclaimed as a sustainable food system, aquaponics is highly adaptable to urban environments (Bernstein 2013; Lennard 2009; Pollard et al. 2017). Another approach is urban beekeeping, facilitated by redesigned accessible beehives, fosters biodiversity and engages communities in sustainable food production (Devany et al. 2006). Again, Shamshiri et al. (2018b) highlighted the potential of adopting modern farming techniques in urban agriculture like greenhouse plant cultivation (GPC) and agricultural robotics technologies (ART) to enhance traditional agriculture, leading to increased yields, improved quality, and lower expenses sustainably. These methods optimize land use through advanced technology and automation while providing plants with controlled microclimate conditions, as supported by various studies (Shamshiri et al. 2018a, b; Lakhiar et al. 2020). Microgreens thrive in small indoor spaces in metropolitan areas, offering fresh, nutrient-dense young plants for cooking. Modern cultivation methods are increasing the appeal of microgreens in cooking, creating a considerable market demand (Kesarwani 2020). In this section, we briefly discussed the cutting-edge technologies to uphold the urban agriculture.

3.3 Aquaponics

Aquaponics is a form of urban agriculture that creates a symbiotic relationship between hydroponics (soilless plant cultivation) and aquaculture (fish farming). This approach has become popular due to its effective and sustainable use of resources, especially in metropolitan settings with limited land (Dos Santos 2016). On the plus side, aquaponics eliminates the need for synthetic fertilizers because fish waste naturally fertilizes plants, uses less water because it recycles water via the fish-plant system, and requires less soil (Pantanella 2018). According to studies, it is effective at yielding large quantities in confined areas, which makes it appropriate for urban settings. The high initial expenses, technical difficulty, and need on consistent water quality and temperature for the health of the fish and plants are some of aquaponics' disadvantages (Dos Santos 2016). New practitioners without specialized understanding may be discouraged by the closed-loop system's sensitivity to imbalances. Although promising, aquaponics is not always accessible in urban areas due to its need for financial commitment and careful supervision (Goddek et al. 2015).

3.4 Rooftop Garden

As a form of urban agriculture, rooftop gardening involves growing plants on building rooftops to provide fresh vegetables and green space in crowded regions. By using underutilized urban spaces, enhancing air quality, and lowering building temperatures through plant transpiration, this strategy helps to mitigate the urban heat island effect (Alhashimi et al. 2018). Additionally, by creating habitats for wildlife, supporting food crops, bees, and other small species, biodiversity can be increased (Riehn et al. 2023). Rooftop gardens can have certain drawbacks, too, such as most rooftop gardeners sometimes incurred greater expenses than returns, lacked skill, and received little financial gain from their gardens (Nitol and Hossain 2021). Extensive structural support is needed, as buildings must sustain the additional weight of water and soil. Installing and maintaining continuing care can be costly, and accessibility issues can arise, especially in high-rise buildings. Additionally, root penetration, chemical corrosion, and aggressive substances in irrigation water can cause problems with concrete (Jittin et al. 2022; Specht et al. 2014). Despite these difficulties, rooftop gardens are a practical means of enhancing the environment and producing food in an environmentally responsible manner in metropolitan areas.

3.5 Community Garden

As a type of urban agriculture, community gardens are open areas where people or organizations work together to grow food, flowers, and other plants. In addition to improving food access in urban settings, these gardens encourage social interaction and give participants a feeling of belonging and purpose (Dhar et al. 2024). Through increasing biodiversity, lowering pollution, and fostering green spaces in urban areas, community gardens benefit the environment. They also give locals educational opportunities to understand gardening techniques and sustainability (Truong et al. 2022). Additionally, it improves mental health by increasing participants' resilience, optimism, and subjective well-being (Koay and Dillon 2020). However, there are a number of obstacles that community gardens must overcome, including the time commitment required for community involvement, the scarcity of resources, and geographical disparities in support. These obstacles frequently reduce the efficacy of assistance initiatives and have an effect on the sustainability of gardens. Funding and ongoing community involvement are also necessary for maintaining these gardens, which can be difficult to maintain over time. Additionally, disagreements over resource allocation and garden management might occasionally arise from different gardeners' objectives and interests (Diaz et al. 2018). Community gardens continue to be a useful tool for advancing food security and urban greening in spite of these obstacles.

3.6 Vertical Farming with LED Technology

Using energy-efficient LED lighting to assist plant growth, vertical farming using LED technology is a cutting-edge urban agricultural technique that involves growing crops in stacked layers, frequently indoors (Panotra et al. 2024). The method uses less water, less area, and less fertilizer and pesticide. Additionally, it shortens the food supply chain, increasing the sustainability and dependability of food production (Oh and Lu 2023). Compared to traditional farming, LED technology provides precise control over light quality, duration, and intensity, enabling year-round crop production and possibly higher yields with less water consumption (Sena et al. 2024; Sumi et al. 2024). Additionally, by reducing transit distances when situated close to metropolitan centers, vertical farming can lessen the environmental effect of food production. However, because of the necessary infrastructure and technology, this approach has certain disadvantages, such as costly initial setup and operating expenses. Despite their efficiency, LED lighting can have high energy requirements, which raise environmental issues if they are not fueled by renewable energy sources. Crop diversity may also be limited by certain crops' poor adaptation to indoor or stacking habitats (Panotra et al. 2024). Although promising, vertical farming using LED technology has environmental and financial issues that must be resolved before it can be used on a wide scale.

3.7 Hydroponics

A type of urban agriculture known as hydroponics uses nutrient-enriched water as the main growing medium rather than soil to grow plants. Because of its effectiveness and adaptability to confined spaces, including rooftops, balconies, and indoor locations, this approach is widely used in urban settings. Because the closed system recycles nutrients and reduces evaporation losses, hydroponics frequently uses less water than conventional soil-based agriculture. Furthermore, because nutrients are delivered straight to roots without the need for soil buffering, hydroponics can lead to quicker plant development and larger yields (Rajendran et al. 2024). The system's disadvantages include expensive setup costs at first and a reliance on constant electricity for indoor artificial lights and water pumps, which leaves it susceptible to power outages. Additionally, hydroponic systems may be more vulnerable to waterborne illnesses that, if left unchecked, can spread quickly (Khatri et al. 2024). All things considered, hydroponics provides a sustainable urban farming method, but its advantages must be maximized via careful management and resource allocation.

3.8 Urban Beekeeping

Urban beekeeping promotes ecological consciousness, pollinates local ecosystems, builds community involvement, and strengthens the bond between city dwellers and the natural world. Furthermore, it can support responsible, self-regulated environmental behaviors in urban areas and advance socioecological justice (Sponsler and Bratman 2021). By producing local honey, wax, royal jelly, propolis, pollen and other bee products and lowering the carbon footprint involved in their transportation, urban beekeeping can support food security (Harianja et al. 2023). It also encourages community involvement and education regarding environmental management and pollinator health (Sponsler and Bratman 2021). The absence of uniform regulations, which differ greatly by area and frequently prioritize safety over biodiversity preservation and apicultural production, is one drawback of urban beekeeping. Furthermore, a lot of regulations are not supported by science, which emphasizes the need for changes to properly regulate urban beekeeping activities (Matsuzawa and Kohsaka 2021). Furthermore, bees in crowded places might cause sting anxiety, particularly in allergy sufferers, which could result in disputes between beekeepers and their neighbors (Charoenwikkai et al. 2024). All things considered, even if urban beekeeping has many advantages for sustainability and community involvement, its success depends on careful consideration of local laws and public opinion.

3.9 Aeroponics

Plants are grown in an air or mist environment using aeroponics, a cutting-edge technique for urban agriculture that eliminates the need for soil or an aggregate medium (Farran and Mingo-Castel 2006). This method can be applied vertically, making the most of the available space, making it especially appropriate for urban environments with limited space. One of the primary benefits of aeroponics is its effective use of nutrients and water; in comparison to conventional farming techniques, the misting system uses a lot less water because it distributes a nutrient solution straight to the plant roots. Aeroponics' ideal oxygenation and nutrient supply can also result in higher yields and faster plant development rates (Sahoo et al. 2024). However, setting up an aeroponic system can be difficult and expensive, and maintaining optimal conditions calls for specialized equipment and ongoing monitoring (Garzón et al. 2023). Aeroponic systems' reliance on technology also leaves them open to problems like pump breakdowns, which can cause stress or even plant death very quickly. Overall, even though aeroponics offers a viable and effective method of urban farming, careful planning and financial support are necessary for its successful use.

3.10 Greenhouse

Because they enable the development of crops in controlled surroundings regardless of the external weather conditions, greenhouses are a vital instrument in modern agriculture. However, heating, cooling, and artificial lighting are the main energy-consuming aspects of greenhouse operation. Because greenhouse agriculture enables precise control over environmental conditions, resulting in quicker growth cycles, it has the potential to improve agricultural productivity and quality (Maraveas et al. 2023; Muñoz-Liesa et al. 2022). Using hydroponic or aquaponic systems, which use less water and use less chemical fertilizers and pesticides, is another way greenhouses may support sustainability. For some urban growers, however, the initial outlay and ongoing expenses of establishing and maintaining a greenhouse can be prohibitive. Additionally, improper management of the energy needs for lighting, heating, and cooling might result in a higher carbon footprint (Maraveas et al. 2023; Otiende et al. 2024). All things considered, greenhouse agriculture has a lot to offer in terms of sustainability and production, but in order to optimize its usage and reduce its negative effects on the environment, careful planning and resource management are necessary.

3.11 Microgreens

Microgreens are edible seedlings of different herbs and vegetables that are picked shortly after the development of the first genuine leaves (Parida 2020). Because it works well in small areas like flats and rooftops, this type of urban agriculture is becoming more popular and accessible to city inhabitants. The ability to produce microgreens quickly—they are usually ready for harvest in 7–21 days—is one of their main benefits. This enables rapid turnover and ongoing output. The appeal of microgreens among health-conscious customers can also be attributed to their high nutrient density, which frequently includes higher quantities of vitamins, minerals, and antioxidants than their mature counterparts (Polash et al. 2019; Sumi et al. 2024; Xiao et al. 2015). There are a few disadvantages to take into account, though. Microgreens are prone to illnesses like damping-off and often need careful control over light and moisture levels. Furthermore, even though microgreens can be grown all year round, market saturation and rivalry from well-established suppliers may pose a threat to their economic viability (Abaajeh et al. 2023). All things considered, microgreens provide a promising future for urban agriculture due to their rapid production cycles and health advantages, but they also demand close attention to market dynamics and cultivation methods.

The aforementioned tactics are frequently employed in urban agriculture to enhance the livelihoods of urban residents. It is obvious that the surge in urban agricultural approaches, from rooftop gardening to aquaponics and vertical farming, not only addresses food security and sustainability but also fosters community engagement and innovation, promising a greener and more resilient urban future. Embracing modern technologies and collaborative community initiatives, urban agriculture holds the key to creating healthier, more self-sufficient cities worldwide.

3.12 Policies of Urban Agriculture

Urban development planning blends ecological models and new urbanism to integrate urban agriculture smoothly. This involves creating supportive policies, enforcing them through official bodies, and implementing programs. It also includes finding suitable land, distributing it according to accessibility rules, and offering incentives like tax breaks to encourage urban agriculture in master plans and zoning regulations (Mubvami et al. 2006; Sarker et al. 2019). In addition, in order to effectively manage public safety, transportation, community health, and land use, a variety of planning techniques are essential. These include site plans, master plans, local plans, neighborhood plans, and subject plans. To safeguard the welfare of both communities and people, they offer agendas for coordinating the use of both public and private land (Mubvami et al. 2006).

Recognizing urban food production as a legitimate land use ensures stability and fair access to resources. This is crucial for marginalized communities. Increasing publicly-owned open spaces for community gardens can help achieve this goal (Howe 2003; Milburn 2010). In addition, incentives are important for promoting urban farming. They help address barriers, involve the community, and make the most of alternative land use like Integrated Rooftop Gardens (Islam et al. 2021). Again, according to reports, water use policy in urban farming is crucial for optimizing benefits and overcoming local limitations and difficulties. It facilitates efficient resource allocation, promotes sustainable water management practices, and fosters synergistic relationships between the urban water sector and urban agriculture (Moglia 2014). Moreover, composting policies in urban farming are crucial for sustainable practices, recycling organic waste and reducing greenhouse gas emissions. Effective policies promote compost use as soil amendment, encouraging eco-friendly farming and potentially enhancing crop nutrition in cities (Ulm et al. 2019).

Food safety rules and policies are crucial for urban farming to produce safe and healthy food. They tackle contamination risks, promote high hygiene standards, and enforce proper storage and handling practices. These measures are vital for safeguarding public health (Kline 2009). Again, in order to reduce environmental contamination, promote public health, and encourage sustainable weed control practices, the significance of pesticide and chemical use policy in urban farming is demonstrated by the varying political interests and regulations of developed nations (Kristoffersen et al. 2008). Moreover, urban agriculture education programs for high school students in the United States show how education and community engagement policies can boost youth participation in addressing social and environmental issues. These programs help students develop leadership skills and a sense of civic responsibility. They play a crucial role in shaping future generations' understanding of sustainable practices and community involvement (Russ and Gaus 2021).

Figure 3.2 presents policies aimed at integrating urban agriculture into urban planning processes and enhancing overall city sustainability. It concludes good long-term city planning is essential for incorporating urban farming into efforts to make cities more sustainable. For urban farming to thrive, it needs official recognition, support, and incentives. Strong rules on water use, composting, pesticide management, and food safety are essential for a resilient urban farming ecosystem. Furthermore, allocating resources to educational programs and community service projects is crucial. These initiatives empower youth and instill a sense of civic responsibility. This commitment ensures a sustainable future for urban agriculture.

Again, in line with sustainable development objectives, policies that promote urban agriculture emphasize accessibility, affordability, environmental sustainability, and community involvement. Subsidies for water use are intended to lower water prices and increase the economic viability of urban farms (Duan et al. 2024). Gardeners' capacity to safely grow food on urban soils should be enhanced by soil testing and information availability (Whitzling et al. 2010). Subsidized testing is available through soil quality testing programs. This guarantees secure growing conditions, which are critical for public health and productivity. Once more, public land leasing at reduced prices improves affordable land access and makes it possible to exploit underutilized places for food production, which is particularly advantageous for low-income communities (Muraoka et al. 2018). Composting incentives

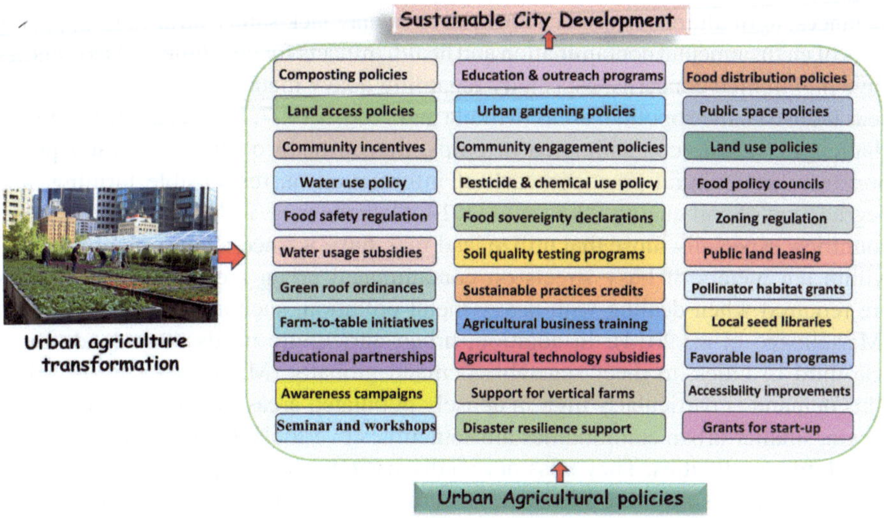

Fig. 3.2 Urban agricultural policies for sustainable city development. Image credit: urban farming in a modern city archive at Freepik (PearPran, Premium license, https://www.freepik.com)

encourage families to contribute organic waste, promoting soil health and waste reduction in order to promote sustainable practices (Widyatmika and Bolia 2023). According to Gonzales-Yanac et al. (2024), farm-to-table programs link urban farms with nearby markets, enhancing local food networks and supporting sustainable, local agriculture. Urban food security and resilience to climatic changes are fostered by grants for climate-resilient crop research, which guarantee that urban agriculture stays flexible (Haysom and Battersby 2023). To support biodiversity, pollinator habitat grants encourage urban farms to create pollinator-friendly areas, vital for plant health and ecosystem stability (Davis et al. 2017). Seed libraries foster crop biodiversity and adaptability, offering free seeds suited to local conditions (Peekhaus 2018).

The aforementioned discussion presents essential aspects of policies in urban agriculture, and the adoption and implementation of these policies can facilitate sustainable urban agriculture.

3.13 Challenges of Urban Agriculture

Urban agriculture, while promising, comes with its own set of challenges due to the unique constraints of urban environments. While, it can help address food shortages and climate change in cities, urban agriculture encounters substantial risks and limitations, mostly with agricultural productivity and site selection. Site selection in urban agriculture has a significant impact not only in environment but also in consumers. For instances, agriculture on former industrial sites may lack safety information, posing risks of environmental contamination and health impacts for consumers. Also, studies have highlighted concerns, especially regarding heavy metal accumulation in crops near industrial metropolitan areas (Mikula and Indeka 1997; Duží et al. 2014). Other challenges include contamination of crops, transmission of diseases, water pollution from agricultural practices, lack of information on responsible farming, and degradation of soil quality (Duží et al. 2014; Galhena et al. 2013). However, several countries, especially those that utilize lead fuel, have warned against planting crops within ten meters of busy streets on contaminated urban area. Special measures are required when dealing with heavy metal pollution, such as lead and cadmium (Martellozzo et al. 2014). In addition, urban agriculture needs more land, about one-third of cities, to meet vegetable demand globally (Martellozzo et al. 2014). This demand is particularly high in densely populated areas facing food insecurity. Again, smaller urban communities are sometimes overlooked, while accounting for two-thirds of all cities. They miss out on discussions about opportunities for urban farming. Addressing these disparities is crucial for sustainable urban agriculture development (Martellozzo et al. 2014). Moreover, urban crop farming's potential to improve food security, waste recycling, and poverty in developing countries is limited by land availability. To boost urban agriculture productivity and profitability, affordability and land tenure security must be addressed (Odudu and Omirin 2012).

Furthermore, urban agriculture faces challenges in managing water supply, particularly in regions undergoing both agricultural expansion and urban growth. Factors like population density in farming areas, crop varieties, and integration with urban water systems affect the sustainability of water resources for urban farming. Addressing these factors is crucial for ensuring the long-term viability of water supply for urban agriculture (Sun and Kato 2022). Other important factors to consider for successful urban farming include resource availability, efficient transportation systems, strong market connections, and effective waste disposal systems (DeKay 1997; Midmore and Jansen 2003; Sarker et al. 2019).

Table 3.1 highlights the challenges associated with integrating urban agriculture into sustainable food systems, categorized across social, economic, human health, and conservation aspects. Social challenges include time and knowledge constraints, perpetuation of inequality, and cultural insensitivities. Economic challenges involve limited crop diversity, underpaid labor, and mixed evidence on its impact. Human health concerns encompass the predominance of non-produce nutrition, pathogenic risks from poultry, and data limitations. Conservation challenges include increased water usage in arid climates, potential nutrient pollution, fear of bees and pathogen spread, emissions, and the offsetting effects of rainwater (Nogeire-McRae et al. 2018).

Nonetheless, its widespread acceptability is hampered by external restraints such as social, economic, and regulatory difficulties, as well as political barriers (Burton et al. 2013; Sarker et al. 2019; Siegner et al. 2018). It's crucial to collaborate and integrate agriculture into urban planning to support sustainable urban growth and overcome challenges. Despite obstacles, urban agriculture innovates, experimenting

Table 3.1 Challenges for urban agriculture as part of sustainable food systems

Aspect	Social	Economic	Human health	Conservation
Challenges	• Time and knowledge constraints • Perpetuates inequality • Cultural insensitivities	• Limited to fruits and vegetables • Underpaid labor supports efforts • Limited impact in some studies	• Primary nutrition isn't produced • Pathogenic risks from poultry • Data constrains	• Extra water in arid climate • Potential for nutrient pollution • Fear of bees and pathogen spread • Rises emissions • Rain water offsets

Source Modified after Nogeire-McRae et al. (2018)

with new methods, crops, and business models to improve efficiency and productivity, fostering resilient and economically viable cities.

3.14 Urban Agriculture in Food Security and Climate Disaster Resilience

Around the world, the integration of agriculture into urban design has gained significant attention in both industrialized and developing nations (Aubry et al. 2012). Urban agriculture offers significant environmental and social benefits, particularly in the realm of food justice. It addresses food access and security issues by increasing availability of fresh produce in urban areas. Additionally, it promotes health by providing nutritious food options and encourages physical activity through gardening (Allen 2008; Reynolds 2015). Urban agriculture creates opportunities for income generation and skill-building among communities, contributing to economic empowerment and self-sufficiency. Furthermore, it fosters community development and connectivity, offering spaces for social interaction and collective action toward addressing structural inequities in food systems (Horst et al. 2017; Tornaghi 2014).

Table 3.2 outlines the potential benefits of integrating urban agriculture into sustainable food systems across four key aspects: social, economic, human health, and conservation. Socially, urban agriculture promotes food access, education, and civic engagement, fostering community involvement. Economically, it reduces food costs, creates job opportunities, and enhances property values. Additionally, it positively impacts human health by contributing to nutrition through access to fresh produce and protein sources, while also conserving land, enhancing nutrient recycling, and fostering biodiversity (Nogeire-McRae et al. 2018).

A 9.3 m^2 garden's nutritional contribution in terms of different minerals and nutrients is shown in Table 3.3, along with the proportion of the daily recommended consumption that the garden alone provides. Vitamins K, C, and A are produced in significant quantities in the garden, albeit they only make up a minor portion of

Table 3.2 Potential benefits for urban agriculture as part of sustainable food systems

Aspect	Social	Economic	Human health	Conservation
Benefits	• Promotes food, education and awareness • Fosters civic engagement	• Cuts food costs • Creation of jobs • Improves property values	• Contributes to human nutrition • Chicken flocks supply protein	• Minimizes land conversion • Enhances nutrient recycling • Boosts biodiversity

Source Modified after Nogeire-McRae et al. (2018)

the daily required consumption. Although they are included in lesser amounts than the required daily intake, other nutrients like riboflavin, protein, phosphorus, and folate are also present. The garden is notably deficient in saturated fatty acids and offers nothing in the way of vitamin D. Yet, exposure to sunlight is a simple way to obtain vitamin D (Nogeire-McRae et al. 2018). Urban agriculture clearly enhances food security, especially for those opting for plant-based diets. Remarkably, just one square kilometer of urban agricultural land has the potential to sustainably feed 1482 individuals adhering to a vegan diet (Sarker et al. 2019).

However, urban agriculture, similar to food security, contributes to the resilience of cities in the face of climate disasters. It plays a vital role in solving the difficulties of climate change mitigation and adaptation, disaster risk management, and building climate resilience. It contributes to environmental sustainability on a global scale through the reduction of greenhouse gas emissions, mitigation of the urban heat island effect, and enhancement of biodiversity within cities (Ebissa et al. 2023). It also plays a crucial role in strengthening the resilience of vulnerable urban areas. It

Table 3.3 Nutritional contribution of a 9.3 m^2 garden

Minerals and nutrients	Total garden produce	Daily recommended intake provided by garden only (%)
Vitamin K (μg)	6763	0.069
Vitamin C (mg)	4571	0.062
Vitamin A (IU)	2,59,052	0.043
Riboflavin (mg)	11	0.005
Protein (g)	220	0.004
Phosphorus (mg)	6044	0.005
Folate (μg)	5171	0.01
Vitamin D (IU)	0	0
Vitamin B6 (mg)	27	0.01
Saturated fatty acids (g)	8	0.0003

Source Modified after Nogeire-McRae et al. (2018)

directly tackling problems like food insecurity and indirectly helping to reduce the negative effects of climate change-induced disasters such as flooding and temperature increases (Veenhuizen and Halliday 2019). Moreover, utilizing open urban spaces for agricultural growth and maintaining agricultural projects as part of urban green infrastructure help lower the hazards of flooding and landslides. In addition, integrating urban agriculture into comprehensive city water management plans fosters sustainable water usage and climate resilience (Dubbeling and De Zeeuw 2011).

From the above information, it is obvious that urban agriculture emerges as a critical strategy for addressing food security, environmental sustainability, and climate resilience in rapidly expanding urban environments worldwide. It makes fresh produce more accessible, lowers greenhouse gas emissions, and lessens the dangers associated with climate change, all of which make urban living more resilient and sustainable. However, there are obstacles to urban agriculture, including space limitations, water management issues, contamination dangers, and socioeconomic limitations. Strong community involvement, encouraging policies, and careful planning are necessary to overcome these challenges. Cities may increase local food access, lessen their environmental effect, and strengthen community bonds by integrating urban agriculture into their design. This strategy helps towns adjust to climate change while simultaneously bolstering local businesses. To reach the full potential of urban agriculture, coordinated efforts at all levels—local, national, and international—are required. By making cities more resilient, sustainable, and self-sufficient, urban agriculture can play a crucial part in giving urban communities a more secure future.

3.15 Conclusion

The rapid expansion of urbanization worldwide has heightened the urgency to address pressing issues such as food security, environmental sustainability, and climate resilience. Urban agriculture emerges as a critical strategy to tackle these multifaceted challenges by promoting local food production, reducing greenhouse gas emissions, and enhancing the resilience of urban communities. Despite facing significant obstacles such as limited space, soil contamination, and regulatory hurdles, urban agriculture offers a plethora of benefits across social, economic, human health, and conservation aspects. Efforts to integrate urban agriculture into urban planning and policy frameworks are essential for creating resilient and sustainable cities of the future. This necessitates the formulation of supportive policies, allocation of suitable land, and implementation of incentives to promote urban farming practices. Moreover, addressing challenges related to water management, soil contamination, and equitable access to resources is crucial for maximizing the potential of urban agriculture in enhancing food security and climate disaster resilience. Urban agriculture promotes community involvement, boosts economic empowerment, and encourages environmental responsibility, leading to healthier and more resilient cities. To take full advantages of its benefits, global, national, and local collaboration is essential in building sustainable urban futures worldwide.

References

Abaajeh AR, Kingston CE, Harty M (2023) Environmental factors influencing the growth and pathogenicity of microgreens bound for the market: a review. Renew Agric Food Syst 38(e12). https://doi.org/10.1017/S174217052300008X.; Akin U (2011) Urban agriculture: an instrument for sustainable urban development in the metropolitan city of istanbul. J Environ Protect Ecol 12:2245–2254; Alhashimi L, Aljawi L, Gashgari R, Alamoudi A (2018) The effect of rooftop garden on reducing the internal temperature of the rooms in buildings. In: The 4th World Congress on Mechanical, Chemical, and Material Engineering, Madrid, Spain. https://doi.org/10.11159/icmie18.114

Allen P (2008) Mining for justice in the food system: perceptions, practices, and possibilities. Agric Hum Values 25(2):157–161. https://doi.org/10.1007/s10460-008-9120-6

Aubry C, Ramamonjisoa J, Dabat M-H, Rakotoarisoa J, Rakotondraibe J, Rabeharisoa L (2012) Urban agriculture and land use in cities: an approach with the multi-functionality and sustainability concepts in the case of Antananarivo (Madagascar). Land Use Policy 29(2):429–439. https://doi.org/10.1016/j.landusepol.2011.08.009

Bernstein S (2013) Aquaponic gardening: a step-by-step guide to raising vegetables and fish together. New Society Publishers, Gabriola Island, Canada

Brown KH, Bailkey M, Meares-Cohen A, Nasr J, Smit J, Buchanan T, Mann P (2003) Urban agriculture and community food security in the United States: farming from the City Center to the urban fringe. Urban Agriculture Committee of the Community Food Security Coalition

Burton P, Lyons K, Richards C, Amati M, Rose N, Des Fours L, Pires V, Barclay R (2013) Urban food security, urban resilience and climate change. Nat Clim Change Adapt Res Facility Gold Coast, 160 pp

Buxton M, Carey R (2014) The use of planning provisions and legislation to protect peri-urban agricultural land. Aust Enviro Rev

Caprotti F (2018) Future cities: moving from technical to human needs. Palgrave Commun 4(1):35. https://doi.org/10.1057/s41599-018-0089-5

Charoenwikkai S, Intapun P, Lao-Araya M (2024) Bee sting injuries in Thailand's high apicultural area: outcome, risk and treatment patterns. Risk Manag Healthc Policy 17:1837–1845. https://doi.org/10.2147/RMHP.S470007

Davis AY, Lonsdorf EV, Shierk CR, Matteson KC, Taylor JR, Lovell ST, Minor ES (2017) Enhancing pollination supply in an urban ecosystem through landscape modifications. Landsc Urban Plan 162:157–166. https://doi.org/10.1016/j.landurbplan.2017.02.011

Deelstra T, Boyd D, Biggelaar M (2001) Multifunctional land use. An opportunity for promoting urban agriculture in Europe. Urban Agricul 4

DeKay M (1997) The implications of community gardening for land use and density. J Archit Plan Res 14:126–149

Dessus S, Herrera S, De Hoyos R (2008) The impact of food inflation on urban poverty and its monetary cost: Some back-of-the-envelope calculations. Agric Econ 39(s1):417–429. https://doi.org/10.1111/j.1574-0862.2008.00348.x

Devany L, Littlewood J, Geens A (2006) Urban beekeeping schemes for sustainable food production and biodiversity

Dhar S, Boruah P, Sarma S, Gogoi S (2024) Urban agriculture and community gardens. In: Vegetable science frontiers: insights of olericulture. Stella International Publication, pp 252–266

Diaz JM, Webb ST, Warner LA, Monaghan P (2018) Barriers to community garden success: demonstrating framework for expert consensus to inform policy and practice. Urban Forest Urban Green 31:197–203. https://doi.org/10.1016/j.ufug.2018.02.014

Dos Santos MJPL (2016) Smart cities and urban areas—aquaponics as innovative urban agriculture. Urban Forest Urban Green 20:402–406. https://doi.org/10.1016/j.ufug.2016.10.004

Duan Y, Zhou S, He J, Bai M (2024) Improving the performance of agricultural temporary water markets: the role of technology-based and transaction-based subsidies. Agric Water Manag 303:109062. https://doi.org/10.1016/j.agwat.2024.109062

Dubbeling M, De Zeeuw H (2011) Urban agriculture and climate change adaptation: Ensuring food security through adaptation. In: Otto-Zimmermann K (ed) Resilient cities. Springer Netherlands, pp 441–449. https://doi.org/10.1007/978-94-007-0785-6_44

Duží B, Tóth A, Bihunova M, Stojanov R (2014) Challenges of urban agriculture: highlights on the Czech and Slovak Republic Specifics, pp 82–107

Ebissa G, Yeshitela K, Desta H, Fetene A (2023) Urban agriculture and environmental sustainability. Environ Dev Sustain. https://doi.org/10.1007/s10668-023-03208-x

Eldridge BM, Manzoni LR, Graham CA, Rodgers B, Farmer JR, Dodd AN (2020) Getting to the roots of aeroponic indoor farming. New Phytol 228(4):1183–1192. https://doi.org/10.1111/nph.16780

Erickson DL, Lovell ST, Méndez VE (2013) Identifying, quantifying and classifying agricultural opportunities for land use planning. Landsc Urban Plan 118:29–39. https://doi.org/10.1016/j.landurbplan.2013.05.004

Farran I, Mingo-Castel AM (2006) Potato minituber production using aeroponics: effect of plant density and harvesting intervals. Am J Potato Res 83(1):47–53. https://doi.org/10.1007/BF02869609

Galhena DH, Freed R, Maredia KM (2013) Home gardens: a promising approach to enhance household food security and wellbeing. Agric Food Sec 2(1):8. https://doi.org/10.1186/2048-7010-2-8

Garzón J, Montes L, Garzón J, Lampropoulos G (2023) Systematic review of technology in aeroponics: introducing the technology adoption and integration in sustainable agriculture model. Agronomy 13(10):2517. https://doi.org/10.3390/agronomy13102517

Gibbon D (2009) Agriculture in urban planning: generating livelihoods and food security. Exp Agric 45(4):516–516. https://doi.org/10.1017/S0014479709990408

Goddek S, Delaide B, Mankasingh U, Ragnarsdottir K, Jijakli H, Thorarinsdottir R (2015) Challenges of sustainable and commercial aquaponics. Sustainability 7(4):4199–4224. https://doi.org/10.3390/su7044199

Gonzales-Yanac T, Nagadeepa C, Jaheer Mukthar KP, Castillo-Picón J, Manrique-Cáceres J, Ramirez-Asis E, Huerta-Soto C (2024) Minimalist farm-to-table practices: connecting consumers with local agriculture. In Alareeni B, Hamdan A (eds) Technology and business model innovation: challenges and opportunities, vol 924. Springer Nature Switzerland, pp 109–122. https://doi.org/10.1007/978-3-031-53998-5_9

Guitart D, Pickering C, Byrne J (2012) Past results and future directions in urban community gardens research. Urban Forest Urban Green 11(4):364–373. https://doi.org/10.1016/j.ufug.2012.06.007

Harianja AH, Adalina Y, Pasaribu G, Winarni I, Maharani R, Fernandes A, Saragih GS, Fauzi R, Tampubolon AP, Njurumana GN, Sukito A, Aswandi A, Kholibrina CR, Siswadi S, Kurniawan H, Hidayat MY, Wahyuni R, Koeslulat EE, Heryanto RB, Kuspradini H (2023) Potential of beekeeping to support the livelihood, economy, society, and environment of Indonesia. Forests 14(2):321. https://doi.org/10.3390/f14020321

Harrison T, Winfree R (2015) Urban drivers of plant-pollinator interactions. Funct Ecol 29(7):879–888. https://doi.org/10.1111/1365-2435.12486

Haysom G, Battersby J (2023) Urban food security and resilience. In: Béné C, Devereux S (eds) Resilience and food security in a food systems context. Springer International Publishing, pp 355–388. https://doi.org/10.1007/978-3-031-23535-1_11

Horst M, McClintock N, Hoey L (2017) The intersection of planning, urban agriculture, and food justice: a review of the literature. J Am Plann Assoc 83(3):277–295. https://doi.org/10.1080/01944363.2017.1322914
Howe J (2003) Growing food in cities: The implications for land-use policy. J Environ Plan Policy Manage 5(3):255–268. https://doi.org/10.1080/1523908032000154188
Islam E, Ahmad Khan S, Khaleque A, Ahammed S (2021) Incentives and barriers of integrated rooftop farming (IRF) in Dhaka city, Bangladesh. Ecol Environ Conserv 27:123–130
Jittin V, Tripti SR, Bahurudeen A, Hammadhu RJ (2022) Cleaner construction of durable green rooftop in residential buildings with municipal water supply. Clean Mater 5:100125. https://doi.org/10.1016/j.clema.2022.100125
Kesarwani A (2020) Microgreens—a millennial super food of urban farming 1:6–7
Khatri L, Kunwar A, Bist D (2024) Hydroponics: advantages and challenges in soilless farming. Big Data Agric 6:81–88
Kline W (2009) Food safety in urban agriculture. HortScience 44:998–998
Koay WI, Dillon D (2020) Community gardening: stress, well-being, and resilience potentials. Int J Environ Res Public Health 17(18):6740. https://doi.org/10.3390/ijerph17186740
Kozai T (2016) Why LED lighting for urban agriculture? In: Kozai T, Fujiwara K, Runkle ES (eds) LED lighting for urban agriculture. Springer Singapore, pp 3–18. https://doi.org/10.1007/978-981-10-1848-0_1
Kristoffersen P, Rask AM, Grundy AC, Franzen I, Kempenaar C, Raisio J, Schroeder H, Spijker J, Verschwele A, Zarina L (2008) A review of pesticide policies and regulations for urban amenity areas in seven European countries. Weed Res 48(3):201–214. https://doi.org/10.1111/j.1365-3180.2008.00619.x
Kulak M, Graves A, Chatterton J (2013) Reducing greenhouse gas emissions with urban agriculture: a life cycle assessment perspective. Landsc Urban Plan 111:68–78. https://doi.org/10.1016/j.landurbplan.2012.11.007
Lakhiar M, Gao J, Naz Syed T, Ali Chandio F, Hussain Tunio M, Ahmad F, Ali Solangi K (2020) School of agricultural equipment engineering, Jiangsu University, Zhenjiang 212013, Jiangsu, China. (2020). Overview of the aeroponic agriculture—an emerging technology for global food security. Int J Agric Biol Eng 13(1):1–10. https://doi.org/10.25165/j.ijabe.20201301.5156
Lennard W (2009) Aquaponics: the integration of recirculating aquaculture and hydroponics. World Aquacult 40:23
Lohrberg F, Licka L, Scazzosi L, Timpe A (2015) Urban agriculture Europe
Lovell ST (2010) Multifunctional urban agriculture for sustainable land use planning in the United States. Sustainability 2(8):2499–2522. https://doi.org/10.3390/su2082499
Mamun SAA, Islam MdM, Okely AD, Hossain MS (2022) Rapid and unplanned urbanization in the least developed districts of Bangladesh: a case study from Jamalpur using geospatial techniques. Discov Sustain 3(1):42. https://doi.org/10.1007/s43621-022-00110-w
Maraveas C, Karavas C-S, Loukatos D, Bartzanas T, Arvanitis KG, Symeonaki E (2023) Agricultural greenhouses: Resource management technologies and perspectives for zero greenhouse gas emissions. Agriculture 13(7):1464. https://doi.org/10.3390/agriculture13071464
Martellozzo F, Landry JS, Plouffe D, Seufert V, Rowhani P, Ramankutty N (2014) Urban agriculture: a global analysis of the space constraint to meet urban vegetable demand. Environ Res Lett 9(6):064025. https://doi.org/10.1088/1748-9326/9/6/064025
Matsuzawa T, Kohsaka R (2021) Status and trends of urban beekeeping regulations: a global review. Earth 2(4):933–942. https://doi.org/10.3390/earth2040054
Midmore DJ, Jansen HGP (2003) Supplying vegetables to Asian cities: is there a case for peri-urban production? Food Policy 28(1):13–27. https://doi.org/10.1016/S0306-9192(02)00067-2
Mikula W, Indeka L (1997) Heavy metals in allotment gardens close to an oil refinery in Płock. Water Air Soil Pollut 96(1–4):67–71
Milburn L-A (2010) Greening cities, growing communities: learning from seattle's urban community gardens; Hou J, Johnson JM, Lawson LJ (2009) University of Washington Press and the landscape architecture foundation 29(2):238–239. https://doi.org/10.3368/lj.29.2.238

Moglia M (2014) Urban agriculture and related water supply: explorations and discussion. Habitat Int 42:273–280. https://doi.org/10.1016/j.habitatint.2014.01.008

Mubvami T, Mushamba S, de Zeeuw H (2006) Integration of agriculture in urban land use planning. Cities Farming Future

Muñoz-Liesa J, Royapoor M, Cuerva E, Gassó-Domingo S, Gabarrell X, Josa A (2022) Building-integrated greenhouses raise energy co-benefits through active ventilation systems. Build Environ 208:108585. https://doi.org/10.1016/j.buildenv.2021.108585

Muraoka R, Jin S, Jayne TS (2018) Land access, land rental and food security: evidence from Kenya. Land Use Policy 70:611–622. https://doi.org/10.1016/j.landusepol.2017.10.045

Nitol N, Hossain M (2021) Problems and contributions of rooftop gardening: a case study of Khulna city 1–12

Nogeire-McRae T, Ryan EP, Jablonski BBR, Carolan M, Arathi HS, Brown CS, Saki HH, McKeen S, Lapansky E, Schipanski ME (2018) The role of urban agriculture in a secure, healthy, and sustainable food system. Bioscience 68(10):748–759. https://doi.org/10.1093/biosci/biy071

Odudu CO, Omirin MM (2012) Evaluating the constraints affecting land access among urban crop farmers in metropolitan Lagos. J Agribus Dev Emerg Econ 2(2):130–146. https://doi.org/10.1108/20440831211272599

Oh S, Lu C (2023) Vertical farming: smart urban agriculture for enhancing resilience and sustainability in food security. J Hortic Sci Biotechnol 98(2):133–140. https://doi.org/10.1080/14620316.2022.2141666

Otiende M, Cheruiyot JK, Opunga J (2024) Evaluating challenges and opportunities in greenhouse farming among smallholder vegetable producers in Kericho county, Kenya. East Afr J Agric Biotechnol 7(1):173–187. https://doi.org/10.37284/eajab.7.1.1866

Panotra N, Belagalla N, Mohanty LK, Ramesha NM, Tiwari AK, Abhishek GJ, Gulaiya S, Yadav K, Pandey SK (2024) Vertical farming: addressing the challenges of 21st century agriculture through innovation. Int J Environ Clim Change 14(4):664–691. https://doi.org/10.9734/ijecc/2024/v14i44150

Pantanella E (2018) Aquaponics production, practices and opportunities. In Hai FI, Visvanathan C, Boopathy R (eds) Sustainable aquaculture. Springer International Publishing, pp 191–248. https://doi.org/10.1007/978-3-319-73257-2_7

Parida S (2020) Innovative farming of edible micro greens at home and their nutritional composition. Test Eng Manag 83:17630–17640

Peekhaus W (2018) Seed libraries: sowing the seeds for community and public library resilience. Libr Q 88(3):271–285. https://doi.org/10.1086/697706

Polash MAS, Sakil A, Sazia S, Hossain A (2019) Selection of suitable growing media and nutritional assessment of microgreens. Agric Res J 56(4):752. https://doi.org/10.5958/2395-146X.2019.00116.9

Pollard G, Ward J, Koth B (2017) Aquaponics in urban agriculture: social acceptance and urban food planning. Horticulturae 3(2):39. https://doi.org/10.3390/horticulturae3020039

Quddus A (2022) Rooftop gardening in the globe: advantages and challenges. Hortic Int J 6(3):120–124. https://doi.org/10.15406/hij.2022.06.00253

Rajendran S, Domalachenpa T, Arora H, Li P, Sharma A, Rajauria G (2024) Hydroponics: exploring innovative sustainable technologies and applications across crop production, with emphasis on potato mini-tuber cultivation. Heliyon 10(5):e26823. https://doi.org/10.1016/j.heliyon.2024.e26823

Reynolds K (2015) Disparity despite diversity: social in justice in New York city's urban agriculture system. Antipode 47(1):240–259. https://doi.org/10.1111/anti.12098

Riehn JK, Fogel NS, Hathaway JN, Camilo GR (2023) Bee diversity on urban rooftop food gardens. Front Sustain Cities 5:1100470. https://doi.org/10.3389/frsc.2023.1100470

Rodríguez-Delfín A, Gruda N, Eigenbrod C, Orsini F, Gianquinto G (2017) Soil based and simplified hydroponics rooftop gardens. In Orsini F, Dubbeling M, De Zeeuw H, Gianquinto G (eds) Rooftop urban agriculture. Springer International Publishing, pp 61–81. https://doi.org/10.1007/978-3-319-57720-3_5

Russ A, Gaus MB (2021) Urban agriculture education and youth civic engagement in the U.S.: a scoping review. Front Sustain Food Syst 5:707896. https://doi.org/10.3389/fsufs.2021.707896

Sahoo SK, Dash S, Satapathy A, Samal AK, Biswal D, Mallik RR, Sahoo JP (2024) Aeroponics: the modern concept of agriculture for global nutritional security. Vigyan Varta Int E-Mag Sci Enthusiasts 5(5):123–129. https://doi.org/10.13140/RG.2.2.23414.41287

Sarker AH, Bornman JF, Marinova D (2019) A framework for integrating agriculture in urban sustainability in Australia. Urban Sci 3(2). https://doi.org/10.3390/urbansci3020050

Sena S, Kumari S, Kumar V, Husen A (2024) Light emitting diode (LED) lights for the improvement of plant performance and production: a comprehensive review. Curr Res Biotechnol 7:100184. https://doi.org/10.1016/j.crbiot.2024.100184

Shamshiri R, Weltzien C, Hameed IA, Yule IJ, I., Grift TE, Balasundram SK, Pitonakova L, Ahmad D, Chowdhary G (2018a) Research and development in agricultural robotics: a perspective of digital farming. Int J Agric Biol Eng 11(4):1–11. https://doi.org/10.25165/j.ijabe.20181104.4278

Shamshiri RR, Kalantari F, Ting KC, Thorp KR, Hameed IA, Weltzien C, Ahmad D, Mojgan Shad Z (2018b) Advances in greenhouse automation and controlled environment agriculture: a transition to plant factories and urban agriculture. Int J Agric Biol Eng 11(1):1–22. https://doi.org/10.25165/j.ijabe.20181101.3210

Siegner A, Sowerwine J, Acey C (2018) Does urban agriculture improve food security? examining the nexus of food access and distribution of urban produced foods in the United States: a systematic review. Sustainability 10(9):2988. https://doi.org/10.3390/su10092988

Specht K, Siebert R, Hartmann I, Freisinger UB, Sawicka M, Werner A, Thomaier S, Henckel D, Walk H, Dierich A (2014) Urban agriculture of the future: an overview of sustainability aspects of food production in and on buildings. Agric Hum Values 31(1):33–51. https://doi.org/10.1007/s10460-013-9448-4

Sponsler DB, Bratman EZ (2021) Beekeeping in, of or for the city? a socioecological perspective on urban apiculture. People Nat 3(3):550–559. https://doi.org/10.1002/pan3.10206

Stewart R, Korth M, Langer L, Rafferty S, Da Silva NR, van Rooyen C (2013) What are the impacts of urban agriculture programs on food security in low and middle-income countries? Environ Evid 2(1):7. https://doi.org/10.1186/2047-2382-2-7

Sumi MJ, Thamid SS, Rabbi RHM, Imran S (2024) Comparative analysis of red and green lettuce microgreens under different artificial LED lighting conditions. Arch Agric Environ Sci 9(2):230–235. https://doi.org/10.26832/24566632.2024.090205

Sun M, Kato T (2022) The effect of urban agriculture on water security: a spatial approach. Water 14(16):2529. https://doi.org/10.3390/w14162529

Tornaghi C (2014) Critical geography of urban agriculture. Prog Hum Geogr 38(4):551–567. https://doi.org/10.1177/0309132513512542

Truong S, Gray T, Ward K (2022) Enhancing urban nature and place-making in social housing through community gardening. Urban Forest Urban Green 72:127586. https://doi.org/10.1016/j.ufug.2022.127586

Ulm F, Avelar D, Hobson P, Penha-Lopes G, Dias T, Máguas C, Cruz C (2019) Sustainable urban agriculture using compost and an open-pollinated maize variety. J Clean Prod 212:622–629. https://doi.org/10.1016/j.jclepro.2018.12.069

Veenhuizen DMR, Halliday JCT (2019) Urban agriculture as a climate change and disaster risk reduction strategy (32p). Retrieved from https://www.institut.veolia.org/en/nos-contenus/la-revue-de-linstitut-facts-reports/urban-agriculture-another-way-feed-cities

Whelan A, Wrigley N, Warm D, Cannings E (2002) Life in a "food desert." Urban Stud 39(11):2083–2100. https://doi.org/10.1080/0042098022000011371

White SA, Hamm MW (2015) Urban agriculture and a planning approach to urban food systems. https://api.semanticscholar.org/CorpusID:8523215

Whitzling L, Wander M, Phillips E (2010) Testing and educating on urban soil lead: a case of Chicago community gardens. J Agric, Food Syst, Commun Dev 167–185. https://doi.org/10.5304/jafscd.2010.012.015

Widyatmika MA, Bolia NB (2023) Understanding citizens' perception of waste composting and segregation. J Mater Cycles Waste Manage 25(3):1608–1621. https://doi.org/10.1007/s10163-023-01636-5
Xiao Z, Lester GE, Park E, Saftner RA, Luo Y, Wang Q (2015) Evaluation and correlation of sensory attributes and chemical compositions of emerging fresh produce: microgreens. Postharvest Biol Technol 110:140–148. https://doi.org/10.1016/j.postharvbio.2015.07.021
Zezza A, Tasciotti L (2010) Urban agriculture, poverty, and food security: empirical evidence from a sample of developing countries. Food Policy 35(4):265–273. https://doi.org/10.1016/j.foodpol.2010.04.007

Chapter 4
Urban Agriculture as an Approach of Urban BGI and Climate Disaster Resilience

Jingyuan Xue, Liuyue He, Jiacheng Xing, Kebei Liu, and Yusen Ye

Abstract Urban Agriculture (UA) is a key strategic approach for sustaining regional ecosystems while offering various direct and indirect benefits to humanity. In the late 1990s, a high-level agricultural summit authorized the United Nations Food and Agriculture Organization (FAO) to promote urban agriculture globally. As the concept of sustainable development gained consensus, urban agriculture emerged as a vital component of urban blue-green infrastructure (BGI). It began to receive recognition for its contributions to reducing greenhouse gas emissions, mitigating the urban heat island effect, ensuring food security, enhancing climate resilience, optimizing spatial layouts, and alleviating poverty. However, the development of urban agriculture is hampered by the absence of policy support, exclusion from urban land-use planning, and the dietary choices of urban residents. Consequently, future research should be extended to investigate UA, as an approach of urban BGI to enhance climate disaster resilience, which will help foster a form of urban agriculture conducive to sustainable urban development.

Keywords Urban agriculture · Climate disaster · Urban resilience · Blue-Green infrastructure · Sustainable development

J. Xue (✉) · J. Xing · K. Liu · Y. Ye
Institute for Disaster Management and Reconstruction, Sichuan University, Chengdu, China
e-mail: jiyxue@scu.edu.cn

L. He
Ocean College, Zhejiang University, Zhoushan, Zhejiang, China

M. G. Kibria et al. (eds.), *Urban Blue-Green Infrastructure Approach for Food Security and Climate Disaster Resilience*, Disaster Risk Reduction,
https://doi.org/10.1007/978-981-96-8746-6_4

4.1 Introduction

4.1.1 Urban Agriculture (UA)

The concept of Urban Agriculture (UA) is driven by social, political, and economic factors (Ren et al. 2019). With the development of urbanization, the role of agriculture within urban settings transformed significantly, mirroring the evolution of UA itself. The accessibility of food has always been a pivotal factor in the establishment and growth of cities. The burgeoning population, coupled with advancements in crop cultivation and storage techniques, led to the emergence of the earliest urban centers. These dense social hubs marked the beginning of urban agriculture, with food cultivation taking place in and around city peripheries (Steel 2013). This era can be seen as the foundational phase of UA, embedding agricultural practices into the core of urban planning. The eighteenth century heralded a dramatic shift with the onset of the Industrial Revolution. This period fundamentally altered the self-contained nature of urban development, effectively severing the direct link between cities and their agricultural activities. Consequently, Urban Agriculture entered a new phase characterized by functional transformation. This shift redefined the relationship between urban spaces and agricultural practices, paving the way for the modern iterations of UA that we see today.

The advent of the First and Second World Wars plunged the world into periods of economic crisis and depression (Andreatta 2015; Bentley 1999), spotlighting the critical issue of food availability. This era marked a significant shift in the function of Urban Agriculture (UA), transitioning from a foundational principle of urban planning to a crucial means of food provision. Japan is often regarded as the birthplace of modern UA. In 1930, the concept of UA was introduced for the first time in the "Osaka Prefecture Agricultural Association Newspaper," which defined it as an agricultural production operation that focused not only on the cultivation of perishable vegetables but also on diversified agricultural activities, including the production of fresh milk and flowers. This definition underscored the adaptability and breadth of UA practices. In 1935, Shiro Aoka, a pioneering Japanese scholar, further elaborated on the concept in his book "Agricultural Economic Geography." Aoka described urban agriculture as specialized forms of agriculture situated in urban industrial, commercial, and residential areas, or on the outskirts of cities. He emphasized that these agricultural practices were inherently dependent on, and directly influenced by, the urban economy. Aoka's work laid the foundation for understanding the dynamic interplay between urban environments and agricultural activities, highlighting the vital role of UA in ensuring food security during times of economic hardship.

Over time, UA has become integral to modern urban planning, sustainable development, and technological innovation. Smit et al. (1992) defined UA as the cultivation of food and fuel within cities, produced directly for the market by urban farmers (Smit and Nasr 1992). In 1995, agricultural economist Zoruji Hashimoto emphasized the importance of preserving rural areas around cities, which benefit from urban infrastructure while being impacted by urbanization. In 1999, the Food and Agriculture

Organization of the United Nations (FAO) highlighted UA's presence in small urban spaces, such as open areas, gardens, and balconies, where crops and small livestock are raised for personal use or neighborhood markets. Mougeot (2000) further expanded the concept, describing UA as industries within urban and suburban areas that produce and distribute food and non-food products, utilizing local resources and serving urban populations. By 2007, the FAO described UA as a dynamic part of the urban socio-economic and ecological system. It uses urban resources, competes for land and water, is shaped by urban policies, and contributes to urban development. This modern understanding underscores UA's critical role in fostering sustainable urban growth and resilience (Table 4.1).

At present, the concept of UA remains varied and debated. While the FAO has provided a general framework, ambiguities persist regarding what qualifies as UA. Some scholars restrict UA to agricultural activities within urban boundaries, while others broaden it to include any agriculture supplying food or ecological services to urban residents, irrespective of location. Given our focus on UA's role in enhancing urban resilience to climate disasters, we define UA as agricultural activities that use limited land and resources in or around urban areas to sustainably produce food, provide ecological services, and offer adaptive solutions to climate challenges. This definition underscores UA's crucial role in mitigating climate change impacts and strengthening urban resilience, providing a clear basis for further research.

4.1.2 BGI (Blue-Green Infrastructure)

Extreme weather is recognized as a component of the human–environment coupling system. However, with climate change, extreme weather events increasingly disrupt food production, water supply, infrastructure, and ecosystems, significantly impacting human health and well-being. To address these challenges, various adaptation measures have been implemented. Blue-Green Infrastructure (BGI) represents a sustainable approach that enhances resilience to climate change while improving human health, social well-being, and environmental quality (Gentle et al. 2001; Sullivan 2008). BGI comprises natural or semi-natural systems such as forests, wetlands, rivers, parks, green roofs, green walls, and bioretention pools. These elements promote ecosystem resilience and human well-being by providing essential ecosystem services (Demuzere et al. 2014; Pauleit et al. 2011). As socio-economic development and climate change drive dramatic shifts in land use patterns, BGI must be integrated at various geographical levels. BGI planning extends beyond urban spaces to include suburban, regional, and rural areas, ensuring comprehensive and adaptive land use strategies.

UA and BGI are closely connected. The concept of UA has evolved over time, paralleling the development of BGI and the growing emphasis on urban sustainability. As UA becomes increasingly recognized as a component of BGI, it plays a crucial role in promoting sustainable urban development. Integrating UA into the construction and management of BGI can achieve multiple benefits, including

Table 4.1 Development of the Urban Agriculture (UA) concept

The development of UA	Influencing factor	Time		Proposers	Highlights
As a basic principle of urban planning	The formation of cities; The need for territorial autonomy; Population growth	Until 17th			As the basic principle of urban planning, it served the basic needs of human survival
As a function	Industrialization; The economic stratification of society; Economic migration to cities	18th–19th			Separated cities from agriculture
	World War; Social-economic depression; Insufficient resources	The first half of the 20th century	1935	Shiro Aoka	The function of UA mainly focused on the supply of food
Renaissance	Sustainable development; Development of science and technology; Changes in social values	The second half of the twentieth century till now	1992	Smit	Based on spatial vision, both urban and suburban
			1995	Zoru Hashimoto	As a dual agricultural front
			1999	FAO	Based on spatial perspective, mainly focused on urban areas
			2000	Mougeot	Focused on interaction between industry human and material resources with the city
			2007	FAO	Integrated urban agriculture with urban resources, policies and planning systems

resource conservation, environmental protection, social advantages, and economic gains. This synergy between UA and BGI underscores their combined potential to enhance urban resilience and sustainability.

4.1.3 Urban Climate Disaster Resilience

The concept of resilience originates from the Latin word "resilio," meaning "to return to the original state." Over time, resilience has been applied across various fields, from industry to medicine (Alexander 2013; Pauleit et al. 2011). In 1973, Holling introduced resilience to ecology, defining it as the ability of a society or ecosystem to absorb disturbances while maintaining its structure and function, self-organize, and adapt to stress and change (Holling 1973). Given that human-nature interactions primarily occur in cities, urban resilience to disasters has become a critical area of study (Farhad 2019). Urban resilience refers to the capacity of urban systems to maintain public safety, social order, and economic activity amidst uncertainties and disturbances (Yiwen and Jiang 2015). Currently, urban resilience theory addresses issues such as disaster and economic resilience, with a particular focus on natural or man-made disasters. As climate change leads to more frequent climate-related disasters, building resilient cities becomes essential for adaptation. The IPCC underscores the importance of creating climate-resilient urban areas to mitigate the risks of climate change (Pachauri 2014). This has led to the concept of urban climate disaster resilience, which emphasizes the long-term impact of urban systems on climate processes and their comprehensive response to extreme climate events. A climate-resilient city exhibits strong adaptability to climate change, resilience to climate disasters, and effective post-disaster recovery.

Scholars have noted that climate change induces phenomena such as rising temperatures and sea levels, leading to severe urban impacts like typhoons, rainstorms, and droughts (Kim and Lim 2016). Studying urban climate disaster resilience enables us to assess and mitigate the risks and vulnerabilities of cities to natural disasters, ultimately enhancing their sustainable development capacity. This approach not only improves resistance to climate disasters but also evaluates and adjusts the effects of environmental changes and human activities on urban systems.

4.1.4 Implications of Studying Urban Agriculture to Enhance the Resilience of Cites to Climate Disasters

With the advent of the Industrial Revolution, cities worldwide have expanded nearly 100-fold, exacerbating the issue of climate change (Wang et al. 2012). The relationship between urban development and climate change is complex. The increase in artificial structures and the impact of human activities on the environment are significant factors. Post-Industrial Revolution, large quantities of CO_2 and other greenhouse gases have been released into the atmosphere, contributing to global warming. Rapid urbanization has led to significant population growth, high consumption of natural resources, and an imbalance in natural ecological regulation, intensifying the greenhouse effect. Urbanization has also transformed cultivated land, forests, and other natural vegetation into construction land, fundamentally altering the thermodynamic properties of the surface. This shift, combined with population growth and city expansion, has resulted in the urban heat island effect—a major problem of 21st-century urbanization and industrialization (Rizwan et al. 2008). Studies indicated that the impact of climate change due to urbanization on urban thermal environments is more severe than the urban heat island effect caused by global climate change (Tam et al. 2015; Zhang et al. 2010). As urbanization progresses, the water resource carrying capacity must also improve (Li et al. 2020). Water, a fundamental resource, is crucial for material circulation and should maintain a dynamic balance with urbanization. However, the current water resource capacity in some regions does not align with urban development, leading to an imbalance between supply and demand. This imbalance can cause numerous climate and environmental problems. The challenge lies in integrating sustainable water resource management with urban development to mitigate the adverse effects of climate change and enhance urban resilience. In summary, the interaction between urban development and climate change is complex and multifaceted. Cities, as hubs of intense human activity, are major sources of greenhouse gas emissions and are significantly affected by climate change. Urban expansion exacerbates climate change through the heat island effect, pressure on water resources, and loss of biodiversity. Conversely, the adverse impacts of climate change—such as sea level rise, extreme weather events, and resource shortages—restrict sustainable urban development and threaten the well-being and safety of urban residents. This dynamic interplay underscores the urgent need for integrated strategies to mitigate climate impacts and promote urban resilience.

Today, cities face numerous disaster risks, prompting scholars to develop assessment systems to analyze the impact of events like earthquakes on urban resilience (Ainuddin and Routray 2012; Dunford and Li 2011). Similar efforts have been made to study urban water system resilience in the context of flood disasters (Kongjian et al. 2015). Given the increasing threat of climate-induced disasters, finding solutions to these challenges is urgent. BGI offers effective strategies for mitigating and adapting to climate change, providing solutions for urban floods, heat waves, and other disasters through Nature-Based Solutions (NBs). As an integral part of BGI, UA not only supports cities in coping with climate disasters but also improves the urban ecological environment and enhances resilience. Therefore, UA is a valuable tool for cities to manage and adapt to climate-related disasters.

4.2 Research on Global Urban Agriculture as a Blue-Green Infrastructure to Enhance Urban Resilience to Climate Disasters

Urban areas, vital to national development, face significant sustainability challenges due to energy-intensive activities, pollution, resource scarcity, and unstable ecosystems. Climate change exacerbates these issues, with urban regions increasingly subjected to frequent and severe climatic disturbances, anticipated to worsen in the subsequent decades (Ebru et al. 2018). Urban Agriculture (UA), a key component of Blue-Green Infrastructure (BGI), offers a comprehensive strategy to enhance urban resilience against climate-induced disasters (Mulya et al. 2023; Valente de Macedo et al. 2021). This chapter synthesizes global research on UA, exploring its contributions to urban resilience in ecological, agricultural, economic, and social dimensions.

4.2.1 Study on Improving Urban Resilience to Climate Disasters in Terms of Ecological Security in Urban Agriculture

4.2.1.1 Research on the Role of Urban Agriculture in Reducing Urban Greenhouse Gas Emissions

Agriculture contributes to greenhouse gas emissions, varying between 0 and 98% across countries, with an average of 30% (Laborde et al. 2021). According to the FAO (2019), UA offers a viable solution to climate change driven by greenhouse gas (GHG) emissions. UA is an effective strategy for reducing GHG emissions and strengthening urban resilience to climate change (Benis and Ferrão 2017; Lacour et al. 2018; Langemeyer et al. 2021). Current research highlights three primary ways UA reduces GHG emissions: (I) Reducing Transportation Emissions: By shortening the distance agricultural products travel, UA decreases carbon emissions (Benis and Ferrão 2017); (ii) Promoting Dietary Changes: UA encourages residents to shift their eating habits, increasing vegetable consumption (Caputo et al. 2021; Cleveland et al. 2017; Lacour et al. 2018); (iii) Advancing Sustainable Practices: UA supports the development of organic agriculture and low-emission animal husbandry (Ahmed et al. 2020; Panchasara et al. 2021). Through these mechanisms, UA not only mitigates GHG emissions but also contributes to the broader goal of creating more sustainable and resilient urban environments.

With population growth and urbanization, the urban–rural divide is becoming more pronounced. The distance between food producers and consumers is increasing, necessitating more extensive infrastructure for food transportation and distribution. As a result, food travels longer distances from production sites to customers' table (Pires 2012). In a study of Lisbon, Portugal, Benis and Ferrão (2017) explored the potential environmental benefits of localizing food production and consumption. They proposed two scenarios: (i) Dietary Change: Residents of Metropolitan Lisbon replace the Current Average Consumption (CAC) with the Recommended Healthy Diet (RHD) as advised by the national health authority; (ii) Local Production: Food currently sourced from abroad or elsewhere in the country is produced within the metropolitan borders of Lisbon. Using a full life cycle assessment, they quantitatively evaluated how these changes in food consumption and production structures could mitigate the environmental impact of Lisbon's food system. Their findings highlight the potential for localized food systems to reduce environmental footprints and improve sustainability in urban areas.

The results indicate that by combining dietary changes and structural changes in the food supply chain, CO_2 emissions can be reduced by 32.5% and land use demand by 33.1%. Specifically, the findings are as follows: (i) Dietary Shifts: Transitioning from Current Average Consumption (CAC) patterns to the Recommended Healthy Diet (RHD) as per national health guidelines can reduce greenhouse gas (GHG) emissions by up to 22%; (ii) Eliminating Waste: Reducing loss and waste across the food supply chain could lower GHG emissions by 8.2% under CAC and by 10.5% under RHD; (iii) Local Production: Producing food locally, with an average transport distance of 30 km, further reduces GHG emissions by 2.6% for CAC and 3.5% for RHD. In the CAC scenario, animal-based products (milk, dairy products, meat, fish, and eggs) constitute 36% of the diet but account for 69.3% of GHG emissions. By contrast, the RHD reduces the consumption of these products to 23%, resulting in their GHG emissions share dropping to 52.8%. Eliminating loss and waste across the entire food supply chain has a more significant impact on reducing GHG emissions than merely decreasing transportation distances. Animal-based products, Animal-based products, which represent 28% of total food supply chain losses, are the largest contributors to this reduction, accounting for a 5.42% reduction in GHG emissions under CAC and 8.84% under RHD.

Vávra et al. (2018) analyzed the rate of household food self-sufficiency in the Czech Republic, and the results indicated that food self-supply has great potential to promote sustainable development. This pattern of food supply helps to reduce GHG emissions by 42–92 kg of carbon dioxide equivalent per person per year (Vávra et al. 2018). This amount represents 2.5–5.4% of food-related GHG emissions and 0.4–0.8% of total national GHG emissions in the Czech Republic (the range depends on including the impact of garden management). Moreover, the positive environmental benefits of autonomous production are not negatively offset by the overuse of industrial fertilizers or the use of cars for transport to the garden. Using land use planning data for the next 20 years in Goiania and Brazil, Marcal et al. (2021) calculated that 10,650 hectares of UA in the Goiania region would store 26 tons of CO_2 per year over the next 20 years (Marçal et al. 2021). Cleveland et al. (2017) conducted a life cycle assessment in Santa Barbara County, California, USA, to model the potential of Alternative Household Vegetable Gardens (AHHVGs) to mitigate global climate change by reducing Greenhouse Gas (GHG) emissions.

The model consists of four key components: (i) Reduction of Lawn Area: Replacing traditional lawns with AHHVG; (ii) Vegetable Production: Growing vegetables in AHHVGs, thereby reducing reliance on vegetables from traditional agri-food systems; (iii) Sewage Irrigation: Utilizing sewage for irrigation, reducing the amount of sewage processed by treatment facilities; (iv) Composting Organic Waste: Using household organic waste for composting, thereby decreasing the amount sent to treatment facilities. The results showed that simulated AHHVGs could reduce CO_2 emissions, by more than 2 kg per kilogram of vegetables grown.

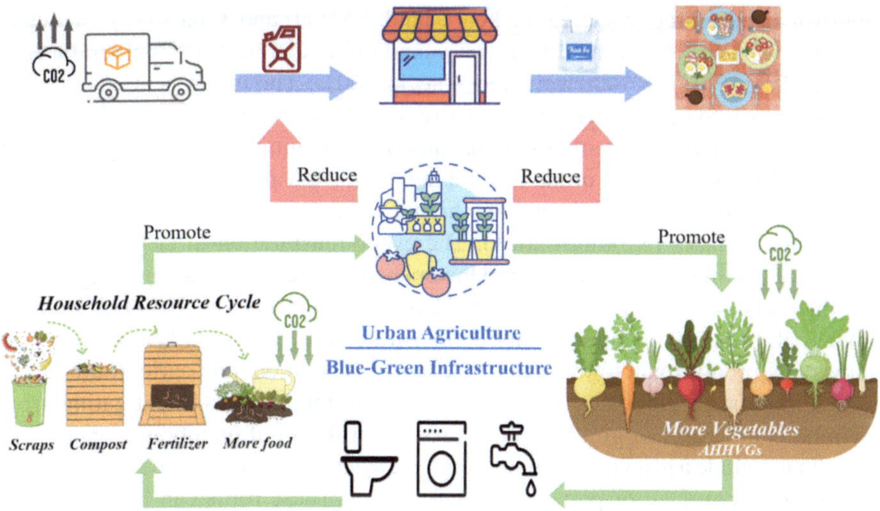

Fig. 4.1 The role of urban agriculture in reducing greenhouse gas emissions

Overall, UA has significant additional potential to reduce the environmental impact of urban food systems by minimizing food loss and waste and reducing transportation distances. However, as Benis and Ferrão (2017) point out, there are areas needing improvement: (i) Land utilization (Fig. 4.1).

Currently, UA supplies only a small portion of the Lisbon metropolitan population due to limited land area. Improving the intensive use of urban land for agriculture is crucial. One promising solution is Integrated Farming in Buildings (BIA), which employs high-performance hydroponic greenhouses on rooftops or building facades (Puri and Caplow 2009). This method not only enhances renewable energy and efficient water use but also supports waste management cycles. BIA systems can reduce fossil fuel consumption by cutting transportation distances and improve building energy efficiency, thereby reducing the overall environmental impact of urban food and energy consumption (Gould and Caplow 2012); (ii) Energy Intensity: High-performance hydroponic greenhouses are energy-intensive. Therefore, it is essential to balance the positive environmental impacts of these technologies with their potential negative effects in practical applications.

4.2.1.2 Research on Alleviating Heat Island Effect of Urban Agriculture

When artificial surfaces like concrete, asphalt, and brick replace natural surfaces that regulate temperature through evaporation and shading, they absorb short-wave solar radiation during the day and re-radiate it as long-wave radiation at night. This process raises the air and surface temperatures of urban areas compared to surrounding suburban and rural regions, creating heat islands (Akbari and Taha 1992). In megacities like New York City, the urban heat island effect can increase temperatures by 3.6–7.2 °F (2–4 °C) (Rosenzweig and Solecki 2001). The urban heat island effect leads to increased energy demand, heat stress, diseases related to air pollution, and excessive emissions of greenhouse gases and other pollutants. Mitigating this effect is essential for sustainable urban development, and UA plays a crucial role in this regard (Ellis 2021). UA surfaces have higher reflectivity of sunlight compared to impervious surfaces (Kirkhorn 2021). The evapotranspiration of crops and soil in UA reduces ambient temperatures and lowers ozone levels in the air, helping to mitigate the global urban heat island effect (Habeeb 2017; Qiu et al. 2013). A study by Koko et al. (2021) in a Nigerian city demonstrated that the urban heat island effect is inversely proportional to the area of wetlands, vegetation, and agricultural land.

Green roofs, a significant form of UA, offer multiple benefits in mitigating urban environmental issues. A green roof consists of a waterproof membrane topped with layers that include a growing medium, drainage system, and root protection, allowing vegetation to cover all or part of the roof surface. These roofs can be categorized into "extensive" green roofs, which are lightweight and planted with hardy, drought-resistant plants to minimize weight, cost, and maintenance, and "intensive" green roofs, which support grass, flowers, trees, shrubs, or crops (Rowe and Whittinghill 2012). Green roofs have been shown to significantly reduce urban temperatures. Large urban green spaces can decrease temperatures by up to 7 °C, while smaller areas, like green roofs, can reduce temperatures by up to 3 °C (Smith and Roebber 2011; Zupancic et al. 2015). Rosenzweig et al. (2006) conducted extensive studies on green roofs in New York City, collecting and modeling data at various scales—from individual buildings to entire neighborhoods. Their research involved developing a methodological framework to support comprehensive data collection and modeling. They established a green roof research station in New York City and used data from Penn State's Green Roof Research Center to calibrate and validate roof energy balance models. These models demonstrated that green roofs significantly reduce the heating and cooling demand of buildings compared to standard roofs. Rosenzweig (2006) also analyzed rainfall and runoff data from Penn State University, comparing it with local weather data from New York to understand green roofs' impact on water management. They employed satellite imagery, GIS software, and local weather data to study the effect of green roofs on the urban heat island effect (Fig. 4.2).

The results suggest that widespread adoption of green roofs could significantly address multiple environmental and human health issues in New York City, such as the urban heat island effect, global climate change, and stormwater runoff. If 50% of the city's roofs were green, average surface temperatures could be reduced by 0.8 °C. This cooling effect mitigates the negative impacts of urban heat islands, including reduced heat stress and heat-related mortality, improved air quality, and lower energy consumption for air conditioning. Santamouris (2014) reviewed studies on improving reflectivity and reducing urban temperatures using green roofs, highlighting their effectiveness on a citywide scale. The findings indicate that for every 0.1 increase in urban albedo, the average ambient temperature can decrease by approximately 0.3 K, and peak ambient temperatures can decrease by about 0.9 K. Modeling

Fig. 4.2 The role of urban agriculture in mitigating heat island effects

studies suggest that green roofs could reduce average ambient temperatures by 0.3–3 K when implemented on a large scale. Green roofs cool surfaces through soil evaporation, plant transpiration and vegetation shading. Simulations indicate that the indirect cooling effect of evapotranspiration is surpasses the direct effect of shading (McPherson et al. 1994). Research by Broadbent et al. (2018) further reveals that temperature reduction primarily depends on species composition and leaf area, and can vary with irrigation intensity—higher irrigation leads to greater cooling effects (Broadbent et al. 2018). This demonstrates that UA can be an effective strategy for energy conservation, improving human health, and mitigating air quality issues (Rosenzweig et al. 2006). However, several challenges remain before green roof technology can be widely incorporated into urban agriculture. These include reducing installation costs, assessing roof carrying capacities, and developing appropriate management methods to ensure sustainable ecological services. Addressing these issues is crucial for the successful implementation and broader adoption of green roofs in urban settings.

4.2.1.3 Research on Urban Agriculture in Flood Control

"Blue and Green" sponge cities offer a promising approach to urban stormwater management, enhancing the resilience of cities to climate change. This strategy combines urban water disaster risk management with reliable, crisis-proof urban water supply services. Key features include improving the water permeability of urban areas, preserving green and blue spaces, enhancing infiltration facilities, and creating green roofs and facades. By providing temporarily available emergency spaces and water emergency routes, sponge cities contribute to the goals of "inclusive, safe, resilient, and sustainable cities" (SDG 11) and "clean water and sanitation" (SDG 6). This approach presents a unique opportunity to advance the United Nations Sustainable Development Goals (SDGs) (Köster 2021).

UA, a key component of urban blue-green infrastructure, provides essential ecosystem services, including flood buffering and regulation (Puppim de Oliveira and Ahmed 2021). Lawson et al. (2014) explored the interdisciplinary aspects of urban blue-green infrastructure design, revealing that urban planners and scientists favor local-scale interventions like urban parks, greenways, and constructed wetlands over traditional drainage systems for effective flood management (Lawson et al. 2014). In Latin America and the Caribbean, "green roofs" are frequently cited for flood and runoff control (Mora-Melià et al. 2018). For instance, the government of Goiania, Brazil, has implemented UA to manage flooding and address the city's water shortages (Marçal et al. 2021) (Fig. 4.3).

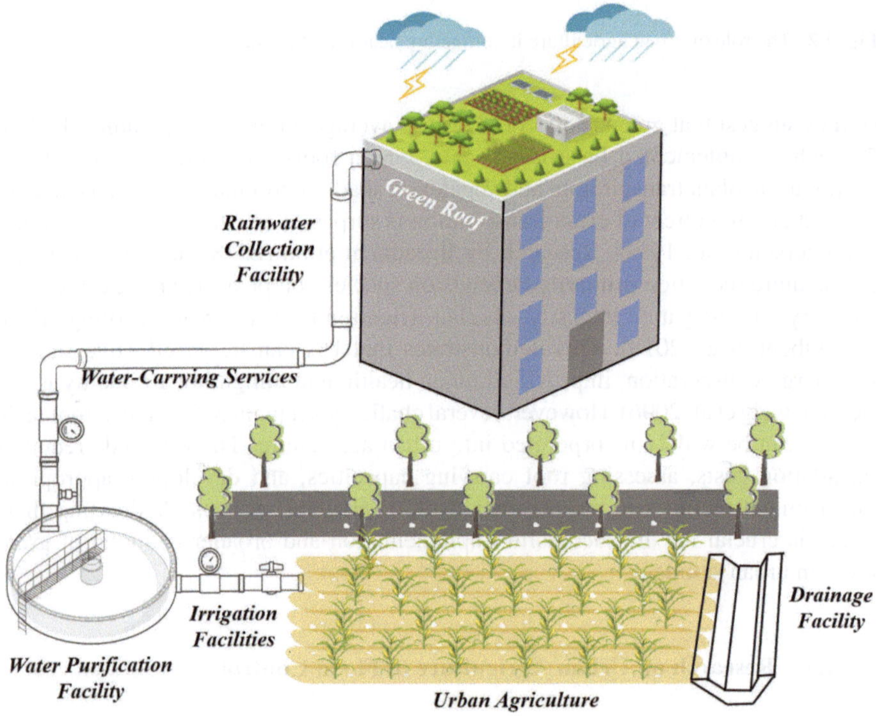

Fig. 4.3 The role of urban agriculture in flood control

Abass et al. (2020) used geographic information technology to analyze land use/cover change, actual flooding, and property damage in the Kumasi region of Ghana from 1986 to 2016. They interviewed nine key informants from various organizations, including the National Disaster Management Organization (NADMO), the City, Municipal and District Councils (MMDAs), the Ministry of Town and Country Planning (T&CPD), the Ghana Meteorological Service (Gmet), the Department of Hydrology (HSD), the Water Resources Commission (WRC), and three traditional authorities. The interviews focused on four main areas: (i) the nature and effects of floods; (ii) the impact of urban expansion and Land Use/Cover Change (LUCC) on flooding; (iii) other flood-causing conditions; and (iv) flood control measures. To analyze rainfall data, the researchers used MAKESENS and time series diagrams, employing the Mann–Kendall test (Z) and Sen's slope (Q) with an α level set to 0.05 to reveal trends and patterns. Their findings indicated a significant increase in urban construction areas from 23.34% in 1986–77.56% in 2016 (an increase of 54.22%). Conversely, green spaces (arable land and vegetation) decreased from 76.66 to 22.44% (a decrease of 54.22%). This expansion of impermeable surfaces at the expense of green spaces has gradually increased the urban flood rate. The study revealed that urban population growth leads many residents to use any available space for residential and commercial activities, including flood plains, swamps, and water-logged areas. These environmentally hazardous areas often lack adequate disaster prevention infrastructure, amplifying the destructive power of floods. The increase in flood rates in Kumasi is thus attributed to a combination of factors, with the reduction in urban green space playing a significant role. To address uncontrolled urban sprawl and its associated flooding challenges, adopting and executing smart growth policies is essential. Effective flood control and mitigation in Kumasi also require the control of land use and the expansion of green areas through legislation and enforcement. Implementing sponge cities requires innovative solutions and extensive stakeholder engagement (Zhou et al. 2019). Public awareness and acceptance are crucial for facilitating stakeholder participation in both the decision-making and implementation processes of blue-green sponge cities (O' Donnell et al. 2018).

4.2.1.4 Research on Biodiversity Enhancement of Urban Agriculture

Biodiversity threats and extinction rates have surged nearly 100-fold since the twentieth century. Maxwell et al. (2016) quantified threats for 8,688 near-threatened or threatened species on the IUCN Red List, revealing that urban sprawl and large-scale agriculture have the most significant impact on biodiversity. UA plays a crucial role in urban biodiversity, encompassing organic foods, plants, animals, and inorganic environments (Contesse et al. 2018; Lin et al. 2017; Zasada et al. 2020). Polling et al. (2016) found that 50% of UAs in 10 European countries are actively promote biodiversity conservation, a practice closely tied to food security.

Montero et al. (2021) studied vegetation and biodiversity changes over the last 20 years by applying landscape ecology indicators to REDD + land cover maps from

1986, 2001 and 2014, and statistically testing these against plant and bird species richness recorded by INBio. The results showed that monoculture and urban expansion fragmented tropical forests, reducing the region's ecological connectivity by 13% and negatively impacting coffee agroforestry. Montero et al. (2021) suggested that enhancing ecological connectivity through agroecological approaches and land use planning, such as preserving green belts in coffee plantations as urban buffer zones, is essential to address these environmental issues. In 2014, Zasada et al. (2020) conducted a survey among 111 gardeners in a Pune, India, using a composite indicator analysis framework. Their findings indicated that protecting and constructing home gardens contribute significantly to urban biodiversity conservation and the aesthetic value of urban green spaces. Consequently, these aspects should be integral to urban planning, design, and management. Salomon et al. (2020) examined soil health and mycorrhizal fungi in urban agriculture systems across 12 sites in the Adelaide metropolitan area, Australia. They analyzed soil samples for chemical and biological parameters, including seven metals, total phosphorus, plant-available phosphorus, and available nitrogen. The study found that all soil metal concentrations were below national guidelines, with plant nutrients derived from urban waste streams supporting plant nutrition and reducing external nutrient inputs. Additionally, plant diversity was high, indicating that UA can support urban biodiversity conservation. The primary goal of urban agriculture is not to enhance biodiversity; however, this provision of ecosystem services for biodiversity conservation is an added benefit (Clucas et al. 2018; Yearbook 2013).

UA contributes significantly to the diversity of plant species, which can enhance the physical health of people, particularly children, and promote healthy living. Many UA farmers prefer to plant exotic vegetation. Colding et al. (2006) found 440 plant species in a 400 m^2 area in Stockholm, while Loram et al. (2008) documented over 1,000 species in 267 private gardens in London. Clarke (2014) identified 278 species in 104 home gardens in Los Angeles, and Lin et al. (2017) noted 16 locally unavailable Asian vegetable crops in Toronto. In Accra, 60% of farmers grow exotic vegetation (Abraham et al. 2006). Urban residents increasingly seek fresh and diverse products from UA (O'sullivan et al. 2019). From a dietary perspective, Luna-Gonzalez and Sørensen (2018) suggested that UA may enhance nutritional diversity and availability, contributing positively to healthy living and improved anthropometry in children. However, they also note that other factors, such as improved sanitation, family planning, nutrition education, and women's empowerment, play significant roles in enhancing diet and nutrition.

Not all studies conclusively show that UA positively affects urban biodiversity. Clucas et al. (2018) defined UA strictly as areas where agricultural products are exclusively grown for human consumption. After reviewing 148 papers from 2000 to 2017, they found only 24 meeting this definition, with 18 addressing both UA and biodiversity. Among these, some studies indicated an increase in biodiversity compared to urban open spaces, while others found no significant difference. The focus of these studies was primarily on plants and invertebrates, and the research was largely concentrated in North America. Clucas et al. (2018) emphasized that the general assumption of UA's positive impact on urban biodiversity requires further investigation. More research is needed across a broader geographical scale, particularly in tropical regions and on a wider range of species and taxa, such as reptiles, birds, and small mammals.

4.2.1.5 Research on Resource Recycling of Urban Agriculture

Resource recycling involves converting waste into new products, playing a crucial role in modern waste reduction and environment sustainability (Kreiger et al. 2014). Currently, cities primarily follow a linear metabolism, characterized by high-quality resource input and low-quality waste output, with limited recycling (Kennedy et al. 2007; Wielemaker et al. 2018). In contrast, circular urban metabolism aims to reduce external resource input and internal waste output by recovering and reusing resources within or between urban functions (Agudelo-Vera et al. 2012). Two emerging strategies in urban resource management are New Sanitation (NS) and UA, both gaining global attention (Metson and Bennett 2015; Mougeot 2006). Integrating these functions could enhance sustainable urban resource recycling, offering a path toward more resilient and eco-friendly cities (Wielemaker et al. 2018).

NS is a framework designed to enhance urban resource management by recovering essential resources from organic waste and wastewater, thereby promoting sustainable food production. Wielemaker et al. (2018) studied Rotterdam using the Urban Harvest Approach (UHA), a type of NS, to assess how well the nitrogen supply from urban water cycles could meet the nutrient demands of urban agriculture (UA), particularly for nitrogen, phosphorus, and organic matter. Their findings indicated that UA could reduce its nutrient and organic matter needs by 65–85%, achieve 100% self-sufficiency in phosphorus, and partial self-sufficiency in nitrogen and organic matter. Weinder and Yang (2020) found that UA significantly relies on solid waste for producing fruits, vegetables, and nutrient supply. In Glasgow and Lyon, UA absorbed 51.7 and 16.9% of food waste, respectively. Another study demonstrated that using solid waste in integrated Rooftop Greenhouse (i-RTG) yielded higher yields of tomatoes and lettuce, especially the former (Manríquez-Altamirano et al. 2020). They recommended using solid waste as ecological materials for such purposes. Rufí-Salis et al. (2020) reported that during a tomato cultivation cycle in an i-RTG in Barcelona, 447 g of phosphorus, 612.8 g of potassium, 5,603.8 g of calcium, and 402.5 g of magnesium were recovered. These nutrients, if not reclaimed, could otherwise contribute to environmental issues like phosphorus-induced eutrophication. Thus, the integration of NS and UA presents a promising approach to achieving urban circular metabolism and sustainable development.

Despite the potential benefits, several challenges need to be addressed to achieve effective resource recycling in urban agriculture. Key concerns including ensuring the, safety and technical feasibility of using recycled products as fertilizers. It is crucial to study and mitigate the risks caused by heavy metals, micropollutants, drugs, and pathogens present in recycled materials (Renner 2000; Singh and Agrawal 2008). These contaminants currently hinder the full realization of circular metabolism in urban settings (de Wilt et al. 2016; Decrey et al. 2011; Uysal et al. 2010). In the

Netherlands, the agricultural use of sewage sludge is restricted due to its heavy metal content (Renner 2000). However, research by Tervahauta et al. (2014) reported that only copper (Cu) and zinc (Zn) levels in blackwater sludge exceed Dutch standards, primarily due to food intake. They concluded that blackwater sludge could be used as fertilizer to support cyclic metal metabolism. To overcome these challenges, it is necessary to develop contaminant removal techniques and establish treatment and reuse protocols that minimize risks (Wielemaker et al. 2018). Since January 2015, the Dutch regulation have permitted the use of struvite, a "recycled phosphate," as a fertilizer, provided it meets guidelines for heavy metals, pathogens, and microcontaminants (van der Grinten et al. 2016).

4.2.2 Research on How Urban Agriculture Can Improve Urban Resilience to Climate Disasters in Terms of Food Security

Currently, 15% of global food production is produced by urban agriculture (Armar-Klemesu 2000; Smit et al. 1996). Despite its limited scale and technology sophistication, UA primarily serves the purpose of self-consumption. It involves the cultivation of various fruits and vegetables, predominantly in gardens and other urban green spaces, and plays a crucial role in improving food security for urban residents, especially in high-poverty areas (Brown and Jameton 2000). In regions such as Bangladesh and India, urban farming, particularly rooftop farming, is increasingly recognized as an effective strategy to address urban land shortages and enhance food security (Safayet et al. 2017; Zinia and McShane 2018).

From mid-November 2011 to early January 2012, Bangkok experienced its worst flooding in nearly 70 years, leading to significant food supply disruptions. Despite Thailand's status as an agricultural exporter, the floods caused short-term food insecurity in the capital. The flooding severely impacted production and transportation chains managed by food companies based in Bangkok. Factories and major roads were flooded, hindering market access for city residents and resulting in higher prices in both modern trade and traditional local markets. The largest vegetable distribution market was also affected, leading to a scarcity of vegetables. In response to these challenges, the citywide UA Policy Network played a crucial role in addressing food insecurity, particularly among the most vulnerable groups. Established in 2010 under the Urban Farm Program, this network aims to create an alternative urban food system in Bangkok. The program is led by the Health Promotion Foundation, a public organization under the Prime Minister's Office, in collaboration with NGOs such as the Sustainable Agriculture Foundation, the Media Centre for Development, and the Food for Change Task Force. It seeks to engage diverse groups, including the urban poor and marginalized citizens. Since 2010, the Urban Farm Project has supported more than 120 community gardens in inner-city Bangkok, Network members have also organized green markets and community-supported farming systems to enhance food security and resilience (Boossabong 2017).

UA plays a crucial role for maintaining health, life safety and social stability of residents, especially during disruptions in external food supply chains. This importance is heightened by global climate change and frequent extreme events. By 2030, nearly 670 million people in sub-Saharan Africa are expected to live in urban areas, increasing the demand for food, water, energy, sanitation, and urban services (UN 2019). Accra, one of Africa's more modern cities and home to 70% of Ghana's industry, illustrates the historical and ongoing significance of UA. During World War II, Accra's UA provided fresh food for the Allied forces. European colonists practiced various forms of farming, planting vegetables in public spaces and gardens (Asomani-Boateng 2002). Post-war, the colonial government restricted on urban agriculture and livestock raising, citing disease concerns—a stance maintained after independence. UA experienced a resurgence during the extreme drought and economic hardships of the 1970s, driven by the "Feed Yourself" initiative (Asomani-Boateng 2002), which aimed to address food shortage in Ghana. In 2006, it was estimated that 1,000 ha. of land in Accra were under integrated agricultural use, mainly producing maize and vegetables, and thousands of backyard gardens, with more than 1,000 farmers completely dependent on the income generated by agriculture, and many more benefiting financially from related supply chain activities (Cofie and Awuah 2008). In the late 1990s, the High-level Conference on Agriculture mandated the Food and Agriculture Organization of the United Nations (FAO) to support the development of urban agriculture worldwide (Taguchi and Santini 2019).

UA holds significant potential to mitigate urban food shortages and reduce vulnerabilities linked to global food supply chains and the concentration of agri-food businesses (Barthel et al. 2019). Artmann and Sartison (2018) highlighted that UA has the ability to provide multiple ecosystem services and can address a variety of urban challenges beyond food provision. Despite this, current urban land use planning often underestimates UA's potential as a multifunctional, nature-based solution. Langemeyer et al. (2021) argued that while sustainable development is a key focus in global urban planning, the complex issues surrounding food supply and the environmental externalities of food imports are frequently overlooked (Seto et al. 2012). Nadal et al. (2018) emphasized the urgent need for a new urban planning paradigm that prioritized unbuilt-up areas. This new approach should be supported by a robust legal framework to promote UA effectively (Chaminuka and Dube 2017).

Urban farming practices have significantly expanded in both scale and scope in recent years, driven by various factors (Chaminuka and Dube 2017). Toriro (1995) argued that the migration of a large rural population to cities and towns has led many unemployed or low-income residents to adopt urban agriculture to enhance family food security. In Zimbabwe, where many low-income urban households struggle to afford basic food items due to high prices, urban agriculture has emerged as a vital coping strategy to fulfill their right to food.

Zimbabwe faces chronic social issues, including drought, an insecure food supply system, and economic poverty (UNDP 2011; Moyo 2011). Chaminuka and Dube (2017) investigated urban agriculture practices among residents in Gweru, focusing on the Mkoba suburb. Their study aimed to identify factors promoting urban agriculture, assess its contributions to urban residents' well-being, review the legal framework governing it, and discuss the challenges urban farmers face. Using purposive sampling, data were collected from 26 participants through interviews and focus group discussions. The results show that the growth of UA is influenced by rapid population growth, high poverty levels, elevated food prices, and high unemployment. UA was shown to enhance food security, creates jobs, and encourage savings. However, *the Municipal Council Act*, Chap. 29, Sect. 15, *the Regional Urban Planning Act*, Chap. 29, Sect. 12, *the Environmental Management Act*, Chap. 20, Sect. 27, *the Public Health Act*, Chap. 15, Sect. 09, *the Forestry Act*, Chap. 19, Sect. 05, and *the Diary Act*, Chap. 18, Sect. 08, impede the practice of urban agriculture. The biggest impediments to UA's current growth include lack of supportive legislation, lack of support from central and local governments, and lack of funding.

By 2030, global food demand in urban areas will increase (Nicholls et al. 2020; Tefft 2017). UA has been identified as part of the solution to several interrelated social, economic and environmental challenges to improve urban food security, reduce poverty, promote urban greening and urban economic development (Davies et al. 2021; Gore 2018; Lee-Smith 2013; Prain 2010).

4.2.3 Research on Urban Agriculture in Terms of Economic Security to Improve Urban Resilience to Climate Disasters

4.2.3.1 Research on Agricultural Tourism of Urban Agriculture

Cities with diversified industrial structures are more resilient to national or local employment shocks (Brown and Greenbaum 2016). However, UA focusing solely on food production has limited impact on urban economic development, job creation, and urban resilience to economic crises (Dieleman 2017). In 1985, Italy passed the first national law to recognize and define agritourism, a tourism industry that attracts people with agricultural scenery and activities, and by 2019, 15% of Italian farms offer agritourism related services. This trend is growing globally (Barbieri and Streifeneder 2019; Lamie et al. 2021).

The development of agritourism offers potential solutions to many challenges faced by urban and rural areas (Yang et al. 2010). Tourism promotes agricultural production, justifies the preservation of arable land in suburban areas, and improves land use efficiency. It creates a lot of high-paying jobs, which contributes to Social Security, especially for low-income and immigrant groups. It has diversified the urban and rural economic structure and brought more business opportunities for local development. In addition, rural cultural traditions are maintained in a way that is important for the rapidly growing urban population. With the development of agritourism, more green Spaces will be created and open to citizens, albeit with a fee. Agritourism is becoming an activity that can be enjoyed by a wide range of the public across different social strata. Yang et al. (2010) discussed the background of suburban agricultural development in Beijing and analyzed the performance of Xiedao Green Resort, a large agritourism enterprise in Beijing, from 2004 to 2008. The results show that this business model combines agricultural production with tourism services, enhancing the supply–demand relationship between urban and rural areas. It improves the quality of agricultural products and services while developing the multiple functions of agritourism with broader economic, environmental, and social benefits, creating opportunities for urban–rural integration and sustainable development.

However, agritourism faces potential limitations. A lack of innovation in agricultural services can hinder business development. In addition, high tourism revenues may drive the expansion of tourism and auxiliary facilities in rural areas, encroaching on farmland. Agritourism also attracts more migrant workers, whose families need more social support. Increasing immigration brings with it problems of integration with local cultures, traditions' and peoples' societies. To a large extent, agritourism is currently in a process of trial and error by farmers to imitate. Governments need more support, regulations, and advocacy, including financial and technical support for agricultural enterprises, land use regulations, and advocacy of successful practices (Yang et al. 2010).

4.2.4 Research on the Improvement of Urban Agriculture's Resilience to Climate Disasters in Terms of Social Security

Human health is increasingly threatened by climate change (Costello et al. 2009). The World Health Organization (WHO) estimates an additional 250,000 deaths per year between 2030 and 2050 due to climate-related impacts (Watts et al. 2015). These impacts include increases in heat-related morbidity and mortality, vector-borne diseases such as dengue fever and malaria, respiratory diseases, and health issues resulting from extreme weather events (Gomez et al. 2017). Secondary effects of climate-related disasters, like rising unemployment, impaired mental health, and increased crime rates, further exacerbate these threats (Dickson et al. 2012; Fernando et al. 2010).

From a social perspective, multifunctional UA offers significant benefits through recreational opportunities. Activities such as riding, running, and walking in the UA landscape provide physical exercise, while nature experiences based on seasonal changes and horticultural participation enhance well-being. Enjoying UA landscapes and engaging in physical activities can help residents relax and escape the stress of city life, thereby improving both physical and mental health (Hawkins et al. 2011; Van Den Berg and Custers 2011).

In general, a key issue in post-disaster and post-conflict policymaking in cities may be determining whether actions promote or inhibit individual and social expression of an urgent pro-life and restored sense of place (Tidball et al. 2014). Tiadball et al. (2014) argued that suppressing this expression may exacerbate disaster or conflict scenarios, and Greening in the Red Zone provides evidence for similar case studies. Cultivating this expression unleashing a cascade of effects that allows humans to re-establish a sense of personal balance, restore and harmonize their place in the ecosystem, re-create a sense of community and place, and take the first steps toward restoring a healthier social ecosystem. UA can create important cultural connections between city dwellers and urban symbols. These relationships have been shown to shape urban people's sense of place and belonging (Okvat and Zautra 2014; Tidball et al. 2014).

With 70% of the global population expected to live in cities by 2050 (UN 2018), fostering these cultural links seems essential in order to increase urban population carrying capacity and build a foundation for change (Theurl et al. 2020). Even small-scale UAs, in the form of schools, community or distribution gardens that make limited or no contribution to food production, can support nature experiences and provide opportunities for environmental education and learning (Beilin and Hunter 2011; Giacchè et al. 2017), which can be seen as a foundation to support the Sustainable Development Goals and potentially promote environmental stewardship for urban people (Langemeyer et al. 2018).

Nowadays, urban agriculture has derived aesthetic, educational, recreational and therapeutic functions from food production (Wu Jianzhai 2020). Velarde et al. (2007) found that compared with urban landscape, urban agricultural landscape has a more positive impact on health. In addition, urban agriculture will be more inclusive to people with disabilities, the homeless, and those released from prison, which can effectively stimulate urban renewal and regeneration (Roth et al. 2015).

However, UA development should also take into account its potential negative social impacts, such as the impact of urban agriculture on women. On the one hand, some research suggested that urban agriculture provides women with the opportunity to earn more economic income, empowers women, and improves the chances of female family advancement (Jacobi et al. 2000). On the other hand, some research reported that urban agriculture will increase the pressure of women's life and may become a trap to restrict women's access to high-paying jobs (Bryld 2003; Smit et al. 1996). In addition, Depietri and McPhearson (2017) argued that in addition to the well-established environmental services provided by GBI, a holistic approach to urban planning needs to consider its potential hazards, such as the spread of infectious disease vectors and safety risks. In some developing countries, urban populations perceive proximity to parks or forest areas as a security risk of physical harm, exposure to wildlife, or criminal attacks (Asibey et al. 2019; de la Barrera et al. 2016).

4.3 Conclusion and Prospect

Amid rapid urbanization and escalating climate challenges, cities face an urgent need for adaptive solutions to enhance resilience and sustainability. Urban Agriculture (UA), once largely a rural practice, is now being integrated into urban landscapes to support food security and broader climate resilience. This chapter explores UA's transformative potential when combined with Blue-Green Infrastructure (BGI), providing cities pathways to mitigate climate impacts while strengthening ecological, economic, and social systems. Research highlights UA's diverse benefits, from reducing greenhouse gas emissions and mitigating urban heat to managing stormwater and promoting biodiversity. Integrated into BGI frameworks, UA offers cities research-driven strategies to address climate challenges dynamically, fostering sustainable urban spaces that deliver measurable ecological and social benefits.

Each section of this chapter has added depth to the understanding of UA's roles in fostering urban resilience. The introduction provided an overview of UA's historical development, tracing its evolution from subsistence agriculture to a scientifically integrated urban strategy aligned with sustainability goals. UA's integration with BGI reveals how elements like green roofs, community gardens, and vertical farms act as natural buffers against extreme weather, reduce pollution, and improve air quality. The second section examined UA's capacity to enhance climate resilience through ecological, food security, economic security, and social lenses, referencing global studies and case examples of successful UA integration in urban planning. This work aims to highlight UA's transformative potential and promote its scalability as a scientifically grounded solution to urban environmental challenges.

4.3.1 UA's Potential

Previous research underscores UA's significant potential to strengthen urban climate resilience by reducing greenhouse gas emissions, mitigating urban heat, managing stormwater, and enhancing flood resilience. UA lowers emissions by shortening food transport distances, promoting plant-based diets, and supporting sustainable agricultural practices, with localized food systems significantly reducing urban carbon footprints. Features like green roofs and vertical farms provide evaporative cooling and shading, effectively mitigating urban heat island effects. Integrated into urban planning, these cooling benefits position UA as a valuable tool in broader climate adaptation strategies. Additionally, UA supports stormwater management with green spaces that absorb and filter runoff. Studies on "sponge cities" illustrate how combining UA with BGI can decrease surface runoff and improve water quality, ultimately boosting flood resilience.

Beyond environmental benefits, UA significantly strengthens social and economic resilience. By securing local food sources, particularly during climate-driven supply chain disruptions, UA provides fresh, accessible foods and shortens distribution chains in dense urban areas. Economic research also shows that UA supports resilience by creating jobs, reinforcing local supply chains, and generating revenue through agritourism, fostering social engagement. UA's social impact extends further by promoting community cohesion and mental well-being, connecting residents with nature, and encouraging sustainable living practices. Community-driven UA initiatives empower urban populations, actively engaging them in sustainability efforts and reinforcing UA as a key element of urban resilience.

4.3.2 Limitations and Challenges

Despite its significant potential, UA faces several scientifically identified challenges that limit its growth and effectiveness in urban contexts. One of the most pressing issues is land scarcity and the competition with other departments, as urban areas often prioritize real estate, infrastructure, and commercial development over agricultural spaces. This limited availability of land constrains UA's scalability and its ability to meet urban food security demands on a larger scale. Similarly, fragmentation and lack of mechanization also limit the scale of UA projects. Most small-scale and fragmented, UA operations are often non-mechanized, making them less efficient than conventional agriculture in terms of productivity. As a result, UA's positively impact on urban food supply and alleviating climate resilience is currently limited.

Additionally, environmental pollution poses significant risks to UA productivity and safety, with air and soil contaminants, such as heavy metals, affecting crop quality. This pollution not only creates barriers to reliable food production but also introduces health concerns associated with consuming produce grown in contaminated soils. Another substantial challenge is UA's sensitivity to climate variability. Extreme weather events—such as floods, droughts, and heatwaves—can significantly disrupt UA productivity, underscoring the need for robust adaptive measures to ensure that UA remains a viable component of urban resilience as climate change intensifies.

Finally, UA's expansion is often hindered by policy and regulatory gaps within urban planning and land-use frameworks. Without supportive policies, including appropriate zoning and land-use regulations, UA is frequently deprioritized in favor of other urban developments. These regulatory challenges limit UA's integration into urban systems, restricting its potential to contribute effectively to urban sustainability. Addressing these interconnected challenges will require coordinated efforts in research, policy development, and urban planning to ensure UA's successful and sustainable integration into city landscapes.

4.3.3 Future Directions and Policy Recommendations

To address these limitations and fully realize UA's potential, a holistic approach is necessary—one that combines technological, policy, and community-driven strategies to establish UA as a sustainable and scalable component of urban resilience. A key step is to prioritize the integration of UA within urban planning frameworks. City planners should emphasize multifunctional spaces that blend residential, recreational, and agricultural uses, fostering a harmonious urban ecosystem where UA contributes to food production, climate adaptation, and social cohesion. Implementing zoning regulations to designate specific areas for UA, especially in underutilized spaces like rooftops, vacant lots, and brownfields, can further support this integration. By creating "agricultural zones," cities can secure accessible land for UA initiatives, facilitating long-term growth and sustainability.

Technological innovation also plays a crucial role in enhancing UA's productivity and sustainability. Tools such as Internet of Things (IoT) sensors, Artificial Intelligence (AI), and remote sensing enable precision agriculture practices on a smaller scale, optimizing resource use, monitoring crop health in real-time, and mitigating adverse conditions in urban settings. Additionally, integrating renewable energy sources like solar-powered greenhouses and hydroponic systems can minimize UA's ecological footprint and improve energy efficiency, aligning UA practices with broader urban sustainability goals and solidifying its role within green infrastructure.

To make UA financially viable and attractive for urban farmers, strengthening policy frameworks and increasing financial support are essential. Governments can provide incentives like tax credits, subsidies, and grants to encourage investment in UA. Partnerships between governments, private companies, and non-profits can pool resources to foster innovation, scalability, and resilience. Such collaborations offer financial backing, expertise, and technological resources, making UA projects more sustainable. Additionally, policies that address climate-related risks, such as crop insurance, adaptive infrastructure, and climate-resilient crop varieties, are necessary to shield UA from unpredictable weather. Establishing risk-sharing mechanisms can further help UA practitioners manage climate variability and economic fluctuations.

Cross-sector and international collaboration is also vital to drive UA innovation and support. Partnerships among governmental bodies, urban planners, scientists, and local communities enable the exchange of best practices from successful UA implementations around the world, enhancing cities' climate resilience and adaptability. Interdisciplinary collaboration, incorporating insights from ecology, urban planning, sociology, and economics, can lead to comprehensive UA models that support job creation, food security, and community well-being. Integrating ecological and socio-economic principles helps cities maximize UA's contributions across multiple dimensions.

Public awareness and community engagement are equally important for UA's success. Raising awareness about UA's benefits can foster greater public support and active engagement. Schools, neighborhood organizations, and non-profits can organize workshops, community gardening sessions, and sustainability programs to encourage participation in UA and educate residents on sustainable food practices. Community-driven UA projects strengthen social bonds and empower residents, reinforcing UA as an essential part of urban life and promoting eco-friendly behaviors and local food security.

Finally, establishing robust monitoring and evaluation frameworks is crucial for measuring UA's impact on urban resilience. Key metrics, such as food production levels, greenhouse gas emission reductions, biodiversity enhancement, and improvements in social well-being, can guide policy adjustments and optimize resource allocation. Ongoing monitoring allows cities to identify challenges, track progress, and refine UA practices over time. By adopting adaptive strategies based on these insights, urban agriculture systems can respond effectively to environmental changes, ensuring their long-term sustainability and success.

In conclusion, UA stands as more than just a tool for localized food production; it serves as a strategic approach for cities seeking resilience in the face of climate change and urbanization. However, maximizing its potential requires concerted efforts across planning, policy, technology, and community engagement. Policymakers are encouraged to integrate UA within urban frameworks, allocate resources, and foster community involvement to create an environment where UA can flourish. Through collaborative action, UA can be transformed into a cornerstone of sustainable urban living, contributing to the ecological, social, and economic resilience of cities worldwide.

References

[UNDP] Z. G. G. U. N. D. P. (2011) Poverty assessment study survey II, main report. Ministry of Public Service, Labor and Social Welfare, Harare

Abass K, Buor D, Afriyie K, Dumedah G, Segbefi AY, Guodaar L, Garsonu EK, Adu-Gyamfi S, Forkuor D, Ofosu A, Mohammed A, Gyasi RM (2020) Urban sprawl and green space depletion: implications for flood incidence in Kumasi, Ghana. Int J Disaster Risk Reduct 51:101915. https://doi.org/10.1016/j.ijdrr.2020.101915

Abraham EM, Drechsel P, Cofie O (2006) The Challenge of urban flood control: the case of Accra's Korle Lagoon. In: 5th worldwide workshop for young environmental scientist in France

Agudelo-Vera CM, Mels A, Keesman K, Rijnaarts H (2012) The urban harvest approach as an aid for sustainable urban resource planning. J Ind Ecol 16(6):839–850

Ahmed J, Almeida E, Aminetzah D, Denis N, Henderson K, Katz J, Kitchel H, Mannion P (2020) Agriculture and climate change: reducing emissions through improved farming practices. McKinsey & Company 23

Ainuddin S, Routray JK (2012) Community resilience framework for an earthquake prone area in Baluchistan. Int J Disaster Risk Reduct 2:25–36. https://doi.org/10.1016/j.ijdrr.2012.07.003

Akbari H, Taha H (1992) The impact of trees and white surfaces on residential heating and cooling energy use in four Canadian cities. Energy 17(2):141–149. https://doi.org/10.1016/0360-5442(92)90063-6

Alexander DE (2013) Resilience and disaster risk reduction: an etymological journey. Nat Hazards Earth Syst Sci 13(11):2707–2716. https://doi.org/10.5194/nhess-13-2707-2013

Andreatta SL (2015) Through the generations: victory gardens for tomorrow's tables. Cult Agric Food Environ 37:38–46

Armar-Klemesu M (2000) Urban agriculture and food security, nutrition and health. Growing cities, growing food: urban agriculture on the policy agenda. A reader on urban agriculture 99–117

Artmann M, Sartison K (2018) The role of urban agriculture as a nature-based solution: a review for developing a systemic assessment framework. Sustainability 10(6):1937. https://www.mdpi.com/2071-1050/10/6/1937

Asibey MO, Yeboah V, Poku-Boansi M, Bamfo C (2019) Exploring the use, behaviour and role of urbanites towards management and sustainability of Kumasi Rattray Park, Ghana. J Urban Manage 8(2):182–194. https://doi.org/10.1016/j.jum.2018.12.003

Asomani-Boateng R (2002) Urban cultivation in Accra: an examination of the nature, practices, problems, potentials and urban planning implications. Habitat Int 26(4):591–607. https://doi.org/10.1016/S0197-3975(02)00027-9

Barbieri C, Streifeneder T (2019) Agritourism advances around the globe: a commentary from the editors. Open Agric 4(1):712–714. https://doi.org/10.1515/opag-2019-0068

Barthel S, Isendahl C, Vis BN, Drescher A, Evans DL, van Timmeren A (2019) Global urbanization and food production in direct competition for land: leverage places to mitigate impacts on SDG2 and on the earth system. Anthr Rev 6(1–2):71–97. https://doi.org/10.1177/2053019619856672

Beilin R, Hunter A (2011) Co-constructing the sustainable city: how indicators help us "grow" more than just food in community gardens. Local Environ 16(6):523–538. https://doi.org/10.1080/13549839.2011.555393

Benis K, Ferrão P (2017) Potential mitigation of the environmental impacts of food systems through urban and peri-urban agriculture (UPA)—a life cycle assessment approach. J Clean Prod 140:784–795. https://doi.org/10.1016/j.jclepro.2016.05.176

Bentley A (1999) Eating for victory: food rationing and the politics of domesticity. Annals Iowa 58(4):444–446. https://doi.org/10.17077/0003-4827.10306

Boossabong P (2017) Floods and food in the city: lessons from collaborative governance within the policy network on urban agriculture in Bangkok, Thailand. In: Allen A, Griffin L, Johnson C (eds) Environmental justice and urban resilience in the global south. Palgrave Macmillan US, pp 215–230. https://doi.org/10.1057/978-1-137-47354-7_12

Broadbent AM, Coutts AM, Tapper NJ, Demuzere M (2018) The cooling effect of irrigation on urban microclimate during heatwave conditions. Urban Clim 23:309–329. https://doi.org/10.1016/j.uclim.2017.05.002

Brown KH, Jameton AL (2000) Public health implications of urban agriculture. J Public Health Policy 21(1):20–39. https://doi.org/10.2307/3343472

Brown L, Greenbaum RT (2016) The role of industrial diversity in economic resilience: an empirical examination across 35 years. Urban Stud 54(6):1347–1366. https://doi.org/10.1177/0042098015624870

Bryld E (2003) Potentials, problems, and policy implications for urban agriculture in developing countries. Agric Hum Values 20(1):79–86. https://doi.org/10.1023/A:1022464607153

Caputo S, Schoen V, Specht K, Grard B, Blythe C, Cohen N, Fox-Kämper R, Hawes J, Newell J, Poniży L (2021) Applying the food-energy-water nexus approach to urban agriculture: from FEW to FEWP (Food-Energy-Water-People). Urban Forest Urban Green 58:126934

Chaminuka N, Dube E (2017) Urban agriculture as a food security strategy for urban dwellers: a case study of Mkoba residents in the city of Gweru, Zimbabwe. PEOPLE: Int J Soc Sci 3(2):26–45

Clarke LW (2014) The biodiversity and direct ecosystem services and disservices of urban gardens. University of California, Riverside

Cleveland DA, Phares N, Nightingale KD, Weatherby RL, Radis W, Ballard J, Campagna M, Kurtz D, Livingston K, Riechers G, Wilkins K (2017) The potential for urban household vegetable gardens to reduce greenhouse gas emissions. Landscape Urban Plan 157:365–374. https://doi.org/10.1016/j.landurbplan.2016.07.008

Clucas B, Parker ID, Feldpausch-Parker AM (2018) A systematic review of the relationship between urban agriculture and biodiversity. Urban Ecosyst 21:635–643

Cofie O, Awuah E (2008) Technology and institutional innovation on irrigated urban agriculture in Accra, Ghana

Colding J, Lundberg J, Folke C (2006) Incorporating green-area user groups in urban ecosystem management. AMBIO: A J Hum Environ 35(5):237–244

Contesse M, Van Vliet BJ, Lenhart J (2018) Is urban agriculture urban green space? a comparison of policy arrangements for urban green space and urban agriculture in Santiago de Chile. Land Use Policy 71:566–577

Costello A, Abbas M, Allen A, Ball S, Bell S, Bellamy R, Friel S, Groce N, Johnson A, Kett M (2009) Managing the health effects of climate change: lancet and University College London Institute for Global Health Commission. The Lancet 373(9676):1693–1733

Davies J, Hannah C, Guido Z, Zimmer A, McCann L, Battersby J, Evans T (2021) Barriers to urban agriculture in Sub-Saharan Africa. Food Policy 103:101999. https://doi.org/10.1016/j.foodpol.2020.101999

de la Barrera F, Rubio P, Banzhaf E (2016) The value of vegetation cover for ecosystem services in the suburban context. Urban Forest Urban Green 16:110–122. https://doi.org/10.1016/j.ufug.2016.02.003

de Wilt A, Butkovskyi A, Tuantet K, Leal LH, Fernandes TV, Langenhoff A, Zeeman G (2016) Micropollutant removal in an algal treatment system fed with source separated wastewater streams. J Hazard Mater 304:84–92. https://doi.org/10.1016/j.jhazmat.2015.10.033

Decrey L, Udert KM, Tilley E, Pecson BM, Kohn T (2011) Fate of the pathogen indicators phage ΦX174 and Ascaris suum eggs during the production of struvite fertilizer from source-separated urine. Water Res 45(16):4960–4972. https://doi.org/10.1016/j.watres.2011.06.042

Demuzere M, Orru K, Heidrich O, Olazabal E, Geneletti D, Orru H, Bhave AG, Mittal N, Feliu E, Faehnle M (2014) Mitigating and adapting to climate change: multi-functional and multi-scale assessment of green urban infrastructure. J Environ Manage 146:107–115. https://doi.org/10.1016/j.jenvman.2014.07.025

Depietri Y, McPhearson T (2017) Integrating the grey, green, and blue in cities: nature-based solutions for climate change adaptation and risk reduction. In: Nature-based solutions to climate change adaptation in urban areas: linkages between science, policy and practice, pp 91–109

Dickson E, Baker JL, Hoornweg D, Tiwari A (2012) Urban risk assessments: understanding disaster and climate risk in cities. World Bank, Washington, DC

Dieleman H (2017) Urban agriculture in Mexico City; balancing between ecological, economic, social and symbolic value. J Clean Prod 163:S156–S163. https://doi.org/10.1016/j.jclepro.2016.01.082

Dunford M, Li L (2011) Earthquake reconstruction in Wenchuan: assessing the state overall plan and addressing the 'forgotten phase'. Appl Geogr 31(3):998–1009. https://doi.org/10.1016/j.apgeog.2011.01.001

Ellis K (2021) The role of ecosystem services in urban agriculture and planning in Austin, Texas (Master's thesis, Iowa State University)

Ebru G, Regina F, Megan L, Xiaoming W, Natenzon CE, Shiraz W, Nivedita M, Maricarmen E, Ibrahim SA, Hori T (2018) Disasters and risk in cities. In: Climate change and cities (ARC 3–2). Second assessment report of the urban climate change research network. Cambridge University Press, pp 61–98

Farhad N, Garg S, Huxley R, Pillay K (2019) Understanding infrastructure interdependencies in cities

Fernando N, Warner K, Birkmann J (2010) Migration and natural hazards: is relocation a secondary disaster or an opportunity for vulnerability reduction? Environ Forced Migr Soc Vulnerability 145–156

Gentle N, Kierce S, Nitz A (2001) Economic costs of natural disasters in Australia. Aust J Emerg Manage 16(2):38–43. https://doi.org/10.3316/informit.378155558898334

Giacchè G, Paffarini C, Torquati B (2017) Cultivating changes: urban agriculture as a tool for socio-spatial transformation. Future Food: J Food, Agric Soc 5:8–20

Gomez J, Richard O, Thompson JD, Candrl C, Chaidez L, Fried E, Gilbert C, Guinane T, Hobson A, Soares J (2017) Climate change: information on potential economic effects could help guide federal efforts to reduce fiscal exposure

Gore CD (2018). How African cities lead: urban policy innovation and agriculture in Kampala and Nairobi. World Dev 108:169–180. https://doi.org/10.1016/j.worlddev.2018.03.011
Gould D, Caplow T (2012) 8—building-integrated agriculture: a new approach to food production. In: Zeman F (ed) Metropolitan sustainability. Woodhead Publishing, pp 147–170. https://doi.org/10.1533/9780857096463.2.147
Habeeb DM (2017) Exploring urban agriculture as a climate change mitigation strategy at the neighborhood scale
Hawkins JL, Thirlaway KJ, Backx K, Clayton DA (2011) Allotment gardening and other leisure activities for stress reduction and healthy aging. Hort Technol 21(5):577–585. https://doi.org/10.21273/horttech.21.5.577
Holling CS (1973) Resilience and stability of ecological systems. Annu Rev Ecol Syst 4(1):1–23. https://doi.org/10.1146/annurev.es.04.110173.000245
Jacobi PR, Amend JR, Kiango SM, Bakker N, Dubbeling M, Gündel S, Sabel-Koschella U, Zeeuw HD (2000) Urban agriculture in dares Salaam: providing an indispensable part of the diet
Kennedy C, Cuddihy J, Engel-Yan J (2007) The changing metabolism of cities. J Ind Ecol 11(2):43–59
Kim D, Lim U (2016) Urban resilience in climate change adaptation: a conceptual framework. Sustainability 8(4):405. https://www.mdpi.com/2071-1050/8/4/405
Kirkhorn P (2021) Urban heat island in Karlstad city-investigating land surface temperature, land cover and vegetation to find vulnerable areas. Population 71:72
Koko AF, Yue W, Abubakar GA, Alabsi AAN, Hamed R (2021) Spatiotemporal influence of land use/land cover change dynamics on surface urban heat island: a case study of Abuja metropolis, Nigeria. ISPRS Int J Geo-Inf 10(5):272
Kongjian YU, Tao XU, Dihua LI, Chunlian W (2015) A review: urban water resilience. Urban Plan Forum
Köster S (2021) How the Sponge City becomes a supplementary water supply infrastructure. Water-Energy Nexus 4:35–40. https://doi.org/10.1016/j.wen.2021.02.002
Kreiger MA, Mulder ML, Glover AG, Pearce JM (2014) Life cycle analysis of distributed recycling of post-consumer high density polyethylene for 3-D printing filament. J Clean Prod 70:90–96. https://doi.org/10.1016/j.jclepro.2014.02.009
Laborde D, Mamun A, Martin W, Piñeiro V, Vos R (2021) Agricultural subsidies and global greenhouse gas emissions. Nat Commun 12(1):2601. https://doi.org/10.1038/s41467-021-22703-1
Lacour C, Seconda L, Allès B, Hercberg S, Langevin B, Pointereau P, Lairon D, Baudry J, Kesse-Guyot E (2018) Environmental impacts of plant-based diets: how does organic food consumption contribute to environmental sustainability? [Original Research]. Front Nutr 5. https://doi.org/10.3389/fnut.2018.00008
Lamie RD, Chase L, Chiodo E, Dickes L, Flanigan S, Schmidt C, Streifeneder T (2021) Agritourism around the globe: definitions, authenticity, and potential controversy. J Agric Food Syst Community Dev 10(2):573–577
Langemeyer J, Camps-Calvet M, Calvet-Mir L, Barthel S, Gómez-Baggethun E (2018) Stewardship of urban ecosystem services: understanding the value(s) of urban gardens in Barcelona. Landsc Urban Plan 170:79–89
Langemeyer J, Madrid-Lopez C, Mendoza Beltran A, Villalba Mendez G (2021) Urban agriculture—a necessary pathway towards urban resilience and global sustainability? Landsc Urban Plan 210:104055. https://doi.org/10.1016/j.landurbplan.2021.104055
Lawson E, Thorne C, Ahilan S, Allen D, Arthur S, Everett G, Fenner R, Glenis V, Guan D, Hoang L (2014) Delivering and evaluating the multiple flood risk benefits in blue-green cities: an interdisciplinary approach. WIT Trans Ecol Environ 184:113–124
Lee-Smith D (2013) Which way for UPA in Africa? City 17(1):69–84. https://doi.org/10.1080/13604813.2012.754177

Li W, Hai X, Han L, Mao J, Tian M (2020) Does urbanization intensify regional water scarcity? evidence and implications from a megaregion of China. J Clean Prod 244:118592. https://doi.org/10.1016/j.jclepro.2019.118592

Lin BB, Philpott SM, Jha S, Liere H (2017) Urban agriculture as a productive green infrastructure for environmental and social well-being. Green Cities Forms Funct 155–179

Loram A, Warren PH, Gaston KJ (2008) Urban domestic gardens (XIV): the characteristics of gardens in five cities. Environ Manage 42:361–376

Luna-González DV, Sørensen M (2018) Higher agrobiodiversity is associated with improved dietary diversity, but not child anthropometric status, of Mayan Achí people of Guatemala. Public Health Nutr 21(11):2128–2141

Manríquez-Altamirano A, Sierra-Pérez J, Muñoz P, Gabarrell X (2020) Analysis of urban agriculture solid waste in the frame of circular economy: case study of tomato crop in integrated rooftop greenhouse. Sci Total Environ 734:139375. https://doi.org/10.1016/j.scitotenv.2020.139375

Marçal D, Mesquita G, Kallas LME, Hora KER (2021) Urban and peri-urban agriculture in Goiânia: the search for solutions to adapt cities in the context of global climate change. Urban Clim 35:100732. https://doi.org/10.1016/j.uclim.2020.100732

Maxwell SL, Fuller RA, Brooks TM, Watson JEM (2016) Biodiversity: the ravages of guns, nets and bulldozers. Nature 536(7615):143–145. https://doi.org/10.1038/536143a

McPherson GE, Nowak DJ, Rowntree RA (1994). Chicago's urban forest ecosystem: results of the Chicago urban forest climate project. https://doi.org/10.2737/NE-GTR-186

Metson GS, Bennett EM (2015) Phosphorus cycling in Montreal's food and urban agriculture systems. PLoS ONE 10(3):e0120726

Montero A, Marull J, Tello E, Cattaneo C, Coll F, Pons M, Infante-Amate J, Urrego-Mesa A, Fernández-Landa A, Vargas M (2021) The impacts of agricultural and urban land-use changes on plant and bird biodiversity in Costa Rica (1986–2014). Reg Environ Change 21(2):48. https://doi.org/10.1007/s10113-021-01767-1

Mora-Melià D, López-Aburto CS, Ballesteros-Pérez P, Muñoz-Velasco P (2018) Viability of green roofs as a flood mitigation element in the central region of Chile. Sustainability 10(4):1130. https://www.mdpi.com/2071-1050/10/4/1130

Mougeot L (2000) Urban agriculture: definition, presence, potentials and risks. Grow Cities, Grow Food Urban Agric Policy Agenda

Mougeot LJ (2006) Growing better cities: urban agriculture for sustainable development. IDRC

Moyo S (2011) Land concentration and accumulation after redistributive reform in post-settler Zimbabwe. Rev Afr Polit Econ 38(128):257–276. https://doi.org/10.1080/03056244.2011.582763

Mulya SP, Putro HPH, Hudalah D (2023) Review of peri-urban agriculture as a regional ecosystem service. Geogr Sustain 4(3):244–254. https://doi.org/10.1016/j.geosus.2023.06.001

Nadal A, Cerón-Palma I, García-Gómez C, Pérez-Sánchez M, Rodríguez-Labajos B, Cuerva E, Josa A, Rieradevall J (2018) Social perception of urban agriculture in Latin-America. A case study in Mexican social housing. Land Use Policy 76:719–734. https://doi.org/10.1016/j.landusepol.2018.02.055

Nicholls E, Ely A, Birkin L, Basu P, Goulson D (2020) The contribution of small-scale food production in urban areas to the sustainable development goals: a review and case study. Sustain Sci 15(6):1585–1599. https://doi.org/10.1007/s11625-020-00792-z

O'sullivan C, Bonnett G, McIntyre C, Hochman Z, Wasson A (2019) Strategies to improve the productivity, product diversity and profitability of urban agriculture. Agric Syst 174:133–144

O'Donnell EC, Lamond JE, Thorne CR (2018) Learning and action alliance framework to facilitate stakeholder collaboration and social learning in urban flood risk management. Environ Sci Policy 80:1–8. https://doi.org/10.1016/j.envsci.2017.10.013

Okvat HA, Zautra AJ (2014) Sowing seeds of resilience: community gardening in a post-disaster context. In: Tidball KG, Krasny ME (eds) Greening in the red zone:disaster, resilience and community greening. Springer Netherlands, pp 73–90. https://doi.org/10.1007/978-90-481-9947-1_5

Pachauri RK, A. M., Barros VR, Broome J, Cramer W, Christ R et al (2014) Climate change 2014 synthesis report. In: Contribution of working groups I, II, and III to the fifth assessment report of the Intergovernmental Panel on Climate Change. IPCC Retrieved from http://www.mendeley.com/research/climate-change-2014-synthesis-report-contribution-working-groups-i-ii-iii-fifth-assessment-report-in-20

Panchasara H, Samrat NH, Islam N (2021) Greenhouse gas emissions trends and mitigation measures in Australian agriculture sector—a review. Agriculture 11(2):85

Pauleit S, Liu L, Ahern J, Kazmierczak A (2011) Multifunctional green infrastructure planning to promote ecological services in the City. In: Niemelä J, Breuste JH, Elmqvist T, Guntenspergen G, James P, McIntyre NE (eds) Urban ecology: patterns, processes, and applications. Oxford University Press, pp 0. https://doi.org/10.1093/acprof:oso/9780199563562.003.0033

Pires I (2012) Do Campo ao Garfo. Desperdício Alimentar em Portugal.

Polling B, Alfranca O, Alves E, Andersson G, Branduini P, Deborah EL, Giacchè G, Heller A, Herkstroter K, Kemper D (2016) Creating added value: societal benefits of urban agriculture. In: Urban agriculture Europe. Jovis, pp 92–100

Prain G (2010) The institutional and regional context. In: Prain G, Lee-Smith D, Karanja N (eds) African urban harvest: agriculture in the cities of Cameroon, Kenya and Uganda. Springer New York, pp 1–11. https://doi.org/10.1007/978-1-4419-6250-8_1

Puppim de Oliveira JA, Ahmed A (2021) Governance of urban agriculture in African cities: gaps and opportunities for innovation in Accra, Ghana. J Clean Prod 312:127730. https://doi.org/10.1016/j.jclepro.2021.127730

Puri V, Caplow T (2009) How to grow food in the 100% renewable city: building-integrated agriculture. In: 100% renewable: energy autonomy in action

Qiu G-Y, Li H-Y, Zhang Q-T, Wan C, Liang X-J, Li X-Z (2013) Effects of evapotranspiration on mitigation of urban temperature by vegetation and urban agriculture. J Integr Agric 12(8):1307–1315

Ren X, Mo D, Storozum M, Lemoine X, Yu Y, Gu W, Lei X, Zhang J, Lü J, Kidder TR (2019) Early urban impact on vegetation dynamics: palaeoecological reconstruction from pollen records at the Dongzhao site, Henan Province, China. Quat Int 521:66–74. https://doi.org/10.1016/j.quaint.2019.07.012

Renner R (2000) Sewage sludge, pros & cons. Environ Sci Technol 34:430A-435A. https://doi.org/10.1021/es0034489

Rizwan AM, Dennis LYC, Liu C (2008) A review on the generation, determination and mitigation of Urban Heat Island. J Environ Sci 20(1):120–128. https://doi.org/10.1016/S1001-0742(08)60019-4

Rosenzweig C, Gaffin S, Parshall L (2006) Green roofs in the New York metropolitan region: research report. Columbia Univ Center Clim Syst Res NASA Goddard Inst Space Stud 59

Rosenzweig C, Solecki WD (2001) Climate change and a global city: the potential consequences of climate variability and change: Metro East Coast: executive summary and recommendations. Columbia Earth Institute

Roth MB, FRixen M, Tobisch C, Scholle T (2015) Finding spaces for urban food production—matching spatial and stakeholder analysis with urban agriculture approaches in the urban renewal area of Dortmund-Hörde, Germany. Future Food J Food Agric Soc 3:79–88

Rowe DB, Whittinghill LJ (2012) The role of green roof technology in urban agriculture. Renew Agric Food Syst 27(4):314–322. https://doi.org/10.1017/S174217051100038X

Rufí-Salís M, Calvo MJ, Petit-Boix A, Villalba G, Gabarrell X (2020) Exploring nutrient recovery from hydroponics in urban agriculture: an environmental assessment. Resour Conserv Recycl 155:104683. https://doi.org/10.1016/j.resconrec.2020.104683

Toriro S (1995) Urban planning and agriculture in Harare: practices and lessons. Urban Agric Mag 1

Safayet M, Arefin MF, Hasan MMU (2017) Present practice and future prospect of rooftop farming in Dhaka city: a step towards urban sustainability. J Urban Manage 6(2):56–65. https://doi.org/10.1016/j.jum.2017.12.001

Salomon MJ, Watts-Williams SJ, McLaughlin MJ, Cavagnaro TR (2020) Urban soil health: a city-wide survey of chemical and biological properties of urban agriculture soils. J Clean Prod 275:122900. https://doi.org/10.1016/j.jclepro.2020.122900

Santamouris M (2014) Cooling the cities—a review of reflective and green roof mitigation technologies to fight heat island and improve comfort in urban environments. Solar Energy 103:682–703. https://doi.org/10.1016/j.solener.2012.07.003

Seto KC, Reenberg A, Boone CG, Fragkias M, Haase D, Langanke T, Marcotullio P, Munroe DK, Olah B, Simon D (2012) Urban land teleconnections and sustainability. Proc Nat Acad Sci 109(20):7687–7692. https://doi.org/10.1073/pnas.1117622109

Singh RP, Agrawal M (2008) Potential benefits and risks of land application of sewage sludge. Waste Manage 28(2):347–358. https://doi.org/10.1016/j.wasman.2006.12.010

Smit J, Nasr J (1992) Urban agriculture for sustainable cities: using wastes and idle land and water bodies as resources. Environ Urban 4(2):141–152. https://doi.org/10.1177/095624789200400214

Smit J, Nasr J, Ratta A (1996) Urban agriculture: food, jobs and sustainable cities. New York, USA 2:35–37

Smith KR, Roebber PJ (2011) Green roof mitigation potential for a proxy future climate scenario in Chicago, Illinois. J Appl Meteorol Climatol 50(3):507–522

Steel C (2013) Hungry city: how food shapes our lives

Sullivan C (2008) Natural hazards in Australia: identifying risk analysis requirements [Book Review]. Aust J Emerg Manage 23:74

Taguchi M, Santini G (2019) Urban agriculture in the Global North & South: a perspective from FAO. Field Actions Sci Rep 20(Special 20):12–17

Tam B, Gough W, Mohsin T (2015) The impact of urbanization and the urban heat island effect on day to day temperature variation. Urban Clim 12:1–10. https://doi.org/10.1016/j.uclim.2014.12.004

Tefft J, Jonasova M, Adjao R, Morgan A (2017) Food systems for an urbanizing world. Retrieved from http://hdl.handle.net/10986/32502

Tervahauta T, Rani S, Hernández Leal L, Buisman CJN, Zeeman G (2014) Black water sludge reuse in agriculture: are heavy metals a problem? J Hazard Mater 274:229–236. https://doi.org/10.1016/j.jhazmat.2014.04.018

Theurl MC, Lauk C, Kalt G, Mayer A, Kaltenegger K, Morais TG, Teixeira RFM, Domingos T, Winiwarter W, Erb K-H, Haberl H (2020) Food systems in a zero-deforestation world: dietary change is more important than intensification for climate targets in 2050. Sci Total Environ 735:139353. https://doi.org/10.1016/j.scitotenv.2020.139353

Tidball KG, Weinstein ED, Krasny ME (2014) Synthesis and conclusion: applying greening in red zones

UN (2018) The world's cities in 2018—data booklet (ST/ESA/SER.A/417)

UN (2019) World urbanization prospects 2018 highlights

Uysal A, Yilmazel YD, Demirer GN (2010) The determination of fertilizer quality of the formed struvite from effluent of a sewage sludge anaerobic digester. J Hazard Mater 181(1):248–254. https://doi.org/10.1016/j.jhazmat.2010.05.004

Valente de Macedo LS, Barda Picavet ME, Puppim de Oliveira JA, Shih W-Y (2021) Urban green and blue infrastructure: a critical analysis of research on developing countries. J Clean Product 313:127898. https://doi.org/10.1016/j.jclepro.2021.127898

Van Den Berg AE, Custers MH (2011) Gardening promotes neuroendocrine and affective restoration from stress. J Health Psychol 16(1):3–11

van der Grinten E, Spijker J, Lijzen J (2016) Hergebruik van grondstoffen uit afvalwater: Belemmeringen en oplossingsrichtingen aan de hand van de cases fosfaat en cellulose

Vávra J, Daněk P, Jehlička P (2018) What is the contribution of food self-provisioning towards environmental sustainability? a case study of active gardeners. J Clean Product 185:1015–1023. https://doi.org/10.1016/j.jclepro.2018.02.261

Velarde MD, Fry G, Tveit M (2007) Health effects of viewing landscapes—landscape types in environmental psychology. Urban Forest Urban Green 6(4):199–212. https://doi.org/10.1016/j.ufug.2007.07.001

Wang H, He Q, Liu X, Zhuang Y, Hong S (2012) Global urbanization research from 1991 to 2009: a systematic research review. Landsc Urban Plan 104(3):299–309. https://doi.org/10.1016/j.landurbplan.2011.11.006

Watts N, Adger WN, Agnolucci P, Blackstock J, Byass P, Cai W, Chaytor S, Colbourn T, Collins M, Cooper A, Cox PM, Depledge J, Drummond P, Ekins P, Galaz V, Grace D, Graham H, Grubb M, Haines A et al (2015) Health and climate change: policy responses to protect public health. The Lancet 386(10006):1861–1914. https://doi.org/10.1016/S0140-6736(15)60854-6

Weidner T, Yang A (2020) The potential of urban agriculture in combination with organic waste valorization: assessment of resource flows and emissions for two European cities. J Clean Prod 244:118490. https://doi.org/10.1016/j.jclepro.2019.118490

Wielemaker RC, Weijma J, Zeeman G (2018) Harvest to harvest: recovering nutrients with new sanitation systems for reuse in urban agriculture. Resour Conserv Recycl 128:426–437. https://doi.org/10.1016/j.resconrec.2016.09.015

Wu Jianzhai ZJ, Jianhua Z, Fantao K (2020) Progress and prospect of research on urban agriculture abroad. Chin J Agric Resour Reg Plan 41:66–75. https://doi.org/10.7621/cjarrp.1005-9121

Yang Z, Cai J, Sliuzas R (2010) Agro-tourism enterprises as a form of multi-functional urban agriculture for peri-urban development in China. Habitat Int 34(4):374–385. https://doi.org/10.1016/j.habitatint.2009.11.002

Yearbook FS (2013) World food and agriculture. Food and Agriculture Organization of the United Nations, Rome, p 15

Yiwen S, Jiang X (2015) Understanding urban resilience: a conceptual analysis based on integrated international literature review. Urban Plan Int

Zasada I, Weltin M, Zoll F, Benninger SL (2020) Home gardening practice in Pune (India), the role of communities, urban environment and the contribution to urban sustainability. Urban Ecosyst 23(2):403–417

Zhang K, Wang R, Shen C, Da L (2010) Temporal and spatial characteristics of the urban heat island during rapid urbanization in Shanghai, China. Environ Monitor Assess 169(1):101–112. https://doi.org/10.1007/s10661-009-1154-8

Zhou M, Köster S, Zuo JE, Che W, Wang X (2019) Cross-boundary evolution of urban planning and urban drainage towards the water sensitive "sponge city". In: Köster S, Reese M, Zuo JE (eds) Urban water management for future cities: technical and institutional aspects from Chinese and German perspective. Springer International Publishing, pp 303–329. https://doi.org/10.1007/978-3-030-01488-9_14

Zinia NJ, McShane P (2018) Ecosystem services management: an evaluation of green adaptations for urban development in Dhaka, Bangladesh. Landsc Urban Plan 173:23–32. https://doi.org/10.1016/j.landurbplan.2018.01.008

Zupancic T, Westmacott C, Bulthuis M (2015) The impact of green space on heat and air pollution in urban communities: a meta-narrative systematic review, vol 61. David Suzuki Foundation Vancouver, BC, Canada

Chapter 5
Blue-Green Infrastructure for Landslide Risk Reduction: A Scenario in Uttarakhand, India

Akshat Vashistha, Srikrishnan Siva Subramanian, and Yunus Ali Pulpadan

Abstract Urban cluster development in steep landscapes has alarmingly increased due to the rapid urbanization of mountainous regions. This trend has made urban areas more vulnerable to mountain hazards like landslides, now referred to as urban landslides. These urban landslides significantly threaten people and property, causing economic disruption. Compared to isolated landslides in remote mountain regions, the financial effects of urban landslides are far more severe. To address this urgent issue, we propose using blue-green infrastructure, which integrates naturally occurring and designed features, including vegetated slopes, drainage systems, and green areas. Blue-green infrastructure makes these locations less vulnerable to landslides, as they absorb excess rainfall, stabilize the soil, and encourage sustainable urban planning. In this context, we evaluate the urban landslides in the Indian state of Uttarakhand during the North India Floods of 2013, emphasizing the need to integrate blue-green infrastructure for improved hazard resilience in hilly urban development.

Keywords Urban landslides · Blue-green infrastructure · Mountainous regions · Hazard resilience · Sustainable urban planning

5.1 Introduction

Urban landslides present a critical and urgent risk in many mountainous regions, with Uttarakhand, India, being a notable example (Pradhan and Siddique 2020; Roberts et al. 2019; Yunus et al. 2021). The region's susceptibility to landslides has been exacerbated by earthquakes, rapid urbanization, deforestation, and the growing impacts

A. Vashistha · S. Siva Subramanian (✉)
Centre of Excellence in Disaster Mitigation and Management, Indian Institute of Technology Roorkee, Roorkee, India
e-mail: srikrishnan@dm.iitr.ac.in

Y. Ali Pulpadan
Department of Earth and Environmental Sciences, Indian Institute of Science and Education Research Mohali, Mohali, India

M. G. Kibria et al. (eds.), *Urban Blue-Green Infrastructure Approach for Food Security and Climate Disaster Resilience*, Disaster Risk Reduction,
https://doi.org/10.1007/978-981-96-8746-6_5

of climate change (Vashistha et al. 2025, Kumar and Anbalagan 2016; Martha et al. 2015; Pradhan et al. 2018; Singh et al. 2013). As urban areas expand and natural landscapes are disrupted, the stability of slopes is increasingly compromised, leading to a heightened likelihood of landslides that can have devastating effects on communities and infrastructure (Alexander 1989; Bhandari and Dhakal 2021; Peethambaran et al. 2020; Petley 2009; Sharma et al. 2023).

In response to these challenges, Blue-Green Infrastructure (BGI) offers a promising solution (Ghofrani et al. 2017). BGI integrates natural processes with engineered systems to manage water flow and vegetation, enhancing slope stability and reducing landslide risks. By incorporating green spaces, vegetation cover, and sustainable water management practices, BGI addresses the immediate risks of landslides and promotes long-term urban resilience and environmental sustainability. This approach represents a forward-thinking strategy to harmonize development with ecological balance, ultimately contributing to safer and more resilient urban environments. As researchers, urban planners, environmentalists, and policymakers, your role in implementing these solutions is crucial.

5.2 Objectives

- Assess Landslide Risks: Identify and map high-risk urban areas in Uttarakhand susceptible to landslides.
- Evaluate BGI Strategies: Analyze the effectiveness of Blue-Green Infrastructure (BGI) techniques in reducing landslide risks.

Uttarakhand, a state located in northern India, is renowned for its rugged, hilly terrain and high rainfall, which significantly influence its landscape dynamics. The state's topography, characterized by steep slopes and fragile geology, makes it particularly susceptible to landslides, especially in rapidly developing urban areas. Cities such as Nainital, Dehradun, and Mussoorie are at the forefront of this vulnerability, facing increased risks due to both natural factors and anthropogenic pressures. The rapid pace of urban expansion in these areas, as well as deforestation and inadequate land use planning, exacerbates the potential for landslides. Heavy monsoon rains further destabilize slopes, making these regions highly prone to landslide events that can severely impact infrastructure and local communities. As such, Uttarakhand represents a critical study area for exploring effective landslide mitigation strategies and implementing sustainable urban planning practices (Fig. 5.1).

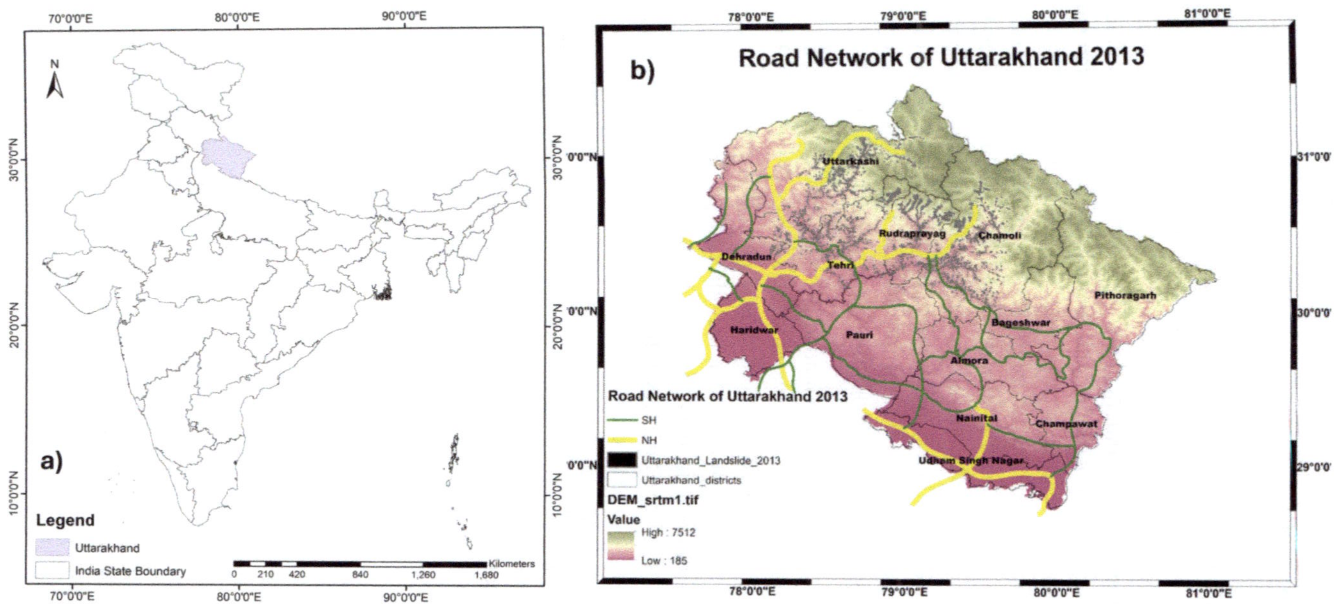

Fig. 5.1 **a** Study area Uttarakhand, India (*Source* Survey of India) and **b** Locations of landslides in Uttarakhand during 2013 (data from Martha et al. 2015) and names of urban (municipal) clusters along with the road network (the road network is digitized using data from the Public Work Department, Uttarakhand, India)

5.3 Methodology

5.3.1 *Risk Assessment*

Data Collection: Gather historical landslide data, topographic maps, rainfall patterns, and urban development plans. Data collection is a critical initial step involving gathering extensive historical landslide data, detailed topographic maps, rainfall patterns, and urban development plans. This diverse dataset provides a foundational understanding of the conditions and factors contributing to landslides. By analyzing historical data, researchers can identify patterns and trends in landslide occurrences, essential for predicting future events. Topographic maps offer valuable information about the terrain, such as elevation and slope, which are key factors in landslide susceptibility. Rainfall patterns are crucial for understanding the trigger mechanisms of landslides, as heavy and prolonged rainfall can destabilize slopes. Urban development plans provide insight into human activities that may increase landslide risk, such as deforestation, construction, and land use changes.

Geospatial Analysis: Use GIS to map landslide-prone areas, incorporating factors like slope, soil type, and land use. Geospatial analysis uses Geographic Information Systems (GIS) to map areas susceptible to landslides. This process involves integrating various factors into a spatial model, including slope steepness, soil type, and land use. Slope steepness is a primary factor in landslide susceptibility, as steeper slopes are more prone to failure. Soil type influences the stability of the ground, with certain soils being more prone to erosion and movement. Land use patterns, such as deforestation or construction, can significantly impact the stability of a slope. Combining these factors in a GIS can create detailed risk maps highlighting the most vulnerable areas. This geospatial analysis is crucial for developing effective risk assessment and early warning systems, ultimately aiding disaster preparedness and mitigation efforts.

5.3.2 *BGI Strategies*

Green Infrastructure: Reforestation, green roofs, and urban green spaces to stabilize soil and reduce runoff. Green infrastructure strategies, such as reforestation, green roofs, and urban green spaces, are essential for stabilizing soil and reducing runoff, thereby mitigating landslide risks. Reforestation helps by restoring vegetation cover, where tree roots bind the soil, preventing erosion and enhancing slope stability. Green roofs, which are covered with vegetation, absorb rainfall, thereby decreasing the volume and speed of runoff, and contribute to cooling urban areas. Urban green spaces, including parks and community gardens, facilitate stormwater management by promoting infiltration and reducing surface runoff. Collectively, these green infrastructure practices not only improve soil stability and water management but also enhance urban biodiversity, aesthetic value, and resilience against climate impacts.

Blue Infrastructure: Rain gardens, permeable pavements, and constructed wetlands to manage stormwater and reduce soil erosion. Blue infrastructure, including rain gardens, permeable pavements, and constructed wetlands, plays a crucial role in managing stormwater and reducing soil erosion. Rain gardens are strategically placed to capture and absorb runoff, utilizing vegetation to filter and retain water, which mitigates flooding and erosion. Permeable pavements allow water to percolate through surfaces like driveways and walkways, reducing runoff and promoting groundwater recharge. Constructed wetlands act as natural water treatment systems, where water is filtered through vegetation and soil, removing pollutants and stabilizing the soil. These blue infrastructure solutions not only manage stormwater effectively but also enhance the resilience of urban areas, improve water quality, and support diverse ecosystems.

5.4 Results and Discussion

The map presented in the study was prepared using the Statistical Approach Frequency Ratio method, meticulously classifying landslide susceptibility into five distinct classes: very low, low, moderate, high, and very high. This classification relied on the Natural Break method, which optimizes data distribution by identifying inherent groupings within the susceptibility values. The analysis uncovered those significant portions of Chamoli, Uttarkashi, Rudraprayag, Bageshwar, and Pithoragarh districts fall into the high and very high susceptibility categories, highlighting areas with a considerable risk of landslides. These regions are characterized by steep slopes, loose soil, and frequent heavy rainfall, which collectively increase the likelihood of landslide occurrences.

However, within these high-risk districts, certain regions were categorized as very low susceptibility zones due to ice cover. The presence of ice stabilizes the terrain, significantly reducing the probability of landslides. Conversely, districts such as Tehri, Pauri, Nainital, Champavat, Almora, and Dehradun were predominantly identified as moderate and high susceptibility areas. These regions exhibit varying degrees of risk, influenced by a combination of natural terrain features and anthropogenic activities, such as deforestation and construction.

In contrast, Haridwar, Dehradun, Champavat, and Udham Singh Nagar districts were found to be predominantly composed of very low and low susceptibility zones. These areas generally have more stable terrain and less severe climatic conditions, resulting in a lower risk of landslides. The comprehensive classification and spatial distribution of susceptibility across these districts provide critical insights for targeted risk management and mitigation strategies. This detailed mapping is essential for informing local authorities and stakeholders, enabling them to prioritize areas for intervention, enhance early warning systems, and implement effective land use planning to mitigate the impact of potential landslides (Fig. 5.2).

The study's findings offer valuable insights into the spatial distribution of landslide susceptibility in the Himalayan region. The identification of high-risk zones can inform targeted mitigation measures, such as slope stabilization techniques, drainage improvements, and early warning systems. By implementing effective land use planning and construction practices, it is possible to minimize human-induced factors that contribute to landslides. Additionally, community-based disaster risk reduction initiatives can empower local communities to take proactive measures to protect themselves and their livelihoods.

However, it is important to note that landslide susceptibility is a dynamic phenomenon influenced by various factors, including climate change, seismic activity, and human activities. Therefore, continuous monitoring and updating of landslide susceptibility maps are essential to adapt to changing conditions. Furthermore, interdisciplinary collaboration between geologists, hydrologists, engineers, and social scientists is crucial for developing comprehensive and sustainable solutions to mitigate landslide risks in the region.

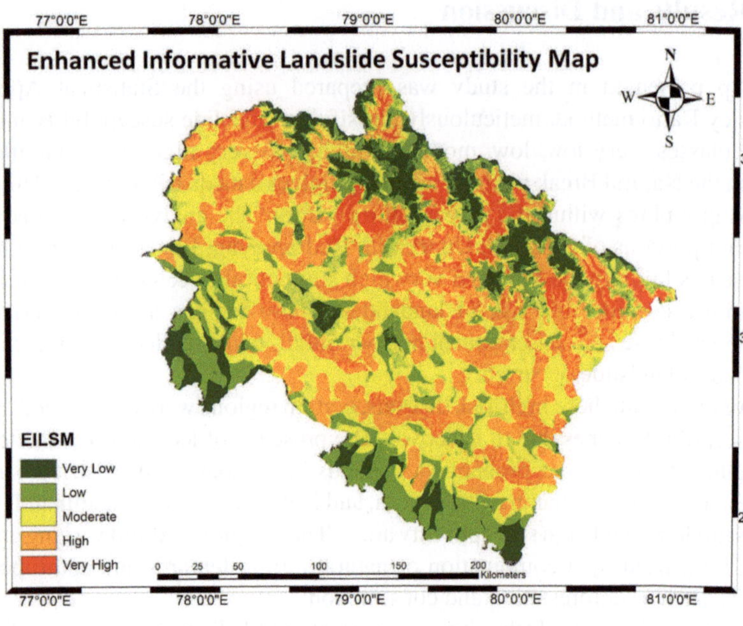

Fig. 5.2 Landslide risk assessment for Uttarakhand, India, considering urban environments

5.5 Conclusion

Implementing BGI in urban areas of Uttarakhand offers a viable solution for reducing landslide risks. Scenario analysis highlights the effectiveness of comprehensive BGI networks, emphasizing the need for strategic planning and investment. By adopting BGI, urban areas can enhance resilience against landslides and achieve sustainable development goals. The conclusions drawn from this study highlight several critical aspects of landslide susceptibility and the factors influencing landslide occurrences in Uttarakhand. The Landslide Susceptibility Mapping (LSM) reveals a high frequency of landslides near roads and streams, pointing to inadequate mitigation measures following hill slope cutting and the encroachment of slopes due to stream flow. The Enhanced Integrated Landslide Susceptibility Mapping (EILSM) further demonstrates that landslides predominantly occur near faults and streams, with approximately 42% of Uttarakhand falling within the very high susceptibility zone as per the Frequency Ratio method.

The study underscores the importance of proper monitoring and mitigation efforts, particularly in areas near the Main Central Thrust (MCT) and Main Boundary Thrust (MBT), which exhibit high susceptibility. Additionally, the study indicates that approximately 13% of Uttarakhand's area (around 6,500 km^2) is prone to landslide damming, with regions near the MCT particularly vulnerable. These areas fall within Zone 5 of the seismic hazard zonation, indicating high seismic activity and necessitating the development of appropriate mitigation strategies. Historical experiences of major magnitude earthquakes triggering numerous landslides underscore the potential impact of future landslides, especially given the significant infrastructure development in the region over the past 25–30 years.

Integrating blue and green infrastructure into land use planning and management is crucial to address these challenges. Implementing green infrastructure measures, such as reforestation, green roofs, and urban green spaces, can enhance soil stability and reduce runoff, mitigating landslide risks. Concurrently, blue infrastructure solutions like rain gardens, permeable pavements, and constructed wetlands can effectively manage stormwater and reduce soil erosion. By incorporating these strategies, it is possible to enhance the resilience of urban areas, improve water management, and reduce the overall landslide risk. Emphasizing these infrastructure improvements, alongside traditional mitigation measures, will contribute to better risk management and greater safety for communities in Uttarakhand.

References

Alexander D (1989) Urban landslides. Prog Phys Geogr 13(2):157–189

Bhandari BP, Dhakal S (2021) A multidisciplinary approach of landslide characterization: a case of the Siwalik zone of Nepal Himalaya. J Asian Earth Sci X 5:100061

Ghofrani Z, Sposito V, Faggian R (2017) A comprehensive review of blue-green infrastructure concepts. Int J Environ Sustain 6(1)

Kumar R, Anbalagan R (2016) Landslide susceptibility mapping using analytical hierarchy process (AHP) in Tehri reservoir rim region, Uttarakhand. J Geol Soc India 87(3):271–286

Martha TR, Roy P, Govindharaj KB, Kumar KV, Diwakar PG, Dadhwal VK (2015) Landslides triggered by the June 2013 extreme rainfall event in parts of Uttarakhand state, India. Landslides 12(1):135–146

Peethambaran B, Anbalagan R, Kanungo DP, Goswami A, Shihabudheen KV (2020) A comparative evaluation of supervised machine learning algorithms for township level landslide susceptibility zonation in parts of Indian Himalayas. CATENA 195:104751

Petley DN (2009) On the impact of urban landslides. Geol Soc, Lond, Eng Geol Spec Publ 22(1):83–99

Pradhan SP, Siddique T (2020) Stability assessment of landslide-prone road cut rock slopes in Himalayan terrain: a finite element method based approach. J Rock Mech Geotech Eng 12(1):59–73

Pradhan SP, Vishal V, Singh TN (2018) Finite element modelling of landslide prone slopes around Rudraprayag and Agastyamuni in Uttarakhand Himalayan terrain. Nat Hazards 94(1):181–200

Roberts NJ, Rabus BT, Clague JJ, Hermanns RL, Guzmán M-A, Minaya E (2019) Changes in ground deformation prior to and following a large urban landslide in La Paz, Bolivia, revealed by advanced InSAR. Nat Hazard 19(3):679–696

Sharma A, Nikam B, Bagri DS (2023) Urbanization trends and challenges: a case study in the asan watershed of uttarakhand on urban growth patterns and sustainable land use planning

Singh R, Umrao RK, Singh TN (2013) Probabilistic analysis of slope in Amiyan landslide area, Uttarakhand. Geomat Nat Haz Risk 4(1):13–29

Vashistha A, Joshi S, Siva Subramanian S (2025) Scenario-based probabilistic risk assessment of earthquake-induced landslides in Uttarakhand, India. Stoch Environ Res Risk Assess. https://doi.org/10.1007/s00477-025-03067-0

Yunus AP, Fan X, Subramanian SS, Jie D, Xu Q (2021) Unraveling the drivers of intensified landslide regimes in Western Ghats, India. Sci Total Environ 145357

Chapter 6
Blue-Green Infrastructure (BGI) in Bangladesh: Present Scenario and Possible Strategies

Mohammad Saidur Rhaman, Jotirmoy Chakrobortty, Sumaiya Farzana, Israt Jahan, Md Moyeed Hasan Talukder, and Md Zayeed Hasan Talukdar

Abstract Bangladesh, a densely populated country with limited land resources, has become more susceptible to several environmental issues due to rapid urbanization and climate change. In this regard, the concept of Blue-Green Infrastructure (BGI) is appeared to be a viable strategy to meet societal and ecological demands. In order to ensure a superior standard of living in an urban environment, it is essential to maintain a harmonious equilibrium between the blue (water), green (vegetation), and gray (built) spaces. Nevertheless, the swift and unregulated progress is significantly modifying the characteristics of the soil, water bodies, and flora in this densely populated, disaster-prone country. Due to the country's rapid and uncontrolled urbanization, Bangladesh is experiencing a decline in the number of blue and green spaces, especially in large cities such as Dhaka, Chittagong, Khulna, and Rajshahi. Dhaka, the capital of Bangladesh, ranks among the world's top 10 megacities as a result of its rapid population increase. Urbanization has led to the conversion of lower lands

---

Mohammad Saidur Rhaman and Jotirmoy Chakrobortty have contributed equally.

M. S. Rhaman (✉)
Department of Seed Science and Technology, Bangladesh Agricultural University, Mymensingh, Bangladesh
e-mail: saidursst@bau.edu.bd

J. Chakrobortty
Department of Soil Science, Khulna Agricultural University, Khulna, Bangladesh

S. Farzana
Graduate Training Institutue, Bangladesh Agricultural University, Mymensingh, Bangladesh

I. Jahan
Department of Soil Science, Bangladesh Agricultural University, Mymensingh, Bangladesh

M. M. H. Talukder
Department of Horticulture and Landscape Architechture, Purdue University, West Lafayette, USA

M. Z. H. Talukdar
Former Consultant, International Finance Corporation (IFC), Dhaka, Bangladesh

M. G. Kibria et al. (eds.), *Urban Blue-Green Infrastructure Approach for Food Security and Climate Disaster Resilience*, Disaster Risk Reduction,
https://doi.org/10.1007/978-981-96-8746-6_6

such as ponds and lakes into reclaimed build-up lands, resulting in a reduction of the blue-green infrastructure. However, in order to enhance the ability of cities to withstand and recover from disasters, the local community and other relevant parties are collaborating with the government and non-governmental organizations to expand the extent of blue-green infrastructure. While BGIs are receiving attention in contemporary urban planning procedures, their balanced implementation into urban planning techniques has not yet been recognized. Consequently, this chapter offers an overview of BGI's overall state in Bangladesh, as well as its effects on Dhaka, the country's most populous metropolis, and potential regulatory improvements.

Keywords Blue-green infrastructure · Climatic hazards · Dhaka · Green spaces · Urbanization

6.1 Introduction

Blue-green infrastructure (BGI) is a concept that integrates both man-made and natural water systems to offer numerous advantages to both humans and the environment in urban and rural planning. The BGI is commonly used as substitute for "Green Infrastructure," a landscape planning term associated with ecological networks (Jongman and Pungetti 2004) and green-ways (Ahern 1995; Ryan and Fabos 2006). Since its introduction in 2000, green infrastructure has been used to design and promote urban green bodies as a cohesive framework for environmental planning. It is believed to be essential for tackling complex socio-ecological challenges by means of strategic network design, management, and planning (Thomas and Littlewood 2010).

Urbanization and rising population densities are global concerns, particularly in Asia and Africa (Liu et al. 2020). According to reports, more than 55% of people on Earth were urbanized in 2020, and by 2050, over 68% of people will be living in cities (Welford and Yarbrough 2021). Urban land has grown globally at a rate of 80% during the previous few decades, compared to a 52% rate of population increase. Developing country cities are pushing hard for the deployment of BGI in response to growing urbanization, with the goal of safeguarding ecosystem services and improving the general quality of life for their citizens (Plieninger et al. 2022). In addition, BGIs are acknowledged as essential for catastrophe risk reduction in addition to their socioeconomic and environmental advantages. Moreover, BGI plays a crucial role in urban planning and design. Its main objective is to produce living spaces that are more sustainable, healthful, and visually appealing. Additionally, BGI methods can help address the threats and difficulties posed by climate change. For example, green areas have emerged as a practical solution for addressing urban flood risks and lessening the impact of urban heat islands in cities (Bai et al. 2018; Kim et al. 2016; Šakić Trogrlić et al. 2018; Liu et al. 2019; Gunawardena et al. 2017). However, fast urbanization is displacing vital green and blue spaces from everywhere (Liu et al. 2020).

Bangladesh has a rapid pace of urbanization compared to other Asian countries, with an annual growth rate of 3.05% as reported by the World Bank (2020). According to Bajracharya and Sultana (2020), it is projected that approximately 57% of Bangladesh's population would live in urban areas by 2050. Bangladesh, the eighth most populous and fifth most disaster-prone country in the world, is experiencing significant changes in areas of bare soil, water bodies, and vegetation due to rapid and unplanned urbanization (Hassan 2017). These changes are enhancing disaster risks and vulnerabilities of the urban areas. In Bangladesh, on the other hand, blue and green spaces are rapidly disappearing as a result of unplanned and rapid urbanization, which poses a variety of threats and hazards, particularly in urban areas. Dhaka, the capital city of Bangladesh, functions as the country's main hub for business and politics. Expanding upon a strong economic base, it has experienced significant growth and established itself as the dominant force in the country's economy. However, the city is currently undergoing uncontrolled and disorganized growth, resulting in the destruction of the city's BGI and negatively affecting the livelihoods of urban people, as well as the quality of the air, land, and water. However, health-conscious people are increasingly turning to BGI for improved opportunities not only in Dhaka, but throughout the country. As a result, this chapter provides a summary of the general condition of BGI in Bangladesh, including its impact on the highly populated metropolis of Dhaka and how regulations might be improved.

6.2 Overall BGI Scenario in Urban Bangladesh

The city's blue, green, and gray infrastructures are essential to its effective operation, and maintaining a proper equilibrium among them is of utmost importance. The majority of Bangladesh's largest cities were constructed along the banks of the Padma, Meghna, Karnaphuli, Bhairab, and Surma Rivers. During the colonial era, cities were founded as hubs for administration and commerce. Over time, cities expanded in line with government objectives and a rise in business transactions. However, as the population grew throughout time, cities emerged as the center of civic amenities, economic opportunities, and other factors that draw people from rural areas. Urban expansion, which takes the form of constructing homes, roads, businesses, industries, etc., is a constant occurrence in the largest cities. Its purpose is to accommodate the expanding population and provide necessary services and facilities. Nevertheless, as a result of this fast urbanization, the availability of blue-green spaces is diminishing steadily. This section provides a concise overview of the decline of BGI in the major urban areas of Bangladesh.

Dhaka, an increasingly overcrowded metropolis, has become gradually uninhabitable day after day. The water bodies in Dhaka City, including ponds, canals, and the flood flow zone, have experienced a significant reduction in size throughout the years. Research indicates that the lowlands had a reduction of 52.5% and water bodies decreased by 32.5% between 1960 and 2008 (Islam et al. 2010). According to the Dhaka Metropolitan Development Plan (DMDP) (RAJUK 2016a), over a span of

10 years, the water bodies dropped from about 8 to 5%, while the agricultural areas had a significant decline from 46 to 30%. The recently revealed draft Detailed Area Plan (2016–2035) for Dhaka City includes a requirement to preserve approximately 30% of agricultural land. This area plays a vital role in the city's green infrastructure, which includes parks, playgrounds, 8% aquatic bodies, and 2.5% open space and forests.

Chittagong is the second largest city in Bangladesh. There are numerous similarities between Dhaka City and Chittagong City. Multiple studies have indicated that the city's green spaces and bodies of water have been steadily diminishing, posing a significant threat. According to the District Fisheries Department, the city had a total of 19,250 bodies of water in 1991. In 2006, the number of water bodies within the Chittagong City Corporation (CCC) territory decreased to 4700, as reported in the CDA (2008). This reduction occurred over a span of 15 years and now accounts for 5.5% of the total land area.

Khulna, situated on the western bank of the Bhairab-Rupsha River, is the third largest city in Bangladesh. The presence of green spaces and water bodies in Khulna City is diminishing steadily, similar to the trends observed in Dhaka and Chittagong. According to a survey conducted in 2012 by the consultants working on the Detailed Area Plan (KDA 2018), the land cover in the Khulna Development Authority area is mostly composed of aquatic bodies (10.50%), urban green space (0.64%), and agricultural land (36%). However, the situation in Khulna City Corporation differs significantly, as there is a relatively little amount of land dedicated to urban green space (0.30%), water bodies (about 6%), and agriculture (13%). According to KDA (2018), between 1998 and 2012, the primary source of green infrastructure, namely, agricultural land, had a decline of more than 49%.

Rajshahi, also called the Silk City or the City of Education, is one of the major metropolitan cities in Bangladesh. The Rajshahi Development Authority (RDA) formulated the inaugural "Rajshahi Metropolitan Development Plan (RMDP), 2004–2024" for the first time in Bangladesh (RDA 2004). They achieved this by employing state-of-the-art planning techniques and tools, such as advanced surveying and mapping technology, covering a territory spanning 364.19 km^2. The plan encompasses an approximate area of 48.06 km^2, specifically the Rajshahi City Corporation. According to the 2003 Physical Feature Survey conducted by the consultant, the RDA planning area consists of around 43% agricultural land, 18% water bodies, and 0.16% open space. In comparison, the city corporation area comprises 19% agricultural land, 11% water bodies, and 1.10% open space. However, still green spaces and water bodies are declining in the city. Based on a study conducted by Kafy et al. (2018), the urban area experienced a loss of 35.64 acres of available land from 2008 to 2018. According to a separate study, the rapid growth of the city corporation area between 1992 and 2017 led to a 14% decrease in aquatic bodies (Kafy et al. 2019).

Mymensingh, previously referred to as Nasirabad, is expanding across the Brahmaputra River and currently ranks as the second most densely populated city in Bangladesh (Sharmin and Perveen 2019). The Mymensingh Strategic Development strategy (MSDP) has formulated a systematic expansion plan based on the analysis and interpretation of field data. This provides a summary of forthcoming advancements that will occur within the next two decades. The objective of this concept is to facilitate the city's forthcoming growth over the river (MSDP 2019). Despite the ongoing loss of green structures in this city, there has been a lack of focused research on BGI.

6.3 Losses of Green and Blue Spaces in Highly Populated Dhaka City

In the current epoch of global warming, the BGI is indispensable not only for enhancing the quality of life but also for facilitating sustainability among the shifting environmental conditions. According to the United Nations Environment Programme (UNEP), a city should possess a minimum of 25% green space in order to enhance the quality of life for its citizens. Moreover, the existence of urban green and blue areas exerts both beneficial and negative effects on the psychological, social, and physical well-being of the residents (Fig. 6.1). Nevertheless, as a result of increasing urbanization, cities in Bangladesh are experiencing a loss of their green spaces and are becoming unsustainable. In this section, we will provide a brief discussion about the present scenario of BGI in the most densely populated city Dhaka.

Over time, Dhaka City's green spaces and water features like ponds, canals, and flood flow zones have become progressively less in number. In the last three decades, the city has experienced a decline of more than 56% of its green spaces. Using GIS and remote sensing, a study examined Dhaka's green spaces during a 30-year period from 1989 to 2020. The results showed that there was a 56% overall decline in

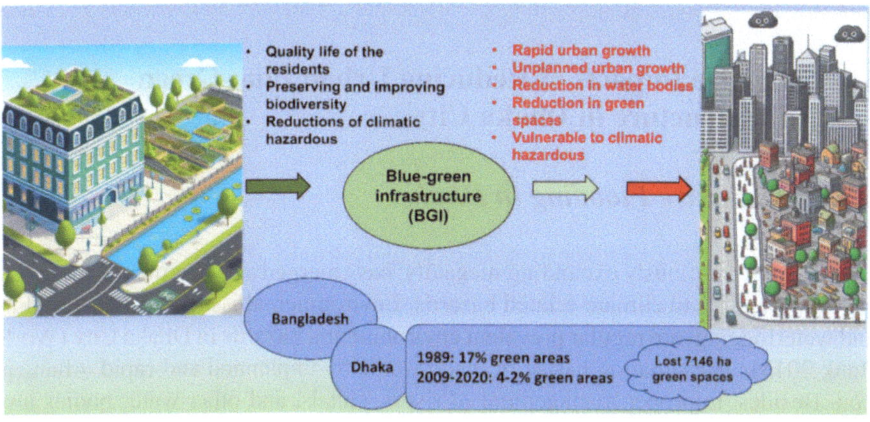

Fig. 6.1 Positive and negative consequences of BGI. Urban green and blue areas continuously diminish and have detrimental impacts on climate and people's physical, mental, and social health. Arrows in shades of green, light green, and red indicate the conversion of BGI to urbanization

vegetation during that time (Nawar et al. 2022). Nevertheless, it was shown that the most significant alterations happened in the overall green areas between 2009 and 2020. In 1989, the healthy green areas occupied a total area of 5202 ha., which accounted for 17% of the total land area. Since 1989, there has been a significant decline of 47% in the presence of healthy green spaces. This decline has resulted in a total of 2754 green spaces, which accounts for around 9% of the original number. Subsequent to the year 2009 and 2020, there were additional reductions of 55.5% and 50% respectively, leading to a final percentage of healthy green areas of 4 and 2% (Nawar et al. 2022). A recent study by Abdullah et al. (2022) revealed that Dhaka witnessed a significant decrease in urban green areas, dropping from 27.46% in 1991 to 10.14% in 2021. Similarly, the city's blue spaces also experienced a fall, decreasing from 20.32 to 13.44% over the same period. That being said, massive urbanization caused Dhaka City to lose 7146 hectares of green area between 1989 and 2020 (Nawar et al. 2022). Apart from green areas, the water bodies in Dhaka City decreased by more than 32.4% between 1992 and 2022 (Shammi et al. 2023).

In 1960, the total area of lowlands and wetlands was 13,527.58 ha. and 2,952.02 ha. respectively. However, the total area decreased to 12,717.73 and 2103.62 ha., respectively, in 1988. The situation deteriorated further, expanding to cover an area of 1990.71 ha. in 2008, indicating a continuous reduction in the size of the lowlands. Therefore, from 1960 to 2008, there was a decrease of 32.57% in lowland areas and a decrease of 52.58% in aquatic bodies (Islam et al. 2010). Moreover, the Bangladesh Institute of Planners (BIP) reports that between 2011 and 2019, at least 3483 acres of lowlands and water bodies in the Dhaka metropolitan area were piled up for infrastructures development (Nabi 2019). Nevertheless, it is imperative to minimize the construction of infrastructure that is unwanted and superfluous in order to preserve the urban settings. The policy recommendation section will offer guidelines on how to improve BGI in metropolitan areas to enhance the quality of life.

6.4 Negative Impacts of Reducing Urban Blue-Green Infrastructure in Dhaka City

6.4.1 Excessive Flooding in Dhaka

Dhaka, an expeditiously expanding megacity, has emerged as a significant urban hub yet is susceptible to climate-related hazards. Environmental issues, such as flooding and waterlogging, are regular prevalent environmental hazards in Dhaka City (World Bank 2015). The situation has been made worse by unplanned and rapid urbanization. Besides, improper management of rivers, canals, and other water bodies also exaggerated the city toward natural hazards. An eminent concern in the city is the occurrence of water logging, which arises from the diminishment of water reservoirs commonly referred to as blue space. The accumulation of water in the soil eventually results in significant floods following a brief period of rainfall (Khan

2001). Insufficient availability of blue spaces results in local floods, which occur when drains become obstructed, leading to streets being submerged in water up to knee or ankle level. The city's problem of waterlogging poses a threat to both the environment and the well-being of the populace (Nishat et al., 2000). For example, during the 2004 flood, around 18% of the western regions of Dhaka and a staggering 94% of the eastern regions were inundated, resulting in a significant impact on roughly 5 million individuals. As a result, there was a severe scarcity of potable water and a widespread breakdown of the sewerage infrastructure (Sayed and Huruyama 2017; World Bank 2017). It is well known that flooding poses health dangers in both the short and long terms, including gastrointestinal ailments, an increase in diseases transmitted by vectors, psychological effects, and even death. For instance, water sources in Dhaka were contaminated by human waste due to problem with drainage congestion issues during the floods of 2007 and 2004 (Islam et al. 2007, 2010). However, reports indicate that the Balu River in the east and the Tongi Khal in the north are the primary causes of 60% of the annual floods in Dhaka City (Gain et al. 2015). Consequently, on an annual basis during the monsoon season, the city encounters severe floods that result in significant detrimental impacts in terms of adverse social, physical, economic, and environmental consequences. In order to maintain biological equilibrium and mitigate the impacts of water logging and abrupt flooding, it is imperative for Dhaka to conserve its remaining wetlands.

6.4.2 Messiness in Dhaka

Dhaka is ideally located in the geographic center of Bangladesh, making it the central hub for social, cultural, economic, and political activities in the country. Additionally, it is the residence of around 10% of the nation's population (Roy et al. 2018; Rahman et al. 2017). Besides, annually, a substantial number of individuals relocate to Dhaka, either on a temporary or permanent basis, in search of employment and other means of sustenance. Weather and climate anomalies are causing people to abandon their rural livelihoods and forcing them to move to metropolitan areas, leading to a rise in population growth in slums and other informal settlements (RAJUK 2016b; Raju et al. 2017). For example, according to data obtained from the Bangladesh Bureau of Statistics (BBS), over 1.8 million individuals, which accounts for about 20% of the entire population, reside in slum areas (Hasan et al. 2021). Moreover, the city's substantial population and plenty of densely constructed buildings, together with the uncontrolled growth of urban development, are posing considerable obstacles to the quality of life in cities. As a result, the city is ranked seventh among the least livable cities globally (The Economist Intelligence Unit 2023).

Due to the city's haphazard urbanization, urban residents are not only losing green and open spaces, but also grappling with social issues such as inadequate affordable housing, urban poverty, income inequality, insufficient health facilities, limited women's empowerment, inadequate public transit, and illiteracy (Satu and Chiu 2019; Yasmin 2019; Barai 2020). Additionally, the haphazard development of

Dhaka has led to a significant increase in pollution levels in the air, soil, and water, reaching a critical point. Recently, Dhaka is ranked second in terms of having the poorest air quality among cities in Asia (WHO 2016). Industrialization, specifically, the textile and garment industries attract a large number of workers from rural areas. However, the unhealthy and overburdened working conditions in this industry lead to pollution. Besides, the exponential increase in population and uncontrolled urban development in Dhaka City have significantly transformed the water ecosystem, resulting in the depletion of groundwater levels and the pollution of river systems. The rivers encircling Dhaka are experiencing rapid degradation, with very low levels of dissolved oxygen and elevated levels of chloride, total dissolved solids, biochemical oxygen demand, and chemical oxygen demand (DOE 2015; WRG 2015). Moreover, severe traffic congestion is a significant impediment for the residents, among other livability difficulties. Although motorized transit, such as buses and minibuses, has experienced significant growth in Dhaka, non-motorized transport, particularly rickshaws, continues to be the prevailing mode of transportation. Furthermore, inadequate garbage management is a significant issue in the city of Dhaka. While the city corporation manages to collect approximately 50% of the waste created in Dhaka City, a significant portion of around 40–60% remains uncollected and is disposed of in an incorrect manner (Prodhan et al. 2019). However, the city is currently grappling with numerous issues stemming from a substantial volume of waste, and it is anticipated that the rate of waste output would rise to 220 kg per person per year by 2025 (Habib et al. 2021). Hence, the uncontrolled and haphazard urban expansion in Dhaka is progressively diminishing open areas, green spaces, water bodies, and community amenities. This poses a significant risk to the sustainable development of Bangladesh.

6.5 Policy Analysis and Recommendations to Enhance BGI

Considering the circumstances of the above-mentioned cities, we may say that the current condition of the BGI in our country's major cities is deplorable. This scenario worsens as the population grows. Therefore, certain policies are needed to implement BGI in large cities. To improve Dhaka City's blue-green infrastructure, policy analysis must strike a balance between ecological sustainability and urban growth, which is crucial considering the city's dense population and its high susceptibility to the effects of climate change. BGI enhances urban resilience and quality of life by integrating green spaces and water management to lower flood risks, improve air quality, and provide recreational areas. In addition to encouraging nature-based solutions, effective policy initiatives should address urban sprawl and land scarcity. These measures could include enhancing regulatory frameworks to protect natural water bodies, incorporating wetland conservation into urban development, and offering financial incentives for green roofs (Kabir and Parolin 2012; Alam et al. 2023). Involving several stakeholders such as local communities, private developers, and government agencies can promote coordinated action and sustainable urban development in Dhaka (Shammi et al. 2023).

Furthermore, improving BGI in Dhaka necessitates a comprehensive strategy that considers environmental, social, and economic aspects. First and foremost, encourage volunteer programs for upkeep and foster a feeling of pride by incorporating local communities in the design and upkeep of green spaces and water features. Secondly, encourage community involvement in conservation efforts and inform the public about the benefits of BGI. Encourage a master plan that incorporates blue-green infrastructure into land use planning as well in order to preserve and incorporate wetlands, parks, and rivers into urban development. However, the BGI framework facilitates the integration of both traditionally separate and autonomous blue and green efforts, with the potential to create a more resilient urban structure in response to climate change. Hence, it is imperative to implement appropriate measures to improve the integration of BGI into urban construction and maintenance. Therefore, following steps can be taken:

- Instead of sticking to rigid urban planning systems, performance-based planning should be adopted, especially in developing areas.
- Developing urban forests and green spaces is a cost-effective and environmentally friendly approach to addressing climate change. Priority should be given to augmenting the proportion of green spaces, growing extensive clusters of trees, and abstaining from intricate-shaped vegetation in order to attain a more efficient cooling impact.
- Increasing the quantity of green areas, such as urban parks, community garden initiatives, agricultural lots, and energy parks, in every area is necessary to effectively control the local temperature.
- An enhanced methodology should be devised to expand vertical greening systems, as they have the ability to reduce temperature increase within buildings and on their exteriors.
- Efficiently coordinated, strategized, and structured urban green spaces and parks should be supervised to optimize the chances for urban inhabitants to engage in a healthy lifestyle. Green parks should be equipped with essential amenities, services, and safety protocols to cater to the needs of park visitors.
- Urban green spaces should be developed by particularly focusing on older individuals who reside in underserved or underprivileged communities, as they have minimum accessibility to green spaces and higher quality parks.
- Concentration is required for the cultivation of street trees, which will aid in reducing humidity, temperature, and atmospheric pollution levels.
- Technology should be developed for the reutilization of treated wastewater in urban agriculture for efficient management of potentially reducing food system energy demand and supporting sustainable development. BGI should be designed to efficiently collect surface runoff, recharge groundwater, and address the conflicting situation of water shortage and urban flooding.

- It is recommended that a technically sound BGI be created in order to effectively use surface runoff or stormwater to the neighboring areas for low-maintenance landscapes and urban rain gardening.
- To enhance environmental resilience and sustainability in urban areas, it is imperative to implement meticulous planning and take decisive action toward the development of waterfronts, restoration of riverbanks, and establishment of water-based recreational amenities.

6.6 Conclusion

Bangladesh is one of the most climate-vulnerable countries globally owing to a combination of frequent natural disasters, high population density, and rapid urbanization. The concept of BGI is increasingly being recognized as an effective solution for addressing urban challenges and building resilience and sustainability. Based on the preceding discussion, it is evident that BGI plays a crucial role in both urban development and quality of life. Additionally, it serves as a means of safeguarding communities against the impacts of climate change. In Bangladesh, the government has implemented regulations and legislation to safeguard water bodies, such as "The Playground, Open Spaces, Parks, and Water Bodies Conservation Act 2000" and "The Public Water Body Management Policy 2009." The Hatirjheel Area Development Project successfully rejuvenated the natural water features in Dhaka North. In addition, authorities in Dhaka North have designated a "chief heat officer" to address the escalating heat strain caused by climate change and the urban heat island phenomenon. However, the policies suggested in this study can be implemented to achieve proper BGI, which can contribute to the sustainability of quality urban life. Moreover, it is essential for both individuals and the government to be conscious and take action to conserve the existing green spaces. Nevertheless, it is imperative to promote widespread consciousness to establish a robust and productive urban existence alongside the BGI.

References

Abdullah S, Adnan MS, Barua D, Murshed MM, Kabir Z, Chowdhury MB, Hassan QK, Dewan A (2022) Urban green and blue space changes: a spatiotemporal evaluation of impacts on ecosystem service value in Bangladesh. Ecol Inform 70:101730

Ahern J (1995) Greenways as a planning strategy. Landscape and urban planning, 33(1–3), 131–155

Alam R, Quayyum Z, Moulds S, Radia MA, Sara HH, Hasan MT, Butler A (2023) Dhaka city water logging hazards: area identification and vulnerability assessment through GIS-remote sensing techniques. Environ Monit Assess 195(5):543

Bai X, Dawson RJ, Ürge-Vorsatz D, Delgado GC, Salisu Barau A, Dhakal S, Schultz S (2018) Six research priorities for cities and climate change. Nature 555(7694):23–25

Bajracharya P, Sultana S (2020) Rank-size distribution of cities and municipalities in Bangladesh. Sustainability 12(11):4643

Barai MK (2020) Conclusion: Bangladesh's development—Challenges to sustainability and the way forward. In: Barai M (ed) Bangladesh's Economic and Social Progress. Palgrave Macmillan, Singapore, pp 383–409

CDA (2008) Preparation of detailed area plan (DAP) for Chittagong Metropolitan Master Plan (CMMP) 1995–2015. Chattogram development authority, Chattogram degradation through hill cutting in Chittagong district of Bangladesh. Int J Nat Soc Sci 1:41–54

DOE (Department of Environment) (2015) River Water Quality Report 2014. Ministry of Environment and Forest, Government of the People's Republic of Bangladesh, Dhaka. https://doe.portal.gov.bd/sites/default/files/files/doe.portal.gov.bd/publications/53bdd241_6208_4848_95ff_4cca3e059033/Water%20Quality%20Report%202014%20(3).pdf

Gain AK, Mojtahed V, Biscaro C, Balbi S, Giupponi C (2015) An integrated approach of flood risk assessment in the eastern part of Dhaka City. Nat Hazards 79(3):1499–1530. https://doi.org/10.1007/s11069-015-1911-7

Gunawardena KR, Wells MJ, Kershaw T (2017) Utilising green and bluespace to mitigate urban heat island intensity. Sci Total Environ 584:1040–1055

Habib MA, Ahmed MM, Aziz M, Beg MRA, Hoque ME (2021) Municipal solid waste management and waste-to-energy potential from Rajshahi City corporation in Bangladesh. Appl Sci 11:3744. https://doi.org/10.3390/app11093744

Hasan M, Khan MSA, Sutradhar I, Hossain MM, Hossaine M, Yoshimura Y, Mridha MK (2021) Prevalence and associated factors of hypertension in selected urban and rural areas of Dhaka, Bangladesh: findings from SHASTO baseline survey. BMJ Open 11(1):e038975

Hassan MM (2017) Monitoring land use/land cover change, urban growth dynamics and landscape pattern analysis in five fastest urbanized cities in Bangladesh. Remote Sens Appl 7:69–83. https://doi.org/10.1016/j.rsase.2017.07.001

Islam AS, Haque A, Bala S (2010) Hydrologic characteristics of floods in the Ganges–Brahmaputra–Meghna (GBM) delta. Nat Hazards 54:797–811

Jongman RH, Pungetti G (eds) (2004) Ecological networks and greenways: concept, design, implementation. Cambridge University Press

Kabir A, Parolin B (2012) Planning and development of dhaka—a story of 400 Years. Int Plan Stud 17(3):289–314

Kafy AA, Faisal AA, Ferdous L (2018) A participatory assessment to identify the causes and impacts of open space reduction in Rajshahi City Corporation (RCC) area, Bangladesh. Int J Environ Plan Dev 4(2):16–28

Kafy A-A, Islam M, Khan AR, Ferdous L, Hossain MM (2019) Identifying most influential land use parameters contributing reduction of surface water bodies in Rajshahi city, Bangladesh: a remote sensing approach. Remote Sens Land 2(2):87–95. https://doi.org/10.21523/GCJ1.18020202

KDA (2018) Preparation of detailed area development plan for Khulna Master Plan (2001) Area

Khan NI (2001) Assessment of water logging conditions using integrated GIS and remote sensing techniques: a study of Dhaka Mega City. Orient Geogr 45(2):41–54

Kim H, Lee DK, Sung S (2016) Effect of urban green spaces and flooded area type on flooding probability. Sustainability 8(2):134

Liu L, Fryd O, Zhang S (2019) Blue-green infrastructure for sustainable urban stormwater management—lessons from six municipality-led pilot projects in Beijing and Copenhagen. Water 11(10):2024

Liu X, Huang Y, Xu X, Li X, Li X, Ciais P, Zeng Z (2020) High-spatiotemporal-resolution mapping of global urban change from 1985 to 2015. Nat Sustain 3(7):564–570

MSDP (2019) Mymensingh strategic development 2011–2035. http://www.msdp.gov.bd/. Accessed 29 Jan 2019

Nabi MS (2019) 36% of Dhaka's water bodies filled up in 9 years, Bangla Tribune

Nawar N, Sorker R, Chowdhury FJ, Rahman MM (2022) Present status and historical changes of urban green space in Dhaka city, Bangladesh: a remote sensing driven approach. Environ Chall 6:100425. https://doi.org/10.1016/j.envc.2021.100425

Nishat A, Reazuddin M, Amin R, Khan AR (2000) The 1998 Flood: impact on environment of Dhaka City. Department of Environment and IUCN Bangladesh, Dhaka, Bangladesh pattern analysis in five fastest urbanized cities in Bangladesh. Remote Sens Appl 7:69–83

Plieninger T, Thapa P, Bhaskar D, Nagendra H, Torralba M, Zoderer BM (2022) Disentangling ecosystem services perceptions from blue infrastructure around a rapidly expanding megacity. Landsc Urban Plan 222:104399

Prodhan AHMSU, Kaeser A (2019) Solid waste management in Dhaka city-a review on the present status and possible solutions, nature study society of Bangladesh. In: Environmental thoughts, part-1. Toyza publications

Rahman MM, Hassan MS, Bahauddin KM, Ratul AK, Bhuiyan MAH (2017) Exploring the impact of rural-urban migration on urban land use and land cover: a case of Dhaka city, Bangladesh. Migr Dev 7(2):222–239

Rajdhani Unnayan Kartipakkha (RAJUK) (2016a) Dhaka structure plan (2016–2035). Rajdhani Unnayan Kartipakkha, Dhaka

RAJUK (2016b) Dhaka Metropolitan Development Plan (DMDP) 2016–2035

Raju D, Kim KY, Nguyen QT, Govindaraj R (2017) Cities, slums, and early child growth: empirical evidence from Bangladesh (Policy Research Working Paper 8094). World Bank Group, Washington, DC

RDA (2004) Rajshahi Metropolitan Development Plan (RMDP) 2004–2024

Roy S, Sowgat T, Ahmed MU, Islam ST, Anjum N, Mondal J, Rahman MM (2018). Bangladesh: national urban policies and city profiles for Dhaka and Khulna. http://www.centreforsustainablecities.ac.uk/wp-content/uploads/2018/06/Research-Report-Bangladesh-National-Urban-Policies-and-City-Profiles-for-Dhaka-and-Khulna.pdf

Ryan RL, Fábos JG, Allan JJ (2006) Understanding opportunities and challenges for collaborative greenway planning in New England. Landsc Urban Plan 76(1–4):172–191

Šakić Trogrlić R, Rijke J, Dolman N, Zevenbergen C (2018) Rebuild by design in Hoboken: a design competition as a means for achieving flood resilience of urban areas through the implementation of green infrastructure. Water 10(5):553

Satu SA, Chiu RL (2019) Livability in dense residential neighborhoods of Dhaka. Housing Stud 34(3):538–559. https://doi.org/10.1080/02673037.2017.1364711

Sayed MB, Haruyama S (2017) Flood risk measuring under the flood protection embankment construction in Dhaka metropolitan zone. J Geosci Geomat 5(2):46–58

Shammi M, Reza F, Sarker AC, Sakib AA (2023) Prospects and challenges of achieving sustainable urban green spaces: a case study of urban greening in Dhaka North City Corporation (DNCC), Bangladesh. PLOS Sustain Transf 2(5):e000006. https://doi.org/10.1371/journal.pstr.0000061

Sharmin F, Perveen M (2019) Evaluating open space in urban fabric on basis of different landscape aspects case study: circuit house park, Mymensingh, Bangladesh. In: Fábos conference on landscape and greenway planning, vol 6, no 1. University of Massachusetts Amherst Libraries

Sirajul Islam M, Brooks A, Kabir MS, Jahid IK, Shafiqul Islam M, Goswami D, Luby S (2007) Faecal contamination of drinking water sources of Dhaka city during the 2004 flood in Bangladesh and use of disinfectants for water treatment. J Appl Microbiol 103(1):80–87

The Economist Intelligence Unit (2023) The global liveability index 2023. https://ethz.ch/content/dam/ethz/special-interest/baug/irl/irl-dam/lehrveranstaltungen/bsc/site-management/Literatur/Economist_Jun-Global-Liveability-Index-2023.pdf

Thomas K, Littlewood S (2010) From green belts to green infrastructure? the evolution of a new concept in the emerging soft governance of spatial strategies. Plan Pract Res 25(2):203–222

Welford MR, Yarbrough RA (2021) Urbanization. Hum-Environ Interact Int 193–214

WHO (World Health Organization) (2016) WHO global urban ambient air pollution database. Geneva: WHO. https://www.who.int/data/gho/data/themes/air-pollution/who-air-quality-database/2016

World Bank (2020) Urban population growth (annual %) | Data. https://data.worldbank.org/

World Bank (2017) Flood risk management in Dhaka: a case for eco-engineering approaches and institutional reform. Unpublished paper, World Bank, Washington, DC

WRG (Water Resources Group) (2015) Consolidation and analysis of information on water resources management in Bangladesh. World Bank, Washington, DC. https://2030wrg.org/wp-content/uploads/2015/11/2030-WRG-Bangladesh-Revised-Final-Report-PwC_Nov15_new cover.pdf
Yasmin F (2019) Identification of the existing social problems and proposing a sustainable social business model: Bangladesh perspective. Int J Finance Econ 11(9):1–81. https://doi.org/10.5539/ijef.v11n9p81

Chapter 7
Global Actions for Climate-Resilient Blue-Green Infrastructure

Sadiya Arefin Juthee, Md. Hosenuzzaman, and Mohammad Golam Kibria

Abstract Natural ecosystems are under increasing pressure due to the shifting global economy, increasing urbanization, natural disasters, and climate change. Blue-Green Infrastructure (BGI) is a collection of nature-based techniques integrating hydrological functions (blue) and vegetated landscapes (green), which offers resilience to climate change and urban multi-functionality. Managing more frequent and intense rainfall events, drier summers, faster urbanization, and fewer public green spaces are global challenges that contribute to water-related issues like flooding, water scarcity, overexploitation of groundwater, water pollution, inefficient drainage systems, and waste of rainwater in urban areas. This chapter will discuss on the global actions for climate-resilient sustainable BGI to overcome the challenges in rapidly urbanized cities.

Keywords Global actions · Global frameworks · Climate resilience · BGI · Sustainable cities

7.1 Introduction

Weather extremes are an accepted component of coupled human–environment systems (IPCC 2014). However, this is unlikely that in the event of the climate changes, we are ready for a potential step change in their frequency and intensity. Extreme weather events in recent years, including heat waves, wildfires, floods, and droughts, have exposed many natural ecosystems and human systems to significant

S. A. Juthee
Department of Crop Botany, Bangladesh Agricultural University, Mymensingh, Bangladesh

Md. Hosenuzzaman
School of Agriculture and Environment, UWA, The University of Western Australia, Perth, Australia

Md. Hosenuzzaman · M. G. Kibria (✉)
Department of Soil Science, Bangladesh Agricultural University, Mymensingh, Bangladesh
e-mail: kibria.ss@bau.edu.bd

M. G. Kibria et al. (eds.), *Urban Blue-Green Infrastructure Approach for Food Security and Climate Disaster Resilience*, Disaster Risk Reduction,
https://doi.org/10.1007/978-981-96-8746-6_7

levels of climatic unpredictability and change. Extremes in the climate can have a variety of negative effects, including disruptions to the supply of food and water, changes to ecosystems, harm to human health and welfare, and destruction of land and infrastructure. These impacts are exacerbated by the lack of awareness and preparedness for existing climate variability in countries at all levels of development (Faggian and Sposito 2009a, b).

Urbanization has prioritized impervious concrete surfaces and high-rise buildings, leading to increased pollutant-laden runoff from city streets (Nguyen et al. 2019). This rapid development has resulted in the destruction of lakes, wetlands, and their ecosystems while also disrupting both natural and historically engineered water systems (Cun et al. 2019; Jiang et al. 2018; Tang et al. 2018; Xu et al. 2018). Additionally, higher flows in municipal sewer outfalls, along with pollutant and nutrient discharge from peri-urban industries and agriculture, contribute to water contamination and pose risks to public health (Guo et al. 2020; Rui et al. 2018).

Climate disruption intensifies the challenges posed by urbanization on urban infrastructure (Duan et al. 2016; Li 2013). Coastal cities face additional threats, including land loss and damage from sea-level rise and coastal erosion in deltas and floodplains (Loftus et al. 2011; Ruckelshaus et al. 2016; Wang et al. 2018). At the same time, rapid urbanization and climate change diminish the quality of life for both humans and urban wildlife. Increasingly extreme weather events and sudden climate fluctuations make daily life more difficult—and at times, even deadly. The expansion of impervious surfaces leads to habitat destruction, weakening urban ecosystems and reducing biodiversity (Guo et al. 2020). Furthermore, the loss of natural spaces fragments communities, erodes human–nature connections, degrades air quality, and exacerbates the urban heat island effect. These factors contribute to mental health issues, respiratory and cardiovascular diseases, heat-related illnesses, and other public health risks (Heckert and Rosan 2016; Kozak et al. 2020; Tzoulas et al. 2007).

We either adapt to and mitigate the effects of climate change, or fail to do so, which will have catastrophic implications. The future will be shaped according to our response to this challenge (Leary 2012). The first step in preparing for future climate change is to reduce susceptibility and exposure to current climatic variability. Flooding and extreme weather events will become more common as a result of climate change, and only long-term solutions like Blue-Green Infrastructure (BGI) will preserve humanity. Strategies such as BGI that include actions with co-benefits for other objectives can increase resilience across a range of possible future climates while helping to improve human health, social and economic well-being, environmental quality, and livelihoods (IPCC 2014; Middlemann and Middlemann 2007).

There is a great demand for anthropogenic environmental alteration in response to climate and weather changes over time, as well as the creation of "sustainable" engineering solutions. BGI, as one of such solutions, provides a solid, long-term foundation for growth that protects against extreme weather events while also providing a variety of other benefits with no drawbacks. Climate change and increased extreme weather require the implementation of systems such as BGI to provide sustainable and resilient communities. In changing climate, the need for resilient and sustainable infrastructure has become paramount. Blue-green infrastructure, incorporating natural and engineered elements, offers a holistic approach to climate resilience. This chapter explores global initiatives and actions aimed at promoting climate-resilient blue-green infrastructure, referencing key strategies and examples.

7.2 Defining Climate-Resilient BGI

The climate-resilient blue-green infrastructure (BGI) integrates water management and ecological considerations into urban planning and development (Fig. 7.1). It encompasses a range of solutions, including green roofs, urban wetlands, sustainable drainage systems, and coastal protection measures. BGI provides multi-functional benefits like less urban water discharge, better climate change adaptation, less water pollution, increased biodiversity, and water storage (Benedict and McMahon 2012; Handley et al. 2007). Green space benefits people's health and well-being as well as their physical and mental development, even on an individual basis. The goal is to building resilience to encounter climate-related challenges, such as extreme weather, water scarcity, and rising sea levels.

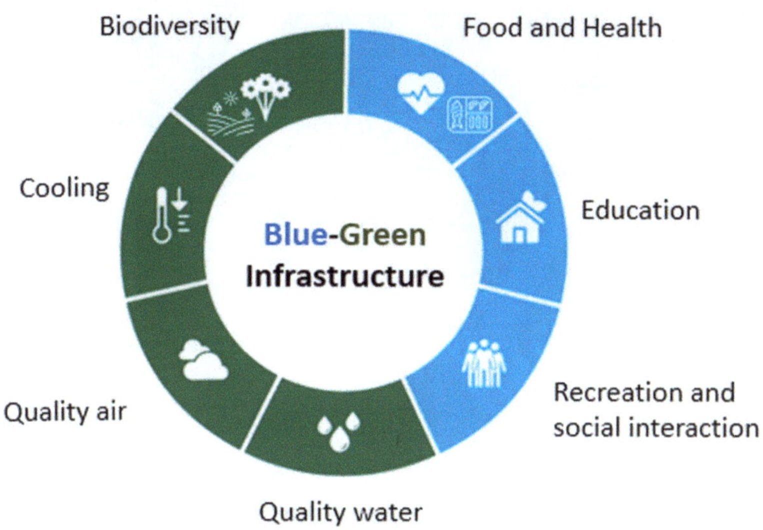

Fig. 7.1 Climate-resilient blue-green infrastructure

7.3 Blue-Green Infrastructure as a Potential Solution for Climate Change Variability and Sustainable Development

Blue-green infrastructure is an interconnected network of water reservoirs, wildlife habitats, wetlands, and other natural areas; parks, greenways, and other conservation lands; ranches, farms, and forests; and wilderness and other open spaces that support species, maintain natural ecological processes, sustain clean air and water resources, and contribute to the health and quality of life (Benedict and McMahon 2002). Hence, BGI is a crucial means for managing extreme weather events (Fig. 7.2).

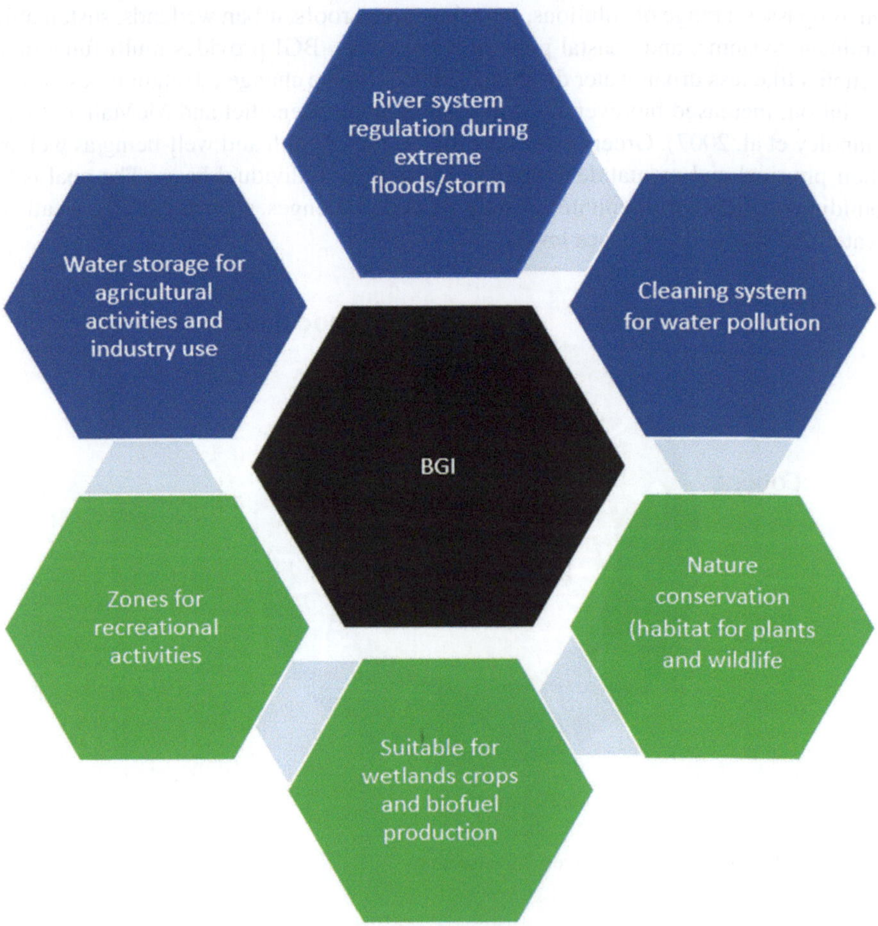

Fig. 7.2 The ways a network of interconnected wetlands, water reservoirs, and their associated (natural) spaces along the rivers in BGI system deal with flooding/extreme weather

The authors of IPCC assessment reports state that BGI is thought of as a win–win approach for urban climate change adaptation and mitigation (Roberts et al. 2012; Bai et al. 2018; Gill et al. 2007; Kazmierczak and Carter 2010; IPCC et al. 2022). BGI concerning urban flood risk reduction and heat mitigation are the most prominent issues studied and implemented to date for adapting cities to extreme weather (Gill et al. 2007; Demuzere et al. 2014). Increasingly, urban areas are incorporating BGI components, i.e., street trees, green roofs, and urban forestry, as ways to counteract the urban heat island effect. These components help cool buildings, provide wind corridors, enhance thermal comfort, and ultimately reduce energy usage (Carter et al. 2015; Moreno-García and Baena 2019; Islam et al. 2024). Temperatures in the vicinity of water bodies might decrease by 2–3 °C when there is a presence of trees (Wu and Zhang 2019; Islam et al. 2024). Research conducted in Manchester, U.K. has found that the presence of mature trees can reduce surface temperatures by up to 15.6 °C (Gill et al. 2007). Furthermore, there have been documented instances in Sacramento, USA, where walls and roofs located near tree canopies saw a decrease in surface temperatures ranging from 11 to 25 °C (USEPA 2008). According to Su et al. (2021), a 0.5 K decrease in the urban heat island effect can result in monthly cooling energy savings of 0.17–1.84 KWh/m^2. However, the degree of cooling and the distance from BGI are not constant; rather, they change according to the kinds of BGI arrangement, climate, topography, urban form, and the kinds of built environments nearby (Bowler et al. 2010; Yu et al. 2020). Urban green spaces like wetlands and woods can act as storm-protection or storm-regulation barriers in addition to being carbon sinks, which lessens the severity of extreme weather.

The changing topography and declining amounts of permeable underlying surface in urban areas brought on by rapid urbanization have enhanced the intensity of disasters related to urban stormwater runoff, as demonstrated by the floods in Petropolis, Brazil in 2022, which claimed hundreds of lives (Feitosa 2022). BGI can be extremely helpful in rainwater collection and storage, which can help mitigate the severe negative water-related effects of climate change, like flooding and landslides, in addition to its potential for cooling and energy savings. In this context, wetlands, woodlands, private gardens, urban parks, riverain areas, playing fields, and other artificial and natural areas in a catchment provide stormwater regulation services, and so help cities mitigate flood risks by retaining a large amount of rainwater for a longer period, so delaying the peak discharge (Shafique et al. 2018; Almeida et al. 2018; Haase 2015).

A larger urban green network can benefit from the presence of green corridors, which are made up of plants and trees that connect open spaces and green areas. These corridors can offer ecosystem services, such as enhanced habitat, rainwater collection, cooling effects, air purification, food production, and flood mitigation, in addition to acting as a catalyst for sustainable mobility networks that improve access and social interaction in urban areas and provide bike and walking paths. Nonetheless, by decreasing the usage of conventional forms of transportation, an increase in the use of alternative modes of transportation combined with the provision of necessary services close to one's home can help lower the ultimate demand for fossil fuels.

7.4 Key Global Initiatives and Frameworks

7.4.1 Sendai Framework for the Disaster Risk Reduction (2015–2030)

The Sendai Framework highlights the importance of ecosystem-based approaches and sustainable land-use planning for disaster risk reduction. It calls for the integration of climate resilience into infrastructure development, aligning with the principles of blue-green infrastructure.

7.4.2 Sustainable Development Goals (SDGs)

Sustainable development goals provide a unified vision for human and environmental peace and development. There are 17 Sustainable Development Goals (SDGs) that include approaches to improving education and health, reducing inequality, spurring economic growth, combating climate change, and preserving our oceans and forests. Among those, SDG 11 focuses on sustainable cities and communities, emphasizing the importance of resilient infrastructure. SDG 13 addresses climate action, recognizing the need for adaptation measures to build resilience against climate impacts.

7.4.3 United Nations Framework Convention on Climate Change (UNFCCC)

The Paris Agreement underscores the significance of adaptation strategies, including the promotion of climate-resilient infrastructure. The Global Climate Action Agenda encourages initiatives that enhance climate resilience through sustainable infrastructure development.

7.4.4 European Green Deal

The European Green Deal (EGD) is the European Union's comprehensive strategy aimed at achieving a climate-neutral continent by 2050, while fostering economic growth, environmental sustainability, and social well-being. A key component of this agenda focuses on enhancing climate-resilient, blue-green infrastructure (BGI)—a framework that integrates natural water management (blue) with vegetation and green spaces (green) to mitigate climate change impacts, particularly in urban areas. By investing in resilient BGI, the EGD aims to mitigate climate risks, protect biodiversity, and promote sustainable urban environments, establishing a foundation for a greener and more climate-adaptive Europe.

7.4.5 Convention on Biological Diversity (CBD)

The **Convention on Biological Diversity (CBD)** is an international agreement aimed at conserving biological diversity, promoting sustainable use of its components, and ensuring the fair and equitable sharing of benefits from genetic resources. The CBD plays a crucial role in supporting climate-resilient blue-green infrastructure, which integrates water management (blue) and vegetation (green) into urban planning to enhance ecological health and reduce climate impacts.

Through CBD guidelines, nations are encouraged to preserve and restore ecosystems—such as wetlands, forests, rivers, and coastal areas—which act as natural buffers against climate impacts like flooding and heatwaves. The CBD framework supports biodiversity-friendly urban designs that improve ecosystem services, enhance urban resilience, and create sustainable habitats within cities. In climate-resilient blue-green infrastructure, CBD principles emphasize the conservation of local flora and fauna, sustainable resource use, and public engagement in biodiversity protection, all of which contribute to robust, adaptive urban landscapes that benefit both biodiversity and human communities.

7.4.6 Global Commission on Adaptation (GCA)

The Global Commission on Adaptation (GCA) is an international initiative aimed at accelerating climate adaptation efforts worldwide, with a focus on sustainable, resilient infrastructure. One of its key areas of emphasis is blue-green infrastructure, which integrates natural water management ("blue" for water) with green spaces, enhancing climate resilience and sustainability. GCA promotes nature-based solutions to manage urban flooding, reduce heat, and protect biodiversity. Examples include green roofs, permeable pavements, urban forests, and restored wetlands, all designed to support climate resilience while providing recreational spaces and improving air quality in cities. The GCA advocates for integrating these solutions into urban planning, especially in vulnerable regions, to reduce climate risks, lower maintenance costs, and enhance community well-being. Through partnerships, research, and funding, the GCA fosters the implementation of blue-green infrastructure as a critical part of climate adaptation strategies worldwide.

7.4.7 Resilient Cities Network

The Resilient Cities Network (RCN) is a global organization focused on helping cities build resilience to climate and other urban challenges, promoting sustainable solutions that enhance social, economic, and environmental well-being. In terms of climate-resilient blue-green infrastructure, RCN advocates for integrating natural systems (green spaces and water management) into urban planning to reduce flooding, manage stormwater, cool urban areas, and improve public health and biodiversity. Overall, RCN's blue-green infrastructure efforts support cities in building resilience against climate impacts while promoting urban environments that are liveable, healthy, and sustainable.

7.4.8 The 100 Resilient Cities Program

The *100 Resilient Cities Program*, pioneered by the Rockefeller Foundation, aimed to bolster the resilience of cities worldwide to physical, social, and economic challenges, including those posed by climate change. A critical focus of the program was the development of *blue-green infrastructure* systems integrating water management ("blue") with vegetation and natural ecosystems ("green") to improve urban resilience. This approach promotes the use of natural solutions, such as urban forests, wetlands, green roofs, permeable pavements, and rain gardens, to manage stormwater, reduce flooding, cool urban areas, and improve air quality. These interventions help cities adapt to climate impacts by enhancing natural ecosystems within urban settings, supporting biodiversity, and improving residents' quality of life.

Participating cities in the program developed resilience strategies tailored to their unique vulnerabilities, incorporating blue-green infrastructure to address issues such as extreme weather, water scarcity, and heat stress. The initiative fostered knowledge sharing across cities, promoted collaboration with local stakeholders, and advanced innovative policies that prioritize sustainability and equity in resilience planning.

7.4.9 Urban Climate Change Resilience Trust Fund (UCCRTF)

The Urban Climate Change Resilience Trust Fund (UCCRTF) is a multi-partner initiative led by the Asian Development Bank (ADB) to support cities in building resilience against climate change through sustainable, climate-resilient infrastructure projects. Established with contributions from governments and institutions, including the UK, Switzerland, and the Rockefeller Foundation, the UCCRTF focuses on blue-green infrastructure-integrating natural and water-based elements, such as wetlands, green corridors, and sustainable drainage, to mitigate flood risks, manage stormwater, and reduce urban heat. Through UCCRTF, cities are implementing projects like flood management systems, green parks, permeable pavements, and restored water bodies, enhancing their adaptive capacity while preserving biodiversity and creating healthier urban environments.

7.4.10 Ramsar Convention on Wetlands

The Ramsar Convention on Wetlands, established in 1971, is an international treaty aimed at the conservation and sustainable use of wetlands. It recognizes wetlands as vital ecosystems that support biodiversity, water resources, and carbon storage, all of which are essential for climate resilience. Wetlands act as natural buffers against floods, droughts, and coastal erosion, making them integral to blue-green infrastructure systems that use natural resources to manage water sustainably and mitigate climate risks.

For climate-resilient blue-green infrastructure, the Ramsar Convention emphasizes protecting and restoring wetlands to enhance ecosystem services, such as water purification, flood control, and carbon sequestration. It promotes international cooperation, encourages sustainable management practices, and designates Ramsar sites—wetlands of international importance. These designated sites help countries create resilient water management strategies that reduce vulnerability to climate impacts, benefiting both people and nature.

7.4.11 The World Bank's Climate Resilience Program

The World Bank's Climate Resilience Program focuses on building sustainable, climate-resilient infrastructure, particularly through "blue-green infrastructure" strategies. These strategies integrate natural (blue) and vegetative (green) elements to create urban spaces that can better withstand and adapt to climate impacts, like flooding, extreme weather, and rising temperatures. The program promotes solutions such as restored wetlands, green roofs, rain gardens, permeable pavements, and urban forests, which work together to manage stormwater, reduce urban heat, and enhance biodiversity. By embedding these nature-based solutions into urban planning, the World Bank aims to protect vulnerable communities, improve urban livability, and support sustainable economic development.

7.4.12 *Natural Capital Protocol*

The *Natural Capital Protocol* is a framework that helps organizations assess their impacts and dependencies on natural capital, supporting the integration of nature-based considerations into business and infrastructure planning. For climate-resilient blue-green infrastructure, the protocol provides guidance on evaluating how green (vegetation-based) and blue (water-based) infrastructure solutions deliver environmental, social, and economic benefits. Using the *Natural Capital Protocol* in blue-green infrastructure planning allows stakeholders to better understand the multifaceted benefits of nature-based solutions, optimize resource use, and contribute to long-term climate resilience.

Table 7.1 summarizes the key global efforts to promote climate-resilient blue-green infrastructure, with a focus on integrating climate adaptation, biodiversity conservation, and sustainable infrastructure development.

7.5 Case Studies and Best Practices

7.5.1 *Singapore's ABC Waters Program*

Singapore's active, beautiful, and clean (ABC) waters program integrates water management with urban development, creating climate-resilient blue-green spaces (Fig. 7.3). ABC waters program was launched in 2006. The ABC water program has gradually developed a pervasive network of about 8,000 km of waterways and 17 reservoirs for our water supply (OPUB 2024). The program includes measures such as rain gardens, bioswales, and the restoration of water bodies for flood control and water quality improvement. It also initiated new recreational spaces, integrated waters with urban spaces. All the water goes to drains and straight into the canals and the water levels rise very quickly. However, ABC water management strategies addressed all these by detaining and treating runoff water on site and the runoff water flow slows down and cleaner water flows into the water reservoirs.

7.5.2 *Netherland's Room for the River Project*

The Netherlands is extremely vulnerable to flooding from the high water levels of the sea and rivers more frequently due to rainwater and meltwater in the changing climate. Hence, the rivers need more space to discharge the flowing water safely from rivers to sea and vice versa. The Room for the River project is a comprehensive approach to flood risk management in the Netherlands. It involves a combination of floodplain restoration, river widening, and sustainable urban development to enhance climate resilience and protect against flooding (Fig. 7.4). BGI is well accepted in The

Table 7.1 Global initiatives and frameworks for climate-resilient BGI

Initiative/ framework	Description	Key focus areas	Organizations involved	Year
Paris Agreement (UNFCCC)	A legally binding international treaty on climate change that focuses on limiting global warming to below 2 °C	Climate resilience, adaptation, emission reduction, and nature-based solutions	United Nations Framework Convention on Climate Change (UNFCCC)	2015
Sendai Framework for Disaster Risk Reduction	A global framework for reducing disaster risks, including those related to climate change impacts and infrastructure	Risk reduction, infrastructure resilience, and sustainable development	United Nations Office for Disaster Risk Reduction (UNDRR)	2015
Sustainable Development Goals (SDGs)	A set of 17 global goals for sustainable development, including clean water, climate action, and sustainable cities	SDG 6 (clean water and sanitation), SDG 13 (climate action), SDG 11 (sustainable cities), and SDG 15 (life on land)	United Nations	2015
European Green Deal	A roadmap for making the EU's economy sustainable by turning climate and environmental challenges into opportunities	Climate neutrality, nature-based solutions, and green infrastructure	European Commission	2019
Convention on Biological Diversity (CBD)	An international treaty aimed at conserving biodiversity, promoting sustainability, and ensuring ecosystem services	Ecosystem-based adaptation, nature-based solutions, and sustainable infrastructure	United Nations	1992
Global Commission on Adaptation (GCA)	An initiative to accelerate adaptation actions in response to climate impacts, including resilient infrastructure	Climate adaptation, resilience in cities, water management, and infrastructure development	Global Center on Adaptation	2018
Resilient Cities Network	A network that focuses on building resilient cities by integrating nature-based and green infrastructure approaches	Urban resilience, green–blue infrastructure, climate risk mitigation	Global Resilient Cities Network	2019

(continued)

Table 7.1 (continued)

Initiative/ framework	Description	Key focus areas	Organizations involved	Year
The 100 Resilient Cities Program	A project of The Rockefeller Foundation aimed at helping cities worldwide become more resilient to physical, social, and economic challenges	Urban climate resilience, infrastructure development, water management, and nature-based solutions	The Rockefeller Foundation	2013
Urban Climate Change Resilience Trust Fund (UCCRTF)	A multi-donor trust fund supporting climate-resilient urban infrastructure in Asian cities	Urban infrastructure resilience, water management, nature-based solutions, and climate adaptation	Asian Development Bank (ADB)	2013
Ramsar Convention on Wetlands	An international treaty that provides the framework for national action and international cooperation for the conservation and sustainable use of wetlands	Wetlands conservation, water management, and biodiversity protection	Ramsar Convention Secretariat	1971
World Bank's Climate Resilience Program	Supports climate adaptation projects, particularly those involving resilient infrastructure and sustainable water systems	Climate-resilient infrastructure, water management, and disaster risk reduction	World Bank	Ongoing
Natural Capital Protocol	A framework to help businesses measure and value their impacts and dependencies on natural capital, including blue-green infrastructure	Sustainable development, corporate responsibility, natural capital, and biodiversity integration	Natural Capital Coalition	2016

Netherlands (Thorne et al. 2018). In particular, climate change impacts on flooding are especially relevant in the Netherlands where about 25% of the land is below mean sea level and over 50% of the country is being protected from flooding by embankments (Koomen et al. 2012; Snepvangers et al. 2011; Sposito et al. 2012; Verburg et al. 2012). More broadly, it has been argued that: "Drastic changes are expected in land-use patterns as a result of socioeconomic developments and climate change. Biodiversity and other land-use functions require larger areas for adaptation to changes in climatic conditions and thus increase competition for available space. This requires a different, more multi-functional type of land-use planning and more

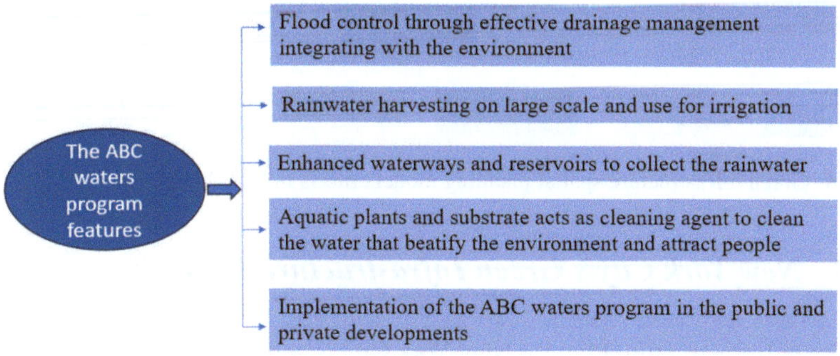

Fig. 7.3 The ABC water program features that improve the quality of stormwater surface runoff while beautifying urban environments

efficient management of resources" (Koomen et al. 2012). Water quality improvements can be ensured by biologically managing this BGI component to create a large natural area along the watercourse. The greater area can absorb far more nutrient and sediment load than the animals and plants along the stream channel, which can become overwhelmed after heavy rains. Therefore, it can enhance the water quality. Moreover, wetlands take a while to fill and empty during a flood, so they help decrease the size of floods for downstream areas (Snepvangers et al. 2011). Different BGI schemes have been applied in different parts of The Netherlands. These approaches combine employing natural and water resources to provide multi-functional climate corridors, which in turn result in efficient nature adaptation, agrobiodiversity, and sustainable farming (Steingrover et al. 2010).

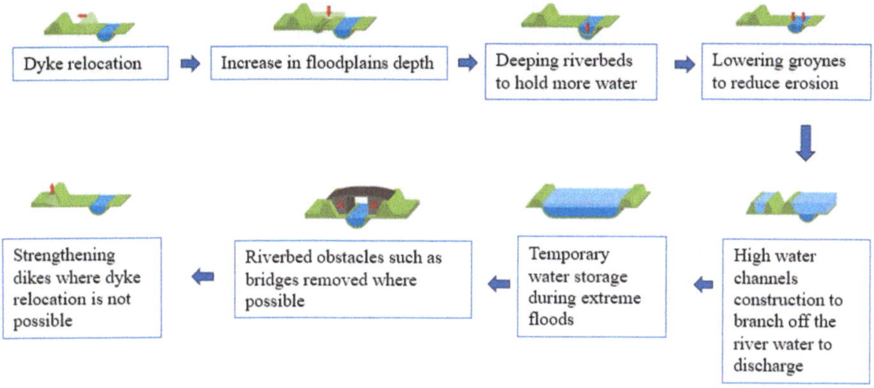

Fig. 7.4 Netherland's room for the river project plan

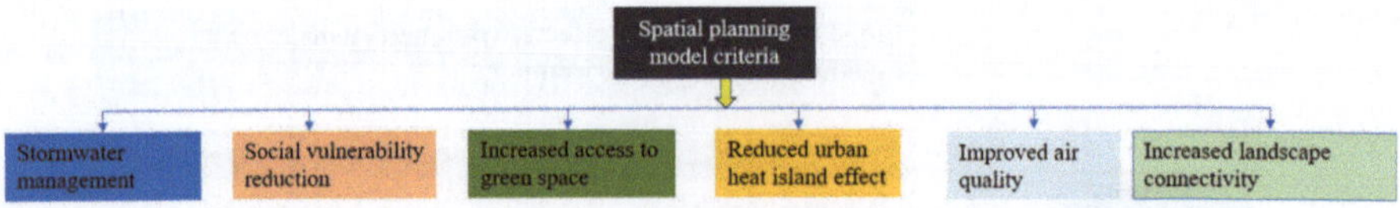

Fig. 7.5 Green infrastructure spatial planning model criteria in New York City

7.5.3 *New York City's Green Infrastructure Plan*

The Green Infrastructure Plan for New York City is centered on enhancing resilience against extreme weather events and managing stormwater through the utilization of natural systems (Fig. 7.5). Initiatives include green roofs, permeable pavements, and the restoration of natural wetlands to reduce flood risk.

7.5.4 *Hoeksche Waard*

In 2010, the Dutch Government conducted a study aiming to design a blue-green network for pesticide control measure in Hoeksche Waard—an agricultural area in the southwest of the Netherlands identified by cultivable fields and a widespread network of dikes, creeks, ditches, and field margin (Steingröver et al. 2010). The objective was to ascertain how local stakeholders (ecologists, farmers, hydrologists, and local policy makers) could apply logical information regarding the relationship between the spatial arrangement of blue-green networks and enhanced natural pest management. The integration of blue-green landscape features has the potential to enhance natural pest management by providing various resources, such as food, shelter, and a favorable microclimate, to encourage the survival of beneficial insects that inhibit agricultural pests. The ultimate goal was thus the development of sustainable farming systems that do not depend on chemical pesticides and are economically viable (Sposito et al. 2012; Steingröver et al. 2010; Alterra 2007). Overall, when developing Hoeksche Waard for pesticide management, three BGI-inspired situations were taken into account and put into action (Fig. 7.6).

7.5.5 *Blue-Green Cities Project in the United Kingdom*

The University of Nottingham, UK, launched a 3-year initiative in 2013 to use blue-green infrastructure to reduce the risk of frequent floods. The project involved institutions working with the government, public, and industry in Newcastle. The project adopted a strategy being the stakeholders at the center, remodeling the drainage systems and minimizing risks by identifying risk factors, studying gaps in the urban

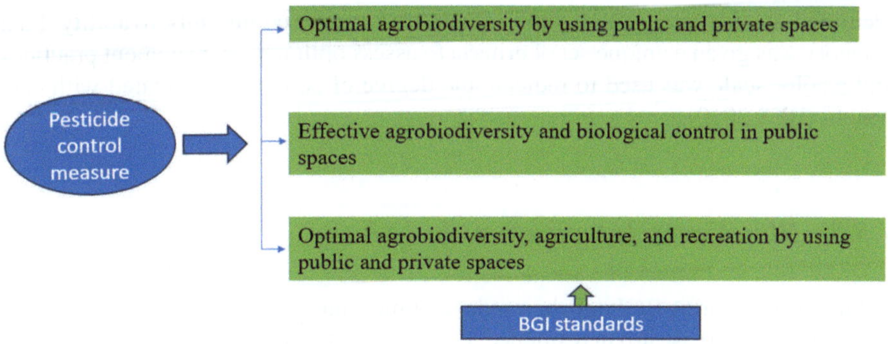

Fig. 7.6 BGI scenarios in Hoeksche Waard

drain networks, working with the communities to develop adaptive strategies, establishing a flood footprint to estimate economic costs, identifying flood prevention barriers, and putting up blue-green infrastructure (Lawson et al. 2014).

7.5.6 Rain City Strategy in Canada

The Vancouver Board of Parks and Recreation approved and formulated the "Rain City Strategy" in 2019. The objectives were to preserve and enhance the quality of urban water, boost urban resilience by managing water sustainably, and promote livability by enhancing natural and urban ecosystems (City of Vancouver 2019). This was accomplished by laying out specific goals, like reducing the amount of pollutants in the air and water, expanding managed impermeable areas and green spaces that absorb runoff, and reducing the impact of the urban heat island effect. Several action plans pertaining to buildings, streets and public spaces, and beaches and parks were implemented in order to achieve the aim of collecting and cleaning 90% of the received rainwater.

7.5.7 Portland's Grey to Green Initiative in USA

The Portland Watershed Management Plan (2005) was made possible by the five-year "Grey to Green" program, which had the overarching objectives of preserving natural resources, reviving important ecosystems, and putting stormwater solutions into place that would blend urban areas with the surrounding environment (BES 2005). In order to accomplish these objectives, the project identified crucial techniques such as culvert removal, revegetation, removal of invasive species, planting, and acquisition of sensitive land parcels. Other features included planting street and yard trees, green streets, and eco-roofs. Three categories of potential advantages were identified: health, energy and carbon sequestration, and community livability. Each category was given a unique set of criteria to assess optimum management practices, and a color scale was used to indicate the degree of certainty associated with each benefit (BES 2010).

7.5.8 *Blue-Green Interventions in India*

India is incorporating BGI into their redevelopment and urban planning interventions in an effort to increase urban resilience. Following extreme weather events, a number of cities have created new resilience programs through their institutions that center on BGI. Despite being a relatively new idea, a number of Indian cities, including Bengaluru, Delhi, Bhopal, Chennai, Madurai, and Kochi, are implementing blue-green elements into their plans in an effort to improve the natural blue systems that already exist in the metropolis and the public areas that surround it by a strategic plan (Table 7.2).

7.5.9 *Wuhan's Sponge City Program in China*

Wuhan, a low-lying district in China's Hubei Province that was formerly known as the "city of a hundred lakes," witnessed rapid urbanization that resulted in the death of numerous lakes. Severe flooding events have damaged transport infrastructure (2012) and caused drainage system collapses (2016), resulting in losses of CNY 2.3 billion (USD 0.32 billion) in terms of lives and property (Jing 2019). In order to solve China's problems with urban water management and restore harmony between the city, the people, and the water, the concept of the "sponge city" was developed in 2013. The national government launched the "sponge city" program in 16 pilot cities in order to provide existing urban infrastructure with additional absorptive capacities. In 2016, the program was expanded to 30 cities with an initial budget of CNY 20.7 billion (USD 2.87 billion) for the duration of 2015–2017. By 2020, the cities were required to provide 20% of their land with sponge characteristics, which could absorb and use up to 70% of the rainfall, and increase that percentage to roughly 80% by 2030 (Peng and Reilly 2021).

The concept of sponge city deals with urban management to resolve urban waterlogging, improve water storage and discharge capacity, improve water quality, and mitigate heat island effects by implementing the concept's six technical measures—"infiltration, retention, storage, purification, utilization, and discharge." The Wuhan Sponge City Project has been a great success so far, showcasing the remarkable power of preventing waterlogging and the possibilities of using natural remedies.

Table 7.2 An overview of the major blue-green infrastructure initiatives in India

City	Area (km^2)	Population (in million)	Responsible authority	Initiative	Purpose
Bengaluru	1307	9.1	Bangalore Water Board and Bangalore Municipal Corporation	Water and Sewerage Masterplan 2050	To maintain ecosystems, transition to a low-carbon economy, ensure resource security, and be resilient to climate change
Bhopal	1017	1.8	Bhopal Municipal Corporation (BMC) and Bhopal Smart City Ltd	Blue-Green Masterplan	To enhance health, build an environmentally sustainable city, and preserve and expand the green space
Delhi	1483	16.7	Delhi Development Authority	Blue-Green Masterplan	To ensure the simultaneous planning of blue and green elements, pollution mitigation, and climate change adaptation
Madurai	148	1.02	Madurai Municipal Corporation	Blue-Green Action Plan	To reduce floods, prepare for it, and boost economic growth through climate-resilient development initiatives

(continued)

Table 7.2 (continued)

City	Area (km^2)	Population (in million)	Responsible authority	Initiative	Purpose
Chennai	426	4.68	Chennai Municipal Corporation	Atal Mission for Urban Rejuvenation and Transformation (2015), Coastal Zone Regulation (CZR 2016), Smart City Mission (SCM 2015), Resilient Chennai Strategy (RCS 2019), Pallikaranai Marshland (2012–2020), Buckingham Canal projects (2019), Master Plan 2026	To improve stormwater and sewerage drainage and adapt to flooding
Kochi	98	0.60	Kochi Municipal Corporation	Kerala State Action Plan on Climate Change (2014), Local Biodiversity Strategy and Action Plan (2020), Mullassery Canal projects (2019), Kerala State Action Plan	To improve drainage and mitigate flooding and sustainable waste management

Source Udas-Manikar and Driver (2021), Sanchez and Govindarajulu (2023)

7.5.10 SWOT Analysis of the Global Actions for Climate-Resilient BGI

A SWOT analysis for global actions aimed at promoting climate-resilient blue-green infrastructure (BGI) can help identify the key strengths, weaknesses, opportunities, and threats in relation to such initiatives (Table 7.3). Blue-green infrastructure refers to natural and engineered solutions that integrate water management (blue) and vegetation or green spaces to mitigate climate impacts, enhance ecosystems, and improve urban environments.

In summary, global actions for climate-resilient blue-green infrastructure have significant strengths and opportunities but face notable challenges, especially in terms of financing, political will, and technical expertise. However, if implemented properly, they can provide long-term resilience to climate impacts and support sustainable development goals globally.

7.6 Recommended Global Actions for Climate-Resilient BGI

Potential global actions for climate-resilient blue-green infrastructure encompass a range of strategies and initiatives aimed at enhancing the resilience of built and natural systems to climate change impacts (Fig. 7.7). While global efforts are underway, challenges remain in scaling up climate-resilient blue-green infrastructure. Financial constraints, policy coordination, and knowledge dissemination are key hurdles. Opportunities lie in leveraging nature-based solutions, engaging local communities, and fostering international collaboration to address shared climate challenges. Implementing climate-resilient blue-green infrastructure requires a coordinated and multifaceted approach at both local and global levels. The following global initiatives can be implemented to encourage climate resilience in blue-green infrastructure.

7.6.1 International Collaboration and Funding

Fostering international cooperation and partnerships to support the development and implementation of climate-resilient blue-green infrastructure projects. This can include financial support from international organizations, such as the United Nations and the World Bank, as well as and public–private partnerships to address the financing gap and scale up investments in resilient infrastructure.

Table 7.3 SWOT analysis: global actions for climate-resilient blue-green infrastructure

Strengths	Weaknesses	Opportunities	Threats
Ecosystem services: Blue-green infrastructure provides multiple benefits like flood mitigation, urban cooling, air and water quality improvements, and habitat preservation **Climate adaptation:** BGI can enhance resilience to climate change by reducing flood risks, managing extreme weather events, and supporting biodiversity **Sustainability:** It is a sustainable approach, promoting the use of natural resources and systems, reducing reliance on hard infrastructure (concrete solutions) **Public health benefits:** Green spaces improve mental and physical health, offering cleaner air, cooling effects, and recreational opportunities **Support for carbon sequestration:** Many BGI solutions (such as wetlands and urban forests) contribute to carbon sequestration, aiding in climate change mitigation. Green spaces can act as carbon sinks, absorbing CO_2 from the atmosphere, contributing to global carbon reduction goals. BGI projects, such as afforestation and urban tree planting, help mitigate climate change while improving air quality **Economic efficiency:** Nature-based solutions can be more cost-effective in the long run compared to traditional infrastructure, particularly when accounting for ecosystem services	**High initial costs:** Although cost-effective over time, BGI projects can require significant upfront investment in planning, design, and implementation. Large-scale BGI projects, such as urban wetlands, green roofs, or permeable pavements, often require significant initial capital investments. This can be a barrier for developing nations or underfunded regions **Knowledge and expertise gaps:** There is still a lack of widespread technical knowledge, training, and expertise in the design, implementation, and maintenance of BGI. Some regions, especially in developing countries, face a shortage of technical expertise and knowledge about designing, implementing, and maintaining BGI systems. Accurate data on local climate impacts, hydrology, and biodiversity are often missing, making it difficult to design effective, site-specific BGI projects **Longer timeframes:** Blue-green infrastructure may take longer to produce visible results compared to traditional infrastructure, making it less appealing to stakeholders seeking immediate solutions. Many stakeholders are reluctant to invest due to the longer time horizons for returns and the difficulty in quantifying benefits like improved ecosystem health or public well-being **Vulnerability to overuse:** If not properly managed, green spaces and water bodies in urban areas can suffer from overuse, degradation, and pollution, undermining their effectiveness **Complexity of integration:** Integrating BGI into existing urban areas and infrastructure networks can be challenging due to space limitations, competing land uses, and regulatory hurdles. Implementing BGI often requires coordination across multiple sectors and governance levels (local, regional, national). A lack of integrated planning and coordination can slow or derail progress. Many countries still lack comprehensive policies that encourage or mandate the integration of BGI into urban planning or climate resilience strategies In dense urban areas, there is competition for land use between development, housing, and green space. Prioritizing BGI can be seen as secondary to economic growth and housing needs. Developers may be resistant to incorporating BGI into projects due to perceived complexity, costs, and the belief that conventional infrastructure is more efficient	**Policy and regulatory support:** Growing international recognition of climate change has led to increased government support, including new policies, funding mechanisms, and regulatory frameworks promoting BGI **Technological advancements:** Innovations in sustainable construction, water management, and landscape design are making blue-green infrastructure more feasible and effective. Integrating BGI with smart city technologies can enhance urban planning, real-time data collection, and predictive climate modeling. For example, sensors in BGI systems can monitor water levels and temperature changes, improving their adaptability to climate risks **Private sector engagement:** There are increasing opportunities for public–private partnerships in financing and implementing BGI projects, as businesses recognize the value of climate resilience. Collaboration between governments, private developers, and civil society offers opportunities for innovative financing mechanisms and policy frameworks that encourage BGI **Community involvement:** BGI offers opportunities for community engagement in local decision-making, fostering a sense of ownership and responsibility for maintaining green spaces and water management systems **Global awareness and collaboration:** International collaborations and knowledge-sharing initiatives can help accelerate the adoption of BGI practices worldwide. Global initiatives, such as the UN's Sustainable Development Goals (SDGs), provide a platform for countries to exchange knowledge, resources, and best practices in implementing BGI There is growing access to climate finance from international bodies like the Green Climate Fund (GCF) and multilateral development banks to support BGI projects, especially in vulnerable regions **Disaster risk reduction:** With climate-induced extreme weather events on the rise, there is an urgent opportunity to promote BGI as a disaster risk reduction strategy	**Climate change acceleration:** As climate change worsens, some BGI projects may become insufficient or overwhelmed by more extreme weather patterns, such as severe droughts or flooding **Political and economic instability:** Shifts in political priorities, economic downturns, or competing development interests can divert attention and resources away from climate-resilient infrastructure projects **Public perception and misunderstanding:** A lack of public understanding of the benefits of BGI can result in resistance to these projects, particularly if their long-term value is not well communicated **Regulatory and bureaucratic barriers:** Navigating the complex web of regulations, especially in urban areas, can slow the adoption of blue-green solutions or lead to ineffective implementation **Funding constraints:** While there are funding opportunities, securing sufficient and sustained financial resources for large-scale BGI projects can be challenging, particularly in developing countries

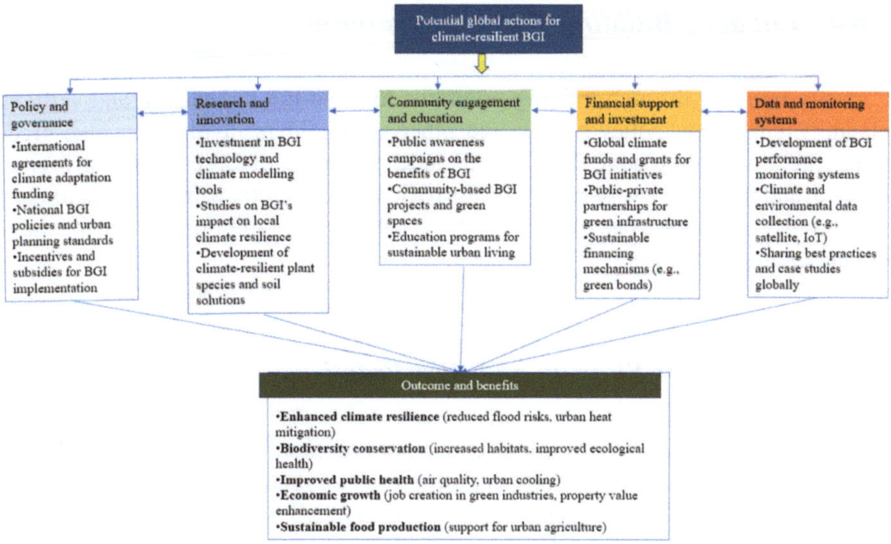

Fig. 7.7 Recommended global actions for climate-resilient BGI

7.6.2 Policy Frameworks and Guidelines

Developing global policy frameworks and guidelines at national, regional, and international levels that promote the integration of climate resilience into blue-green infrastructure planning and design. These frameworks should emphasize nature-based solutions, ecosystem services, and adaptive management approaches to enhance resilience to climate change impacts in urban planning, water management, and land-use policies.

7.6.3 Research and Innovation

Funding on research and innovation to advance the understanding of climate change impacts on blue-green infrastructure and developing innovative technologies and approaches to enhance resilience. This includes research on green infrastructure design, ecosystem-based adaptation strategies, and climate modeling to assess future risks.

7.6.4 Capacity Building Through Training

Capacity building through training programs to build the knowledge and skills of planners, engineers, policymakers, and other stakeholders involved in blue-green infrastructure development. This can include exchange of best practices, lessons learned, and innovative approaches across regions and sectors, workshops, seminars, and technical assistance on climate resilience, nature-based solutions, and sustainable urban planning practices.

7.6.5 Knowledge Sharing and Best Practices

Facilitating knowledge sharing and exchange of best practices among countries and regions to promote learning and replication of successful climate-resilient blue-green infrastructure projects. This can be done through platforms such as international conferences, online portals, and peer-to-peer networks.

7.6.6 Public Awareness and Engagement

Public awareness raising about the importance of climate-resilient blue-green infrastructure for climate change adaptation and mitigation. Engaging local communities, organizations, and other stakeholders in the process of planning, designing, and implementation of blue-green infrastructure projects to ensure their sustainability and effectiveness.

Ensuring the meaningful participation of local communities, indigenous peoples, and marginalized groups in the planning, design, and implementation of blue-green infrastructure projects. Empowerment of communities to contribute local knowledge, priorities, and needs, and ensure equitable access to benefits and decision-making processes.

7.6.7 Nature-Based Solutions and Ecosystem Restoration

Promoting the ecosystem-based approaches and use of nature-based solutions in blue-green infrastructure projects. Restoration and conservation of natural ecosystems such as wetlands, mangroves, and forests to enhance resilience to climate change impacts, reduce flood risks, and improve water quality.

7.6.8 Innovative Financing Mechanisms

Exploring cutting-edge funding options likc climate funds, green bonds, and public–private partnerships to encourage funding for climate-resilient blue-green infrastructure. Development of blended finance approaches that leverage public, private, and philanthropic resources to maximize impact and sustainability.

7.6.9 Awareness Raising and Advocacy

Raising awareness of policymakers, stakeholders, and the general public on the importance of climate-resilient blue-green infrastructure for building resilience, reducing disaster risk, and promoting sustainable development. Advocate for greater investment, policy support, and action to mainstream climate resilience into infrastructure planning and development processes.

7.6.10 Monitoring and Evaluation

Establishment of monitoring and evaluation mechanisms to track the performance and effectiveness of climate-resilient blue-green infrastructure initiatives over time. This includes monitoring key indicators such as flood resilience, water quality improvement, biodiversity conservation, and social co-benefits to inform adaptive management and decision-making. Use of data-driven decision-making and adaptive management approaches to assess project outcomes, identify lessons learned, and make necessary adjustments over time.

By taking these global actions, countries can enhance the resilience of blue-green infrastructure to climate change impacts, promote resilience and sustainable development, and safeguard the well-being of communities and ecosystems globally.

7.7 Conclusion

Blue-green infrastructure can operate across jurisdictional lines and exist at different geographic levels (such as region, urban, peri-urban, river basin, catchment, and watershed). Consequently, BGI is not restricted to urban areas, and its planning can be taken into account in a variety of planning settings and at different levels, including urban, peri-urban, regional, and rural planning. Although there are now very few BGI projects that have been completed successfully or that are in the planning stages, this number will rise in the coming years. The promotion of climate-resilient blue-green infrastructure on a global scale is essential for building a sustainable and resilient future. By learning from successful case studies, aligning with international frameworks, and addressing challenges collaboratively, countries can enhance their capacity to adapt to climate change. This chapter underscores the urgency of incorporating climate-resilient blue-green infrastructure into global policy agendas and development initiatives to create a more resilient and sustainable world.

References

Almeida CMVB, Mariano MV, Agostinho F, Liu GY, Giannetti BF (2018) Exploring the potential of urban park size for the provision of ecosystem services to urban centres: a case study in São Paulo, Brazil. Build Environ 144:450–458

Alterra research institute (2007) Green-blue veining: agro-biodiversity as innovation for sustainable agriculture. Technical report, VROM Dutch Ministry of Housing, Spatial Planning and the Environment, Den Haag

Bai X, Dawson RJ, Ürge-Vorsatz D, Delgado GC, Salisu Barau A, Dhakal S, Schultz S (2018) Six research priorities for cities and climate change. Nature 555(7694):23–25

Benedict MA, McMahon ET (2002) Green infrastructure: smart conservation for the 21st century. Renew Resour J 20(3):12–17

Benedict MA, McMahon ET (2012) Green infrastructure: linking landscapes and communities. Island Press

BES (Bureau of Environmental Services) (2005) The City of Portland Oregon, Portland Watershed Management Plan, Portland Oregon

BES (Bureau of Environmental Services) (2010) The City of Portland Oregon, Portland's Green infrastructure: quantifying the health, energy, and community livability benefits, Portland Oregon

Bowler DE, Buyung-Ali L, Knight TM, Pullin AS (2010) Urban greening to cool towns and cities: a systematic review of the empirical evidence. Landsc Urban Plan 97(3):147–155

Carter JG, Cavan G, Connelly A, Guy S, Handley J, Kazmierczak A (2015) Climate change and the city: Building capacity for urban adaptation. Prog Plan 95:1–66

Cun C, Zhang W, Che W, Sun H (2019) Review of urban drainage and stormwater management in ancient China. Landsc Urban Plan 190:103600

del Carmen Moreno-García M, Baena I (2019) The microclimatic effect of green infrastructure (GI) in a Mediterranean city: the case of the urban park of Ciutadella (Barcelona, Spain). Arboric Urban Forest (AUF) 45(3):99–107

Demuzere M, Orru K, Heidrich O, Olazabal E, Geneletti D, Orru H, Faehnle M (2014) Mitigating and adapting to climate change: multi-functional and multi-scale assessment of green urban infrastructure. J Environ Manage 146:107–115

Duan W, He B, Nover D, Fan J, Yang G, Chen W, Liu C (2016) Floods and associated socioeconomic damages in China over the last century. Nat Hazards 82:401–413

Faggian R, Sposito VA (2009) Systemic regional development—a systems thinking approach. In: Proceedings of the 53rd annual meeting of the ISSS-2009, Brisbane, Australia

Faggian R, Sposito VA (2009) Systemic regional development-a systems thinking approach. In: Proceedings of the 53rd annual meeting of the ISSS-2009, Brisbane, Australia

Feitosa A (2022) Ocupação irregular de áreas em Petrópolis mais do que dobrou entre 1985 e 2020. G1 News. https://g1.globo.com/rj/regiao-serrana/noticia/2022/02/23/ocupacao-irregular-de-areas-em-petropolis-mais-do-que-dobrou-entre-1985-e-2020.ghtml. Accessed 23 Feb 2022

Gill SE, Handley JF, Ennos AR, Pauleit S (2007) Adapting cities for climate change: the role of the green infrastructure. Built Environ 33(1):115–133

Guo Y, Wu Y, Wen B, Huang W, Ju K, Gao Y, Li S (2020) Floods in China, COVID-19, and climate change. Lancet Planet Health 4(10):e443–e444

Haase D (2015) Reflections about blue ecosystem services in cities. Sustain Water Qual Ecol 5:77–83
Handley J, Pauleit S, Gill S (2007) Landscape, sustainability and the city. In: Benson J, Roe M (eds) Landscape and sustainability. Taylor & Francis, London, pp 183–211. https://doi.org/10.4324/9780203962084
Heckert M, Rosan CD (2016) Developing a green infrastructure equity index to promote equity planning. Urban Urban Green 19:263–270
Intergovernmental Panel on Climate Change (IPCC) (2014) Climate change 2014: impacts, adaptation, and vulnerability. Cambridge University Press Cambridge, New York, NY
IPCC (2022). In: Portner H-O, Roberts DC, Tignor M, Poloczanska ES, Mintenbeck K, Alegría A, Craig M, Langsdorf S, Loschke S, Moller V, Okem A, Rama B (eds) Climate change 2022: impacts, adaptation and vulnerability. In: Contribution of working group ii to the sixth assessment report of the intergovernmental panel on climate change. Cambridge University Press
Islam MR, Talukdar S, Rihan M, Rahman A (2024) Evaluating cooling effect of blue-green infrastructure on urban thermal environment in a metropolitan city: using geospatial and machine learning techniques. Sustain Cities Soc 105666
Jiang Y, Zevenbergen C, Ma Y (2018) Urban pluvial flooding and stormwater management: a contemporary review of China's challenges and "sponge cities" strategy. Environ Sci Policy 80:132–143
Jing L (2019) Inside China's leading 'sponge city': Wuhan's war with water. The Guardian. https://www.theguardian.com/cities/2019/jan/23/inside-chinas-leading-sponge-city-wuhans-war-with-water
Kazmierczak A, Carter J (2010) Adaptation to climate change using green and blue infrastructure. A database of case studies. Technical report, University of Manchester
Koomen E, de Moel H, Steingröver E, van Rooij S, van Eupen M (2012) Land use and climate change; synthesis report. Clim Changes Spat Plan
Kozak D, Henderson H, de Castro Mazarro A, Rotbart D, Aradas R (2020) Blue-green infrastructure (BGI) in dense urban watersheds. The case of the Medrano stream basin (MSB) in Buenos Aires. Sustainability 12(6):2163
Lawson E, Thorne C, Ahilan S, Allen D, Arthur S, Everett G, Wright N (2014) Delivering and evaluating the multiple flood risk benefits in blue-green cities: an interdisciplinary approach. WIT Trans Ecol Environ 184:113–124
Leary N (2012) Climate change and adaptation, Earthscan
Li B (2013) Governing urban climate change adaptation in China. Environ Urban 25(2):413–427
Loftus A, Anton B, Philip R (2011) Adapting urban water systems to climate change: a handbook for decision makers at the local level. ICLEI-EU, IWA, UNESCO-IHE. ICLEI European Secretariat Gmbh
Middlemann MH, Middelmann M (2007) Natural hazards in Australia: identifying risk analysis requirements. Geosci Aust
Nguyen TT, Ngo HH, Guo W, Wang XC, Ren N, Li G, Liang H (2019) Implementation of a specific urban water management-Sponge City. Sci Total Environ 652:147–162
OPUB (2024) About ABC waters programme. Singapore's National Water Agency. Accessed from: https://www.pub.gov.sg/Public/Places-of-Interest/Our-Reservoirs-and-Waterways/ABC-Waters/About
Peng Y, Reilly K (2021) Using nature to reshape cities and live with water: an overview of the Chinese Sponge City programme and its implementation in Wuhan. Report for the EU Project GROWGREEN—Green Cities for Climate and Water Resilience, Sustainable Economic Growth, Healthy Citizens and Environments (Grant Agreement No 730283)
Roberts D, Boon R, Diederichs N, Douwes E, Govender N, Mcinnes A, Spires M (2012) Exploring ecosystem-based adaptation in Durban, South Africa:"learning-by-doing" at the local government coal face. Environ Urban 24(1):167–195
Ruckelshaus MH, Guannel G, Arkema K, Verutes G, Griffin R, Guerry A, Rosenthal A (2016) Evaluating the benefits of green infrastructure for coastal areas: location, location, location. Coast Manag 44(5):504–516

Rui Y, Fu D, Do Minh H, Radhakrishnan M, Zevenbergen C, Pathirana A (2018) Urban surface water quality, flood water quality and human health impacts in Chinese cities. What do we know? Water 10(3):240

Sanchez FG, Govindarajulu D (2023) Integrating blue-green infrastructure in urban planning for climate adaptation: lessons from Chennai and Kochi, India. Land Use Policy 124:106455

Shafique M, Kim R, Kyung-Ho K (2018) Green roof for stormwater management in a highly urbanized area: the case of Seoul, Korea. Sustainability 10(3):584

Snepvangers J, Kamerling-Baake A, Rooij SAM, Steingröver EG (2011) Plannen met multifunctionele groenblauwe netwerken in Salland. Technical report, Ecologische Netwerken

Sposito VA (2012) Multifunctional agriculture and regional development. Technical report, Department of Primary Industries, Melbourne

Steingröver EG, Geertsema W, van Wingerden WK (2010) Designing agricultural landscapes for natural pest control: a transdisciplinary approach in the Hoeksche Waard (The Netherlands). Landsc Ecol 25:825–838

Su MA, Ngarambe J, Santamouris M, Yun GY (2021) Empirical evidence on the impact of urban overheating on building cooling and heating energy consumption. Iscience 24(5)

Tang YT, Chan FKS, O'Donnell EC, Griffiths J, Lau L, Higgitt DL, Thorne CR (2018) Aligning ancient and modern approaches to sustainable urban water management in China: Ningbo as a "Blue-Green City" in the "Sponge City" campaign. J Flood Risk Manage 11(4):e12451

Thorne CR, Lawson EC, Ozawa C, Hamlin SL, Smith LA (2018) Overcoming uncertainty and barriers to adoption of Blue-Green Infrastructure for urban flood risk management. J Flood Risk Manage 11:S960–S972

Tzoulas K, Korpela K, Venn S, Yli-Pelkonen V, Kaźmierczak A, Niemela J, James P (2007) Promoting ecosystem and human health in urban areas using Green Infrastructure: a literature review. Landsc Urban Plan 81(3):167–178

Udas-Mankikar S, Driver B (2021) Blue-green infrastructure: an opportunity for Indian cities. ORF Occas Pap (317)

USEPA (2008) Reducing urban heat islands: compendium of strategies. State and Local Branch Climate Protection Partnerships Division, Washington, DC, pp 1–32

Verburg PH, Koomen E, Hilferink M, Pérez-Soba M, Lesschen JP (2012) An assessment of the impact of climate adaptation measures to reduce flood risk on ecosystem services. Landsc Ecol 27:473–486

Wang J, Banzhaf E (2018) Towards a better understanding of green infrastructure: a critical review. Ecol Ind 85:758–772

Xu YS, Shen SL, Lai Y, Zhou AN (2018) Design of sponge city: lessons learnt from an ancient drainage system in Ganzhou, China. J Hydrol 563:900–908

Yu Z, Yang G, Zuo S, Jørgensen G, Koga M, Vejre H (2020) Critical review on the cooling effect of urban blue-green space: a threshold-size perspective. Urban Fores Urban Green 49:126630

Part III
Urban Resource Management and Coordination for BGI, Food Security and Climate Disaster Resilience

Chapter 8
Urban Housing, Poverty, and Informal Sector: Asian Dilemma Toward Livable Blue-Green Cities

Irtija Alam, Gulsan Ara Parvin, and Mohammad Golam Kibria

Abstract Asia is the hub of urban growth in twenty-first century. Climate change and environmental stresses are the main push factors that aggravate rapid rural–urban migration in Asian cities. Rural poverty is shifting to cities and a large percentage of these poor live in informal settlements in low-lying and flood-prone areas on steep slopes, with limited or no access to basic services and viable livelihoods. For employment and livelihood urban poor depend of informal sectors. However, due to slum eviction, and eviction of informal markets and on-street business urban poor are in constant struggle to secure their basic rights. On the other hand, combating with urban challenges like growth of informal and illegal settlements, water scarcity, waste management, transportation, rapid reduction of green spaces and frequent disasters is the key concern of urban governance. This paper aims to highlight the Asian dilemma toward keeping cities livable with blue-green infrastructures by minimizing urban challenges.

Keywords Urban · Informal sector · Housing · Asian cities · Blue-green infrastructure

I. Alam
Department of Urban and Regional Planning (URP), Khulna University of Engineering & Technology (KUET), Khulna, Bangladesh

G. A. Parvin (✉)
College of Policy Science, Ritsumeikan University, Osaka-Ibaraki Campus, Osaka, Japan
e-mail: niruurp@yahoo.com

M. G. Kibria
Department of Soil Science, Bangladesh Agricultural University, Mymensingh, Bangladesh

M. G. Kibria et al. (eds.), *Urban Blue-Green Infrastructure Approach for Food Security and Climate Disaster Resilience*, Disaster Risk Reduction,
https://doi.org/10.1007/978-981-96-8746-6_8

8.1 Asian Cities, Poverty, and Slums

According to World Cities Report 2022, "Urban poverty and inequality remain one of the most intractable and highly complex problems confronting cities." From developed to developing countries, homelessness, housing crisis, overcrowded and unhygienic slums, and poverty have become key policy concerns of most of the cities, especially in Asia. World cities report 2020 claims that 1.6 billion people or 20% of the world's population live in inadequate housing and among them 1 billion reside in slums and informal settlements. This number will reach to 3 billion by 2030 and it is needed to build 96,000 new affordable homes every day to meet this demand (UN-Habitat 2020). It is claimed that the world is facing a critical crisis, which is "Global Urban Housing Affordability Crisis," since in many urban centers the trend of housing-related household expenses is rising faster than salary and wage increases (Wetzstein 2017). Especially after the coronavirus disease (COVID-19) pandemic, along with geopolitical tensions and war, rampant consumption, speculation, financialization, and hunger for land, has triggered an economic downturn, escalated inflation, energy shortages, surging food prices, and heightened job insecurity.

All these crises have multifold economic, ecological, social, and political crises which have had a profound impact on millions of individuals striving to meet their basic needs, including housing expenses (UN 2023). All these crises are also the driving forces for rural-to-urban migration and shifting multidimensional deprivation and poverty from rural-to-urban areas, especially in Asian cities. In Asian cities, affordability of housing has been an acute crisis that is leading these migrants to live in slums. Spatial and socio-economic marginalization (informal/makeshift settlements, illegal occupation of private/public land, gated communities, and peri-urbanization) can be translated by the symbolic term "slum," (Bolay 2020). United Nations also claims that the main drivers of slum growth in developing countries include rapid urbanization, ineffective planning, lack of affordable housing options for low-income households, and poverty. Though the overall poverty trend is declining (Fig. 8.1) in different countries in Asia, the number of challenges for urban poor is aggravating. Among these challenges having affordable housing is the most crucial one.

Despite poverty is decreasing in different countries of Asia, the absolute number of people living in slums or informal settlements grew to over 1 billion (in 2018), with 80% attributed to three regions: Eastern and South-Eastern Asia (370 million), sub-Saharan Africa (238 million), and Central and Southern Asia (227 million). Figure 8.2 reveals that in different countries of Asia there are one-third to more than half of the people who are living is slum areas.

Different parts of Asia are having remarkable economic gains. Further, in case of income-related urban poverty the levels are low in Asia. However, urban poor are constantly combating with different forms of deprivation and challenges in most megacities in South Asia, such as Dhaka, Bangladesh. In different Asian cities, the spatial concentration of deprivation is embedded in the daily lives of the urban poor (UN-Habitat 2022) and this deprivation is reflected in their inhuman living conditions of slums. Climate migrants, refugees, women, the elderly, and others who are

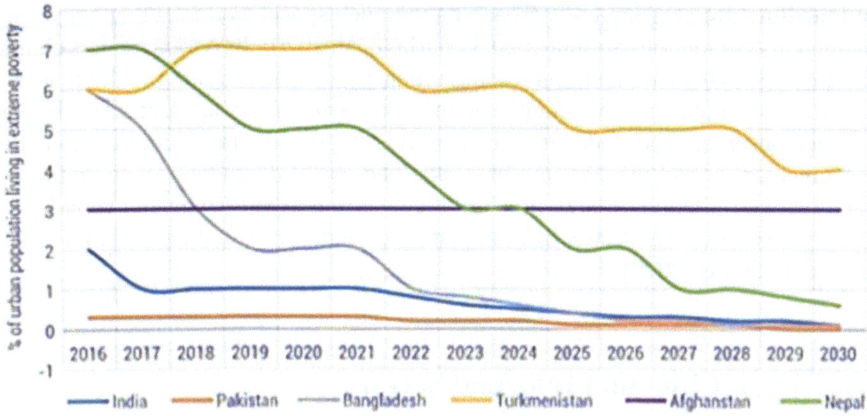

Fig. 8.1 Proportion of urban population living in extreme poverty in selected Asian countries (2016–2030). *Source* World cities report, 2022, UN-Habitat (2022)

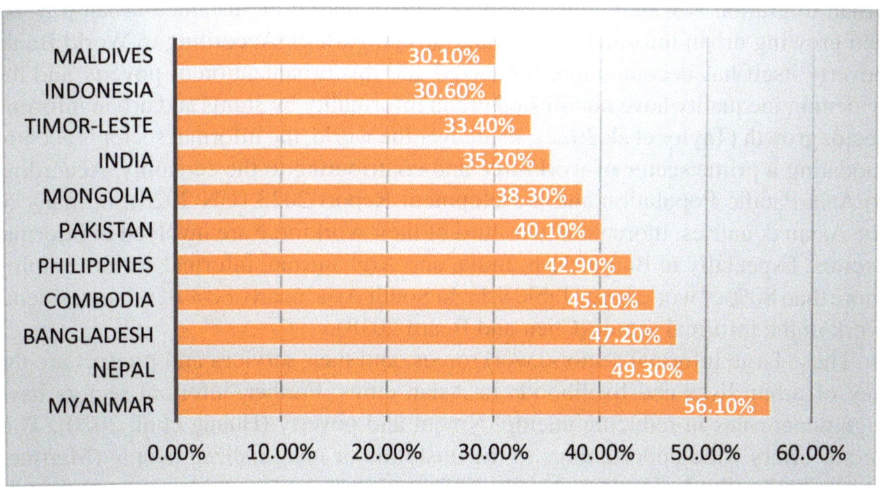

Fig. 8.2 Share of slum population in different countries of Asia and Pacific in 2018. *Source* Adopted from https://www.statista.com/statistics/691746/asia-pacific-share-of-urban-population-living-in-slums-by-country/, Published by Leander von Kameke, September 12, 2022, accessed on September 2, 2024

the most vulnerable populations among urban poor live in these overcrowded and risk-prone informal slum settlements (Cities Alliance 2022). Moreover, most of these vulnerable people need to depend on informal low-paid income opportunities, which do now allow them to afford for basic services, including decent accommodation. In different cities of Asia, slums or informal settlements and related informal services and facilities are growing by encroaching the open space, wetlands that cause environmental degradation (Sinthia 2020; Legarias et al. 2020). In Africa too it is claimed that rapid growth of slums is decreasing urban green spaces. Along with the growth of informal settlements and slums, uncontrolled structural change of a city by integrating non-urban landscape into urban areas is responsible for declining urban blue and green space and various environmental and social degradations. In addition to the direct assimilation of land, process of converting urban fringe and rural land to urban land uses causes destruction of open space, natural vegetation, and loss of farm lands (Mukherjee 2020).

8.2 Asian Cities and Informal Sector

During the pandemic COVID-19, the world has realized the importance of informal sectors and better recognized informal sector workers for their legitimate contribution to urban economy (UN-Habitat 2022). Like the growth of slums rural-to-urban migration is widely considered as a main reason of sustained urban poverty and growing urban informality (Das and Susanto 2022). According to World Bank, poverty itself has become more urbanized and this urbanization of poverty and the widening inequality have manifested urban informality by slums and urban informal sector growth (Taylor et al. 2022). In all over the world, the informal sector is accommodating a prime sector of workforce and contributing to the economy. According to Asia-Pacific Population and Development Report 2023 (UN 2023), in many of the Asian countries, more than two-third of their workforce are involved in informal sectors. Especially in Bangladesh, India, and Afghanistan, informal sectors employ more than 80% of workforce (Table 8.1). In South Asia, nearly 80% of urban residents work in the informal sector (Chen and Beard 2018).

These large informal sectors, workforces, and their services and product are the key of urban lives and livelihoods in Asian cities. Further, informal sectors have significant roles in reducing unemployment and poverty (Huang et al. 2020). This sector offers wide opportunities to the unskilled or marginalized people (Martínez et al. 2017). Study implies that the informal sector plays a detrimental role in the sustainable development of developing countries while economic growth and economic freedom contribute positively (Sultana et al. 2022). In fact, informal sectors directly contribute to a number of sustainable development goals, especially SDGs 1–5 (No Poverty, No Hunger, Good Health, Quality Education, Gender equity). SDG 8 is very clear, namely, to "Promote sustained, inclusive and sustainable economic growth, full and productive employment and decent work for all." Similarly, SDG 11 ensures *making cities and human settlements inclusive, safe, resilient, and sustainable.*

Despite being essential parts of urban lives and economy, informal sectors and the associated labor force are largely excluded from public infrastructure and services, public space, and public procurement contracts (Chen and Beard 2018). Therefore, most often, by illegal and temporary occupancy, the informal sectors' employee

Table 8.1 Informal employment in non-agricultural employment—percentage of total employment, by Asia-Pacific country

Subregion	Country	Year	Percentage
East and Northeast Asia	Mongolia	2022	25.8
	Republic of Korea	2019	23.1
Southeast Asia	Brunei Darussalam	2021	33.5
	Cambodia	2019	83.1
	Indonesia	2022	74.4
	Lao People's Democratic Republic	2017	75.1
	Myanmar	2020	79.4
	Thailand	2018	51.9
	Timor-Leste	2021	70.5
	Viet Nam	2022	53.3
South and Southwest Asia	Afghanistan	2021	82.4
	Bangladesh	2017	91.3
	India	2021	81.2
	Maldives	2019	40.9
	Nepal	2017	77.4
	Pakistan	2021	75.2
	Sri Lanka	2020	58.1
	Türkiye	2022	17.5
North and Central Asia	Armenia	2021	21.0
	Georgia	2020	25.3
	Kyrgyzstan	2021	56.7
Pacific	Australia	2021	25.2
	Cook Island	2019	27.7
	Fiji	2016	24.6
	Kiribati	2020	43.5
	Marshall Island	2021	32.0
	Nauru	2021	47.8
	Samoa	2017	19.4
	Tonga	2021	96.1
	Vanuatu	2019	50.5

Source Adopted from International Labour Organization, ILOSTAT explorer, see https://ilostat.ilo.org/data/. Adopted from UN (2023)

continues activities in public spaces, such as roadside, sidewalks, parks, stadiums, or squares, and create problems in and around cities' public spaces, parks, and playground (Widjajanti et al. 2019). This illegal and unplanned occupancy imposes problems and challenges to urban blue and green infrastructures and eventually the urban environment. It is stated that "Informality is a complex and heterogeneous intrinsic urban phenomenon that must be considered by local authorities and incorporated within sustainable urban development programs and strategies in the context of rapid urbanization manifested in poverty, vulnerability, and marginalization, creating conditions for informal manifestations (Davidescu et al. 2022)."

8.3 Urban Poor, Informal Sectors, and Urban Challenges: Case Studies

Asia is experiencing urban growth rate of 2.16% (Statistica 2023) which is the home of 54.3% of the world's urban population and holds 69% of the world's total urban poor (UN-Habitate 2024). Due to massive industrialization, better employment and income opportunities, and the consequence of natural calamities, the rate of rural-to-urban migration rate is quite high in Asian cities (Gero et al. 2017). Due to this high rate of rural-to-urban migration, cities are experiencing extra rate of growth resulting in rapid urbanization in the developing nations of Asia like Bangkok (Thailand), Dhaka (Bangladesh), and Manila (Philippines) (Mathur 2013). Most of the migrated people are poor, unskilled, and unemployed and often choose the low wage informal sector as the last option for their involuntary growth as they do not get enough job opportunities in the formal sector and fall in the poverty trap which increases urban poverty (Khan and Kraemer 2014).

Due to their low-income scope urban poor people cannot afford adequate houses and housing environments and often choose informal settlement to reside (Hasam et al. 2017). The informal settlements are built through forced encroachment of environmentally sensitive areas or in the private lands developed by land speculators destroying blue and green infrastructures (Begum 2007). These informal settlements and residences are deprived from access to basic services such as water, sanitation, solid waste management, education, health, and so on (Gero et al. 2017). Being ignored by city authority and unaware of their unsustainable activities, dangerous settlement locations and living conditions due to which they are making themselves exposed to hazard increases disaster risk (Hongtong 2019). Thus, the developing nations of Asia are facing challenges to ensure liable, equitable, and sustainable cities balancing urbanization, environment, and informality. These problems and challenges of informality are acute in large cities of Asia, especially in Bangkok, Dhaka, Manila, Mumbai, Jakarta, and Hanoi.

8.3.1 *Informal Sector and Housing Environment of Asian Cities—Highlighting from Bangkok, Dhaka, and Manila*

Dhaka

In the rapidly urbanizing country like Bangkok, Dhaka, and Manila, land management for housing is a big issue. The land price in Dhaka City is too high to develop sustainable housing for lower middle and low-income people as well as the housing rent. Thus, 37% of the total city population is forced to live in 4% land with scarce household resources, limited provisional public facilities, and overcrowded and poor environmental conditions in the fear of eviction (Baser et al. 2020). Housing environment of the informal settlements (Fig. 8.3) is severely degraded because of open waste dumping, insanitary disposal of human excreta, lack of drainage condition, air pollution, and often face water logging even in low precipitation (Dewan et al. 2012). These houses are often the result of unplanned and rapid urbanization, unequal resource distribution, manipulations of the housing market, and proper institutional concern. The number and area of slums is expanding with an increased number of rural migrants every year.

The residents of informal settlements are also deprived of the opportunity to get formal employment due lack of proper skill and education. The major occupation of the informal sector in Dhaka City is daily labors, rickshaw and van drivers, construction workers, waste pickers, repair service, street vendors, hawkers, maids, sweepers, and transport workers (Amin 2016). Although some of them are involved in illegal works, the informal sector has made significant contributions to uplift the national economic growth (Hassan 2018). These informal houses and informal sector activities are encroaching, as well as polluting urban low-lying areas, open spaces, and water bodies and posing critical challenges on protecting and maintaining urban blue-green infrastructures to keep cities livable for current and future generation.

Fig. 8.3 Informal settlements of Dhaka City. *Source* Open Source (https://www.shutterstock.com/image-photo/korail-one-largest-slums-bangladesh-covers-1956499717); Author

Fig. 8.4 Poor housing condition of Manila slum dwellers. *Source* Open source (https://www.alamy.com/stock-image-typical-poor-housing-area-buildings-made-of-light-materials-jutting-169985154.html; https://www.alamy.com/stock-photo/manila-slum.html?sortBy=relevant)

Manila

The situation of metro Manila is quite similar to Dhaka City. Manila is often facing hydro-meteorological hazards due to its geography and the city is facing rapid growth of urban population who reside in informal settlements. Informal settlements in Manila are mostly growing along the rivers, canals, garbage disposal sites, public or private vacant lands, swamp areas, and along transportation lines in both urban centers and peri-urban areas. The employment of the Manila slum people is similar to Dhaka City. But in recent years the poverty level of the slum people is getting much worse that started illegal economic activities (Boquet 2017). Due to a massive urban renewal program the informal settlements of the urban core are undergoing eviction by fear and fire. Due to city-wide hardscape development, slum people face flooding where the water is full of solid and organic waste, human waste, and industrial waste (Fig. 8.4). The building structure of the slum is so poor that life and property loss in heavy participation and wind storms. Slums of both Dhaka and Manila Cities are like most dangerous, unhealthy, and threatening places where people are forced to live and have significant influence on their physical and mental health, nutrition, and quality of life (Yulu 2021).

Bangkok

Growth slum in Bangkok City has been triggered due to massive urbanization and economic growth, higher wage, urban poverty, and natural population not for the rural migrants. With the greatest income inequality Bangkok is the home of 62% slum people in Thailand (Fig. 8.5). The living environment of these slums is unsafe, overcrowded, risky, and polluted. This is the result of failure of different housing programs and slum improvement interventions. Due to eviction, resettlement, and housing programs, the number of slums in Bangkok is decreasing every year.

However, the housing resource and public facility access are getting better than that of Dhaka and Manila. Likewise, in other countries of Asia, the location preference of the slum people totally depends on the employment location. Where there is industry there is slum which mostly developed along the water bodies (Chayakul 2015). Because of the high rate of industrial development in Bangkok, people of

Fig. 8.5 Slums of Bangkok City. *Source* Open source (https://www.alamy.com/stock-photo/bangkok-slum.html?sortBy=relevant); (Khiri 2022) (permission for the picture has been taken)

informal settlement are the source of labor in those industries and that is why the unemployment rate in Thailand is low where slum people of Dhaka and Manila struggle to get better income opportunities. So, the informal sector plays a very significant role in the Thai economy (Thaiprasert et al. 2020). However, the high wage has no positive effect in the housing market. Still 80% people in Bangkok are unable to pay for cheapest house, which is an indicator of never-ending barrier toward adequate housing (Chayakul 2015).

8.3.2 Environmental Poverty and Impact

The population share of informal settlements in all Asian countries is much higher than those who live in formal residence (Howlader and Sarkar 2020). Rapid growth of urban poor population and scarcity of housing land creating pressure on housing sector and public facilities which ended up creating unsuasible informal settlements is deteriorating the urban environment from time to time in Dhaka, Manila, and Bangkok Cities.

The private land speculator buys agricultural land in the peri-urban areas with relatively low prices and opens up informal settlements developed on rent. People of the informal settlement start living in either environmentally sensitive areas destroying blue or green infrastructure or poor environmental areas like garbage dump sites, under the bridge, etc. Most of the informal settlements of Bangkok and Manila Cities developed along the water bodies, swamp areas, and wetlands that disturb the inland ecosystem.

When they have developed the land, they evict the slum people by fire instead of dealing with them, maybe for fear of compensation. This happens in the urban cores of Bangkok and Manila Cities also (Kamalipour 2016; Yulu 2021; Boquet 2017). The informal settlers of Dhaka City often forcefully encroach the public lands, vacant private lands, railway and highway sides, river banks, forest lands, etc. (Begum 2007; Sultana et al. 2020) (Fig. 8.8). Once the urban poor settle in the slums or squatters then start to degrade the surrounding environment more and more through their unhealthy and unsustainable habits and lifestyles. Turns out they are the exposure of environmental hazards which they have created by themselves. This situation often triggers by lack of provisional public facilities and amenities, failure in housing programs, and lack of proper institutional interventions.

Unsustainable traffic vehicles, industries, and informal residences are the main contributors of air pollution in Bangkok (Khan 1994). Most of the industries of Bangkok are coal based. On the other hand, people of informal settlements are deprived of gas and proper waste management facilities use coal, wood for cooking, and burn wastes that significantly contributes in air pollution. The informal workers who work close to the transportation corridors, industries, and other informal sectors are exposed to air pollution reporting for health symptoms like sneezing, headaches, burning or irritated eyes, hoarse throat, and nasal congestion. Open waste dumping (domestic waste, human waste, and solid waste) directly in the water bodies and open spaces is polluting water and soil severely in Bangkok. People are having water-borne diseases and skin diseases as they are using polluted water for bathing and cleaning both in normal time and during the flood. Due to waste dumping the water flow gets clogged and disrupts the drainage system and water vehicles. Waste dumping and burning of waste pollute the soil which is unable to grow any vegetation, even capable of damaging existing vegetation also.

The scenario of Dhaka is not different from Bangkok. Slum dwellers are being deprived of sanitation facility, setting open latrine above encroached water bodies or disposing of human waste in surrounding open spaces resulted in water and soil pollution which spreads diseases (Chowdhury and Amin 2006; Ishtiaque et al. 2014). People throw organic and inorganic waste into the water bodies and make them unable to flow which turns into the home of mosquitoes and spreads dengue, malaria, and other water-borne diseases in Dhaka City. Informal settlers of Dhaka City have no access to gas service and they cannot afford cylinder due to poverty. They use wood, coal, charcoal, and other materials that make smoke and contribute to air pollution in the city that causes respiratory diseases (Biplob et al. 2011; Rahman and Rabbani 2007). The waste generated and thrown by slum dwellers in Manila creates flooding in heavy participation and overflows the septic tanks and pollutes river water. Dump side and coastal informal settlement is the exposure of air pollution, sanitation and flooding due solid waste (Siddique 2017). The people living on the roadside are exposed to air and noise pollution (Ballesteros 2010).

The LULC map of Bangkok, Manila, and Dhaka Cities (Figs. 8.6, 8.7 and 8.8) also visualize that the city is expanding both in the center and outskirt destroying the blue and green infrastructure. Where there is industrialization, there is urbanization, most importantly growth of informal settlement who works in the industries and related informal sectors. The land use change pattern has been shown in Fig. 8.9, where green means no change, yellow means change from waterbody to light vegetation, and red means the land converted into built-up area. If we rank the change pattern Dhaka City is in worse condition than quite good change patterns of Manila and Bangkok Cities. The above also admits that the environmental condition of Dhaka City and Manila City is experiencing massive destruction of green and blue infrastructure and going far behind to achieve the SDG 11. However, to make cities livable it is high time to check the conversion of land uses into built-up areas by any means.

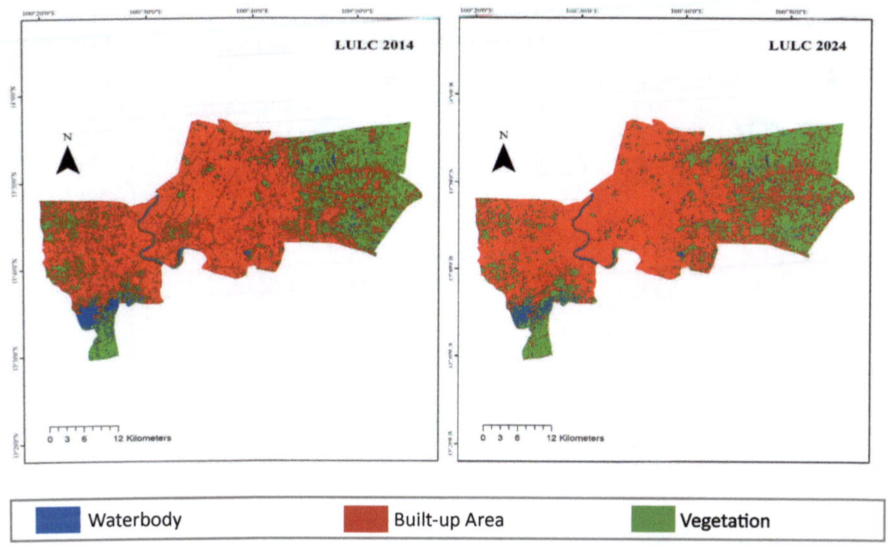

Fig. 8.6 LULC change of Bangkok City from 2014 to 2024. *Source* Author

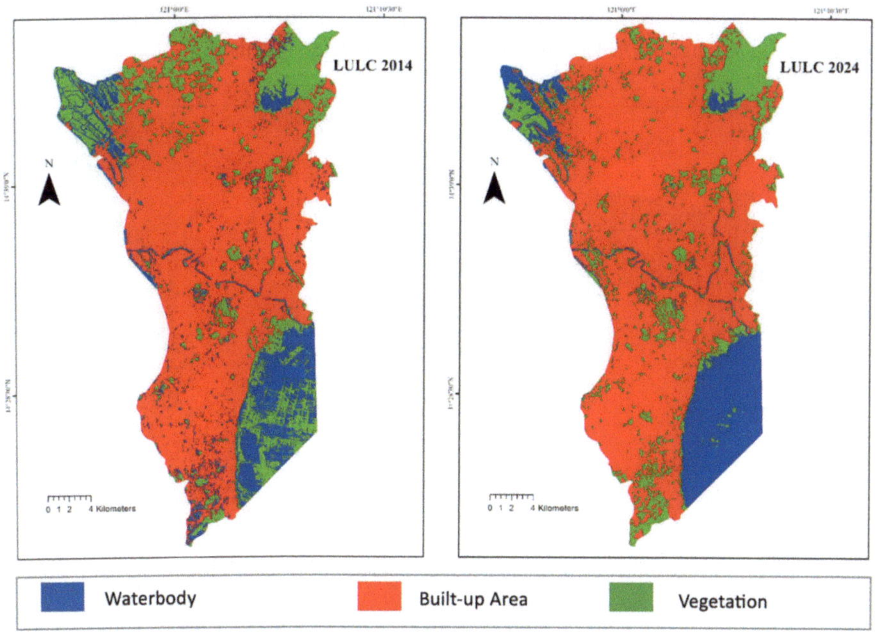

Fig. 8.7 LULC change of Manila City from 2014 to 2024. *Source* Author

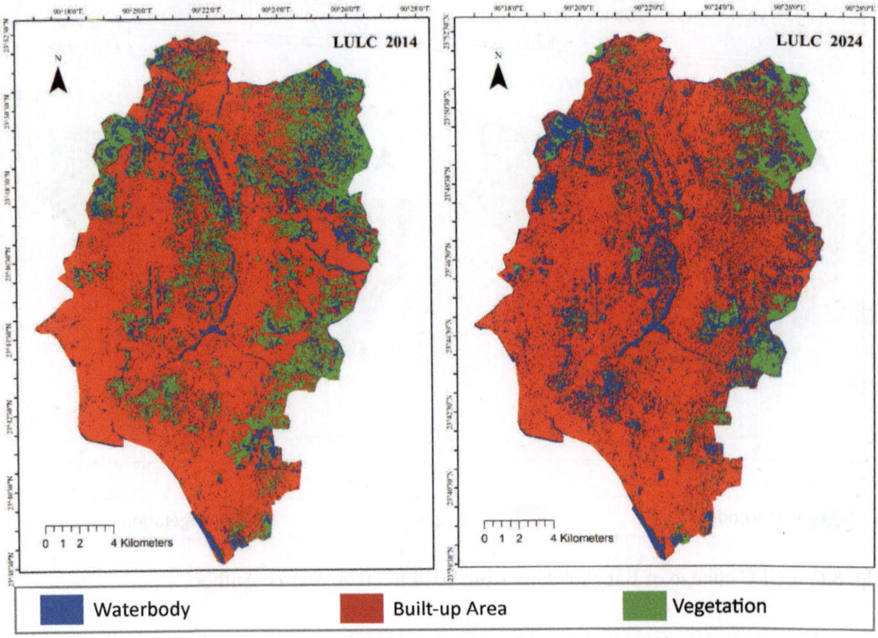

Fig. 8.8 LULC change of Dhaka City from 2014 to 2024. *Source* Author

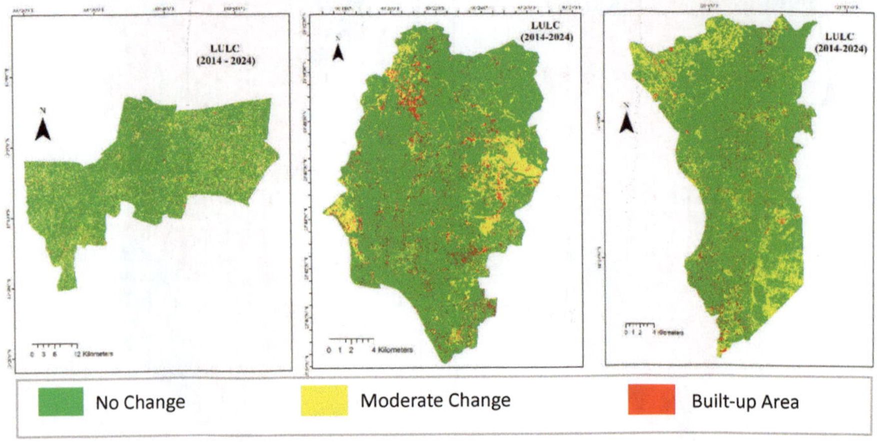

Fig. 8.9 LULC change direction map of Bangkok, Dhaka, and Manila Cities from 2014 to 2024. *Source* Author

8.3.3 Dilemma Toward Sustainable City—Highlighting from Bangkok, Dhaka, and Manila

Urban poor people reside in environmentally poor and sensitive areas, make the living environment worse, find themselves exposed to environmental pollution and hazard, and face the cyclic consequence (Cobbinath et al. 2021).

The slum dwellers who reside in the lowlands, wetlands, flood plains, river banks, river islands, beside the canals are more exposed to primary and secondary hydro-meteorological hazards. The effect of unplanned urbanization, threats of climate change, poverty, growth of informal communities in dangerous locations, socio-economic and physical vulnerability, and weak disaster management capacity increasing risk of disaster and damage. Moreover, those aspects are the major dilemmas toward keeping Asian cities livable.

Bangkok, Dhaka, and Manila Cities are adorned with many rivers, tributaries, and other types of water bodies. The cities are very prone to flood, cyclone, wind storm, heat wave, and other hydro-meteorological hazards. The building structure of most slum dwellers in Manila is not capable of tolerating the wind storm and flooding. How they will be capable of protecting people in the time of a cyclone. Though slum dwellers of the Dhaka City do not face the effect of cyclone but due to poor drainage condition they face flooding, even in a regular amount of precipitation. The Bangkok City slum dwellers are also the same. These slum dwellers have been falling into a cyclical trap. Due to poor housing and facilities, they degrade the environment and then face environmental hazards that make them poorer and they cannot get rid of the informal sectors anymore or not even from the vulnerability and disaster risk (Berquist et al. 2015; Yasmin 2016; Balgos 2016). To solve the dilemmas, it is crucial to analyze the root causes by the government and other concerned authorities and solve the problem through whole society approaches (Okopi and Mohammed 2020).

8.4 Challenges in BGI with the Informality in Cities

Urban blue-green infrastructure (BGI) presents a multifaceted approach to addressing the challenges posed by urbanization, climate change, and environmental degradation. However, its implementation is fraught with various challenges that can hinder its effectiveness and adoption in urban planning. Some of the worth mentioning challenges in BGI with the informality in cities are listed hereunder.

Lack of Understanding about Economic Value of BGI

One of the primary challenges in the implementation of BGI is the limited understanding of its economic value and multifunctionality. Many urban planners and decision-makers still prioritize traditional gray infrastructure due to a lack of awareness regarding the multiple benefits that BGI can provide, such as enhanced stormwater management, improved air quality, and increased biodiversity (Ncube and Arthur 2021; Sehrawat and Shekhar 2024). This oversight is particularly evident in cities where economic considerations dominate urban planning, leading to a preference for conventional infrastructure solutions that may not offer the same ecological benefits (O'Donnell et al. 2017). Furthermore, the integration of BGI with existing gray infrastructure often faces resistance due to the perceived complexity and cost associated with transitioning to more sustainable systems (Kapetas and Fenner 2020). Due to inappropriate planning and lack of urban control, city growth and informal development often destroy BGI and hinder natural drainage system, which is the main cause of urban flooding and consequent economic loss in different Asian cities.

Lack of awareness, Coordination, and Budget Allocation

Another significant barrier is the public awareness and acceptance of BGI. Many communities are not fully informed about the benefits of BGI, which can lead to skepticism regarding its implementation. This lack of awareness can result in insufficient public support for BGI projects, making it challenging to secure the necessary political backing and funding for such initiatives (Thorne et al. 2015). Additionally, the fragmented nature of urban governance complicates the coordination required for successful BGI implementation and maintenances multiple agencies may need to collaborate across various sectors (O'Donnell et al. 2017). Further, BGI systems demand continuous care and management to ensure their functionality and longevity (Gomes et al. 2021). This requirement can strain municipal budgets, particularly in cities facing financial constraints. Informal housing, informal trade and commerce, services and economic activities, and informal land development are not monitored and protected in a coordinated way in many countries that enhance loss of BGI in cities (Kamjou et al. 2024).

New Forms of Social Inequalities

Cities' informal settlements often experience environmental inequalities such as the unjust distribution of green areas and vulnerability to environmental hazards (Kamjou et al. 2024). Recently different studies focus on a new form of inequality in informal settlements associated with green developments. Studies argued that greening gentrification and climate change adaptation policies may exacerbate inequality and social inequalities in cities, through replacement of low-income residents and reducing their access to a safe and protected environment (Anguelovski et al. 2019; UN-Habitat 2020). The irregular distribution of BGI can lead to social inequities, where certain communities may not receive the same level of access to green and blue spaces as others, further complicating the case for BGI (Sowińska-Świerkosz et al. 2021). It is claimed that the growth of informal settlements in urban fringes not only resulted in urban sprawl into conservation areas but also drove social challenges (Macedo et al. 2021).

While blue-green infrastructure offers promising solutions to urban challenges, its successful implementation is hindered by lack of efficient planning and control of city's growth, effective valuation of BGI, and lack of awareness, budget allocation, and coordination to manage cities' informal growth, settlement, and informal economic activities. Addressing these challenges requires a concerted effort to enhance public awareness, improve inter-agency coordination, and develop robust management strategies that can sustain BGI systems over time.

8.5 Livable Blue-Green Cities with Poverty, Urban Slums, and Informal Sectors: Conflict and Consensus

The concept of livable blue-green cities is increasingly relevant in the context of urban slums, poverty, and informal sectors. These cities integrate green and blue infrastructure to enhance urban resilience, improve living conditions, and address the challenges posed by rapid urbanization, particularly in low-income areas. The interplay between urban planning, poverty alleviation, and environmental sustainability is critical in shaping these urban landscapes.

Urban slums often emerge as a consequence of inadequate urban planning and socio-economic disparities. Urban planning has historically contributed to the proliferation of slums, necessitating a fundamental review of planning practices to effectively address urban poverty and slum conditions (Abu-Salia et al. 2015). In India, it was highlighted that policies aimed at creating slum-free cities frequently overlook the broader spectrum of urban poverty, indicating a need for a paradigm shift in how urban poverty is conceptualized and addressed (Biswas 2023). It emphasizes the importance of including informal settlements in urban planning to achieve sustainable development goals (Satterthwaite et al. 2020).

The integration of blue-green infrastructure (BGI) is pivotal in enhancing urban livability. BGI not only mitigates the effects of climate change but also provides essential ecosystem services that can significantly benefit slum dwellers. For instance, it was discussed how green infrastructure can improve the lives of slum residents by enhancing their access to green spaces and natural ecosystems, which are crucial for their well-being (Adegun 2021). Furthermore, intensified measures to alleviate poverty in urban slums have been advocated, emphasizing the need for improved access to education and family planning as part of a comprehensive strategy to enhance living conditions (Htwe et al. 2017).

The design and planning of residential facilities in poor areas are also crucial for reducing slum conditions and argue that improving residential facilities can significantly impact the quality of life in these neighborhoods (Tutuko et al. 2018). This aligns with the findings of Parikh et al. (2013), who demonstrate that community-led initiatives focusing on water and sanitation can lead to substantial improvements in housing and education, ultimately contributing to poverty alleviation. Moreover, the role of green spaces in urban areas cannot be overstated; they provide psychological and physical health benefits, as highlighted by Vaeztavakoli et al. (2018), who note the therapeutic effects of blue and green spaces on urban populations. In addressing the challenges of urbanization, it is essential to recognize the spatial dynamics of green development. Research indicates that the efficiency of green development can have significant spatial spillover effects, impacting the resilience of urban human settlements (Wang 2023). This suggests that a coordinated approach to urban planning that considers the interconnectedness of different urban areas is necessary to avoid negative externalities.

However, the development of livable blue-green cities in the context of urban slums and informal sectors requires a multifaceted approach that integrates effective urban planning, poverty alleviation strategies, and the enhancement of green and blue infrastructure. By prioritizing these elements, cities can not only improve the living conditions of their most vulnerable populations but also foster sustainable urban environments that are resilient to the challenges of climate change and rapid urbanization.

8.6 Conclusion

Addressing the Asian dilemma of urban housing, poverty, and the informal sector while moving toward livable blue-green cities requires a holistic approach that combines economic, social, and environmental policies. The interplay between urban housing, poverty, and the informal sector in the context of developing livable blue-green cities in Asia presents a multifaceted challenge. The concept of blue-green cities integrates water management and green infrastructure to enhance urban livability while addressing pressing socio-economic issues such as poverty and informal settlements. This synthesis draws upon various studies to elucidate the complexities and potential pathways toward achieving sustainable urban environments. Firstly, the integration of blue-green infrastructure (BGI) is essential for enhancing urban livability. BGI not only addresses flood risks but also contributes to the aesthetic and ecological quality of urban spaces, thereby improving the overall quality of life for residents. Moreover, the socio-economic implications of urban greening initiatives cannot be overlooked. While BGI can enhance livability, it may also trigger gentrification, leading to the displacement of low-income communities. In addition, the role of social housing as a strategy to mitigate poverty and enhance livability is critical. By addressing the housing needs of disadvantaged groups, such initiatives can contribute to reducing poverty and enhancing social cohesion within urban areas.

In conclusion, the path toward livable blue-green cities in Asia is fraught with challenges related to urban housing, poverty, and the informal sector. However, by integrating blue-green infrastructure with social housing initiatives and prioritizing social justice in urban planning, cities can create inclusive environments that enhance livability for all residents. The successful implementation of these strategies requires a collaborative approach that engages various stakeholders, including government agencies, community organizations, and residents, to ensure that the benefits of urban development are equitably distributed.

References

Abu-Salia R, Osmannu I, Ahmed A (2015) Coping with the challenges of urbanization in low-income areas: an analysis of the livelihood systems of slum dwellers of the Wa Municipality, Ghana. Curr Urban Stud 03(02):105–118. https://doi.org/10.4236/cus.2015.32010

Adegun O (2021) Green infrastructure can improve the lives of slum dwellers in african cities. Front Sustain Cities 3. https://doi.org/10.3389/frsc.2021.621051

Amin AN (2016) The informal sector and urban poor. Dhaka: Bangladesh Rural Advancement Committee (BRAC), Dhaka

Anguelovski I, Irazábal-Zurita C, Connolly JJ (2019) Grabbed urban landscapes: socio-spatial tensions in green infrastructure planning in Medellín. Int J Urban Reg Res 43(1):133–156

Balgos BC (2016) Securing the safety of informal settler families along waterways in Metro Manila, Philippines: government-civil society organisation partnership. In: Disaster governance in Urbanising Asia, pp 177–193

Ballesteros MM (2010) Linking poverty and the environment: Evidence from slums in Philippine cities (No. 2010-33). PIDS Discussion Paper Series

Baser S, Nahar S, Rahman MM (2020) Inequality in household resources and amenities in Dhaka City: the case of Dhanmondi, Pallabi, and Kallayanpur. Pallabi, and Kallayanpur

Begum A (2007) Urban housing as an issue of redistribution through planning? The case of Dhaka city. Soc Policy Adm 41(4):410–418

Berquist M, Daniere A, Drummond L (2015) Planning for global environmental change in Bangkok's informal settlements. J Environ Plan Manag 58(10):1711–1730

Biplob P, Sarker DC, Sarker RC (2011) Assessment of water supply and sanitation facilities for Korail slum in Dhaka City. Int J Civ Environ Eng 11(5):115–128

Biswas D (2023) Has the geography of deprivation changed in post-reform urban India? A look into relative poverty and access to basic services. Geoscape 17(1):1–19. https://doi.org/10.2478/geosc-2023-0001

Bolay JC (2020) Urban facts. In: Urban planning against poverty. Future City, vol 14. Springer, Cham. https://doi.org/10.1007/978-3-030-28419-0_2

Boquet Y (2017) Managing Metro Manila. In: The Philippine archipelago, pp 567–615

Chayakul J (2015) Informal landscape architecture: a tool to improve water quality for informal settlements along waterways in Bangkok. Louisiana State University and Agricultural & Mechanical College

Chen M, Beard VA (2018) Including the excluded: supporting informal workers for more equal and productive cities in the global South. In: Working paper: towards a more equal city. World Resources Institute, Washington, DC. http://www.wri.org/publication/towards-more-equal-city-including-the-excluded?utm_source=twitter.com&utm_medium=wricities&utm_campaign=socialmedia

Chowdhury FJ, Amin AN (2006) Environmental assessment in slum improvement programs: some evidence from a study on infrastructure projects in two Dhaka slums. Environ Impact Assess Rev 26(6):530–552

Cities Alliance (2022) Annual Report 2022. https://www.citiesalliance.org/resources/publications/annual-reports/annual-report-2022

Cobbinath PB, Asibey MO, Zuneidu MA, Erdiaw-Kwasie MO (2021) Accommodating green spaces in cities: perceptions and attitudes in slums. Cities 111:103094

Das A, Susantono B (2022) Informal Services in Asian Cities: lessons for urban planning and management from the COVID-19 pandemic. Asian Development Bank Institute

Davidescu AA, Petcu MA, Curea SC, Manta EM (2022) Two faces of the same coin: exploring the multilateral perspective of informality in relation to sustainable development goals based on bibliometric analysis. Econ Anal Policy 73:683–705

Dewan AM, Kabir MH, Nahar K, Rahman MZ (2012) Urbanisation and environmental degradation in Dhaka Metropolitan area of Bangladesh. Int J Environ Sustain Dev 11(2):118–147

Gero A, Kohlitz J, Willetts JR (2017) Informal settlements in the Pacific and links to sustainable development. Dev Bull

Gomes MVR, Battemarco BP, Guimarães LF, Oliveira AKBd, Rutigliani VdA, Cabral FM et al (2021) The use of blue-green infrastructure as a multifunctional approach to watersheds with socio-environmental vulnerability. Blue-Green Syst 3(1):281–297. https://doi.org/10.2166/bgs.2021.119

Hasam A, Arafin S, Naznin S, Mushahid M, Hossain M (2017) Informality, poverty and politics in urban Bangladesh: an empirical study of Dhaka city. J Econ Sustain Dev 8:158–182

Hassan H (2018) Determinants and effects of the informal economy: Bangladesh perspective. Janata Bank J Money Financ Dev 5:197–208

Hongtong T (2019) Klong Toey slum's days numbered. Bangkok Post

Howlader D, Sarkar B (2020) Exploring the applicability of sustainable development goals in fringe areas of fast growing metropolises. In: Shaping urban change–Livable City Regions for the 21st Century. Proceedings of REAL CORP 2020, 25th international conference on urban development, regional planning and information society. CORP-Competence Center of Urban and Regional Planning, pp 1107–1111

https://www.habitatforhumanity.org.uk/blog/2018/09/urbanisation-slum-housing/

Htwe T, Oo W, Lwin N, Win K, Dar H (2017) Poverty among households living in slum area of Hlaing Tharyar township, Yangon city, Myanmar. Int J Res Med Sci 5(6):2497. https://doi.org/10.18203/2320-6012.ijrms20172435

Huang G, Xue D, Wang B (2020) Integrating theories on informal economies: an examination of causes of urban informal economies in China. Sustainability 12(7):2738

Ishtiaque A, Mahmud MS, Rafi MH (2014) Encroachment of canals of Dhaka City, Bangladesh: an investigative approach. GeoScape 8(2):48–64

Kamalipour H (2016) Urban morphologies in informal settlements: a case study. Contour J 1(2)

Kamjou E, Scott M, Lennon M (2024) Green infrastructure inequalities in informal settlements. Habitat Int 147:103058

Kapetas L, Fenner R (2020) Integrating blue-green and grey infrastructure through an adaptation pathways approach to surface water flooding. Philos Trans R Soc A: Math Phys Eng Sci 378(2168):20190204. https://doi.org/10.1098/rsta.2019.0204

Khan MMH, Kraemer A (2014) Are rural–urban migrants living in urban slums more vulnerable in terms of housing, health knowledge, smoking, mental health and general health? Int J Soc Welf 23(4):373–383

Khan SA (1994) Attributes of informal settlements affecting their vulnerability to eviction; a study of Bangkok. Environ Urban 6(1):25–39

Khiri (2022) A Warm Welcome in the Slums of Bangkok. https://khiri.com/bangkok-slums-khlong toey/

Legarias TM, Nurhasana R, Irwansyah E (2020) Building density level of urban slum area in Jakarta. Geosfera Indonesia 5(2):268–287

Macedo LSV, Picavet MEB, de Oliveira JAP, Shih WY (2021) Urban green and blue infrastructure: a critical analysis of research on developing countries. J Clean Prod 313:127898

Martínez L, Short JR, Estrada D (2017) The urban informal economy: street vendors in Cali, Colombia. Cities 66:34–43

Mathur OP (2013) Urban poverty in Asia. Asian Development Bank, Metro Manila, Philippines, 1–122

Mukherjee K (2020) Delineating fringe by rough-set theoretic approach: a case study on Barasat city, India. J Phys: Conf Ser 1579(1):012026. IOP Publishing

Ncube S, Arthur S (2021) Influence of blue-green and grey infrastructure combinations on natural and human-derived capital in urban drainage planning. Sustainability 13(5):2571

O'Donnell E, Lamond J, Thorne CR (2017) Recognising barriers to implementation of blue-green infrastructure: a newcastle case study. Urban Water J 14(9):964–971. https://doi.org/10.1080/1573062x.2017.1279190

Okopi UM, Mohammed MUKTAR (2020) Spontaneous settlements at the peri-urban fringe: the benefits of adopting the principles of sustainability. Covenant J Res Built Environ
Parikh P, Parikh H, McRobie A (2013) The role of infrastructure in improving human settlements. Proc Inst Civil Eng Urban Des Plan 166(2):101–118. https://doi.org/10.1680/udap.10.00038
Rahman MO, Rabbani KA (2007) Megacities: Slums, pollution, and Ill health: the Case of Dhaka, Bangladesh. Population
Satterthwaite D, Archer D, Colenbrander S, Dodman D, Hardoy J, Mitlin D, Patel S (2020) Building resilience to climate change in informal settlements. One Earth 2(2):143–156. https://doi.org/10.1016/j.oneear.2020.02.002
Sehrawat S, Shekhar S (2024) Potential sites for blue-green infrastructure in Gurugram, India; a multicriteria analysis. Geojournal 89(1):36
Siddique A (2017) Bangladesh's urban underbelly a cause for concern. Dhaka Tribune
Sinthia SA (2020) Analysis of urban slum: case study of Korail slum, Dhaka. Int J Urban Civil Eng 14(11):416–430
Sowińska-Świerkosz B, Michalik-Śnieżek M, Bieske-Matejak A (2021) Can allotment gardens (AGS) be considered an example of nature-based solutions (NBS) based on the use of historical green infrastructure? Sustainability 13(2):835. https://doi.org/10.3390/su13020835
Statistica (2023) Urban population growth rate Asia 2015–2050. https://www.statista.com/statistics/1379556/asia-urban-population-growth-rate/#:~:text=Between%202015%20and%202020%2C%20the,percent%20between%202045%20and%202050
Sultana N, Rahman MM, Khanam R (2022) The effect of the informal sector on sustainable development: evidence from developing countries. Bus Strategy Dev 5(4):437–451
Sultana R, Birtchnell T, Gill N (2020) Urban greening and mobility justice in Dhaka's informal settlements. Mobilities 15(2):273–289
Taylor et al (2022) Nourishment at the Margins of a Megacity: How the Poor Eat in Dhaka, Bangladesh. In Das and Susantono (eds) Informal services in Asian Cities: lessons for urban planning and management from the COVID-19 pandemic. Asian Development Bank Institute
Thaiprasert N, Leurcharusmee S, Jatukannyaprateep P, Sirisrisakulchai J (2020) Determinants of labor force participation and wages in Thailand: what is the role of the informal sector? Asian Econ J 34(3):301–326
Thorne CR, O'Donnell E, Ozawa CP, Hamlin S, Smith LA (2015) Overcoming uncertainty and barriers to adoption of blue-green infrastructure for urban flood risk management. J Flood Risk Manag 11(S2). https://doi.org/10.1111/jfr3.12218
Tutuko P, Subagijo E, Aini N (2018) The planning and design of residential facilities in poor areas with limited land. Int Rev Spatial Plan Sustain Dev 6(3):22–36. https://doi.org/10.14246/irspsd.6.3_22
UN-Habitate (2024) Asia and the Pacific Region. https://unhabitat.org/asia-and-the-pacific-region#:~:text=Urbanization%20continues%20to%20be%20a,an%20additional%201.2%20billion%20people
UN-Habitat (2022) World Cities Report 2022, Chapter 3: Poverty and Inequality: enduring Features of an Urban Future? United Nations Human Settlements Programme. https://unhabitat.org/sites/default/files/2022/07/chapter_3_wcr_2022.pdf. Accessed 1 Sep 2024
Un-Habitat (2020) World Cities Report 2020: The value of sustainable urbanization. UN
UN (2023) Chapter 2: Poverty Eradication and Employment, Asia-Pacific Population and Development Report 2023 (ST/ESCAP/3112), United Nations, Economic and Social Commission for Asia and the Pacific (ESCAP)
Vaeztavakoli A, Lak A, Yiğitcanlar T (2018) Blue and green spaces as therapeutic landscapes: health effects of urban water canal areas of Isfahan. Sustainability 10(11):4010. https://doi.org/10.3390/su10114010
Wang Z (2023) Spatial spillover and threshold effect of green development efficiency on urban human settlement resilience in Yangtze River delta urban agglomeration. Plos One 18(10):e0292230. https://doi.org/10.1371/journal.pone.0292230
Wetzstein S (2017) The global urban housing affordability crisis. Urban Studies 54(14):3159–3177

Widjajanti R, Dewi SP, Khadiyanta P (2019) Mapping of space compatibility for Street Vendors in Urban Public Space, Taman Tirto Agung, Banyumanik, Semarang. IOP Conf Ser: Earth Environ Sci 313(1):012017 (IOP Publishing)
World Cities Report 2020, Chapter 1: Urbanization and Cities, Trends of a New Global Force, United Nations Human Settlements Programme (UN-Habitat) 2022. https://unhabitat.org/sites/default/files/2020/10/wcr_2020_report.pdf. Accessed 2 Sep 2024
Yasmin S (2016) Impact of climate change on urban slum dwellers in Bangladesh: a case study of four selected slums in Dhaka City. Arch Curr Res Int 4(3):1–15
Yulu A (2021) Slums of Southeast Asia: the case of Manila, Philippines. Türk Coğrafya Dergisi (77):171–182

Chapter 9
Land Use Management, Planning, and Control for BGI, Food Security, and Climate Disaster Resilience

Md. Esraz-Ul-Zannat

Abstract This book chapter explores the critical role of land use management in addressing the complex challenges posed by urbanization, including food security and climate change. By examining the intricate relationship between land use patterns, blue-green infrastructure (BGI), and urban development, this chapter seeks to provide insights into effective strategies for integrating BGI into urban environments. The study investigates the impacts of urbanization on land use, food production, and disaster risk while highlighting the potential of BGI to mitigate these challenges. It delves into land use planning and policy frameworks, land acquisition, and case studies to illustrate best practices and lessons learned. Ultimately, this chapter emphasizes the importance of effective land use management for achieving food security, climate resilience, and sustainable urban development through BGI implementation.

Keywords Urbanization · Climate change · Blue-green infrastructure · Food security · Urban Resilience

9.1 Introduction

Urbanization is a significant global trend with profound implications for society and the environment. The process of urbanization has accelerated over the years, with half of the world's population currently residing in cities (Liu and Li 2017) and projections indicating a rise to 68% by 2050 (Humbal et al. 2023). Urbanization is interconnected with climate change, mutually reinforcing processes. Climate change impacts such as increased rainfall intensity, storm surges, and temperature fluctuations are anticipated to affect urban health and infrastructure globally (Debangshi et al. 2022). The urban

Md. Esraz-Ul-Zannat (✉)
Griffith School of Engineering & Built Environment, and Cities Research Institute, Griffith University - Gold Coast Campus, Southport, Australia
e-mail: md.esraz-ul-zannat@griffithuni.edu.au

M. G. Kibria et al. (eds.), *Urban Blue-Green Infrastructure Approach for Food Security and Climate Disaster Resilience*, Disaster Risk Reduction,
https://doi.org/10.1007/978-981-96-8746-6_9

heat island effect and extreme weather events caused by climate change can further influence urbanization by altering population distribution and urban layout (Chai et al. 2022). Moreover, urbanization can exacerbate climate change by changing land use and cover, potentially modifying precipitation patterns and increasing the risk of urban flooding (Koukoula et al. 2024). These factors collectively influence population distribution, urban infrastructure, and overall urban resilience. However, understanding the multifaceted effects of urbanization and climate change is crucial for sustainable development and effective policymaking in an increasingly urbanized world.

The integration of blue-green infrastructure (BGI) plays a crucial role in fostering sustainable urban development by addressing various environmental challenges and enhancing urban resilience. BGI encompasses natural and artificial features that can mitigate issues like urban water pollution, climate hazards, and declining green cover (Cavadini et al. 2024). In other words, BGI can be a strategy for climate change adaptation and disaster risk reduction, addressing impacts such as heat waves, and floods. It incorporates green infrastructure elements such as green roofs, rain gardens, urban greenspaces, urban trees and forests, and swales, along with blue infrastructure like basins, ponds, channels, detention basins, retention ponds, and permeable surfaces (EEA 2021). European Commission defined BGI as "A strategically planned network of natural and semi-natural areas with other environmental features, designed and managed to deliver a wide range of ecosystem services such as water purification, air quality, space for recreation and climate change mitigation and adaptation" (European Commission 2019, p. 5). "A strategically planned network" refers to the spatial and functional connection between green and blue infrastructures through an integrated planning process to harness maximum benefits. The "natural and semi-natural areas" emphasizes the need for a healthy ecosystem with multiple services and benefits to achieve biodiversity, essential to the BGI strategy. This ecosystem should encompass various landscape elements at different scales for biodiversity conservation. "Designed and managed to deliver" signals maintaining and enhancing ecosystem services to manage the blue-green network effectively. BGI can effectively reduce Combined Sewer Overflows (CSOs), improve air quality, provide psychological relaxation, and promote physical activity among urban residents (Cavadini et al. 2024). This approach not only helps in managing stormwater but also contributes to creating more sustainable and resilient cities by enhancing ecological balance, reducing pollution levels, and promoting community well-being with numerous benefits and co-benefits (Santiago 2024).

The integration of blue elements like water bodies with green elements such as parks and forests in urban area is crucial for enhancing climate resilience, promoting equity, and catalyzing sustainable urban development. Research emphasizes the significance of BGI in countering climate hazards, fostering resilience, and supporting economic transitions for sustainable urban futures. In other words, BGI plays a crucial role in urban land use management, planning, and control by integrating natural and semi-natural areas with built environments to mitigate environmental challenges including managing stormwater, reducing heat island effects, and enhancing urban resilience. Studies emphasize the importance of integrating

BGI elements like bioretention cells, permeable pavement, green roofs, and detention ponds to reduce CSOs and enhance stormwater management (Cavadini et al. 2024; Wu et al. 2024). Implementing BGI strategies can help break the 'control paradox' cycle by making more room for water and shifting flood risk management discourse towards planning for resilience (Dyca et al. 2024). Overall, a comprehensive approach that combines BGI with effective land use planning is essential for creating resilient and sustainable urban environments.

Urban Blue-Green Infrastructure (UBGI) emerges as a vital component in fostering urban food security while promoting overall sustainability. BGI significantly contributes to food security by supporting sustainable urban agriculture and improving ecosystem services. It enhances local food production, optimizes nutrient management, and fosters sustainable urban environments through various mechanisms including urban agriculture, green spaces, and water management systems. Research indicates that integrating green spaces and water bodies within urban environments can increase local food production, thereby reducing reliance on external food sources and enhancing food accessibility to fresh produce for urban populations (Pinto et al. 2023) and enhancing the resilience of urban food systems against climate change impacts. People increasingly recognize the importance of BGI in addressing urban flooding, mitigating heat stress, preserving agricultural land for food security, and providing open spaces for physical and mental well-being (Parvin et al. 2022). The integration of BGI not only improves urban landscapes but also connects agricultural areas to urban settings, thereby facilitating local food production and enhancing ecological stability (Pagac et al. 2023). Moreover, UGBI is crucial in the context of the food-energy-water (FEW) nexus, as it can provide essential resources while also posing challenges related to resource management. Effective management of UGBI can mitigate negative impacts, such as excessive water use and pollution from agricultural practices, thereby promoting sustainable urban food systems (Shah et al. 2021). However, a holistic approach is essential to balance the benefits and trade-offs associated with UBGI, ensuring that food security is achieved without compromising environmental health (Shah et al. 2021). Overall, the integration of blue-green infrastructure in urban planning and development is a promising approach to enhancing urban food systems and resilience.

Effective land use management, planning, and control are crucial for enhancing disaster resilience. Several effective land use management strategies and practices for enhancing climate induced disaster resilience are in practice around the world. Local government perceptions of disaster risks significantly influence resilience planning, as evidenced by studies showing that managers' hazard-specific concerns shape their adaptation priorities across various hazards, such as flooding and hurricanes (Kim et al. 2024). Implementing climate-resilient land use strategies such as floodplain and wetlands restrictions, tree protection and water conservation, design standards and zoning, is vital for enhancing disaster resilience. Studies also highlight the importance of identifying critical conflict patches in urban areas to optimize land use and avoid hazard risks, thereby promoting sustainable urban growth (Wang et al. 2024). Integrating interdisciplinary approaches, such as the 4 Point Approach (4PA)

(It's a strategic framework primarily used in the context of urban flood management, which extends the traditional Three Point Approach (3PA) by incorporating the concept of design for failure) strategy, emphasizes the need for collaboration among governing bodies to optimize flood management and land-use planning, thereby addressing climate change uncertainties (Huang and Wang 2024). Besides, a comprehensive approach that combines local knowledge, regulatory frameworks, and adaptive planning is vital for mitigating climate-related disasters (Malmström et al. 2024). Furthermore, local disaster planning must leverage intergovernmental resources and collaborative strategies to effectively manage risks, particularly in multijurisdictional contexts (Nolon 2022). Risk Sensitive Land Use Planning (RSLUP) is an essential framework for integrating disaster risk management into land use strategies, particularly in areas vulnerable to natural hazards. RSLUP involves assessing and mapping risk sensitivity to inform decision-making processes. For instance, studies have developed GIS-based multi-criteria zoning approaches to evaluate earthquake risk sensitivity, identifying zones from high to low risk, which aids policymakers in determining where interventions are necessary (Adem Esmail and Geneletti 2018; Barua et al. 2023). In regions like Nepal, implementing Risk Sensitive Land Use Plans (RSLUP) is essential for identifying hazard-prone areas and guiding sustainable urban development, thus enhancing community resilience (Sukhwani and Shaw 2022). By integrating proactive strategies and good practices around the world into urban planning and development, communities can better prepare for and respond to the challenges posed by climate change, ultimately safeguarding lives, property, and the environment.

This chapter explores the vital role of land use management in tackling the challenges of urbanization, such as food security and climate change. By examining the relationship between land use patterns, BGI, and urban development, it provides insights into effective strategies for integrating BGI into urban settings. The chapter covers topics including Urban Land Use Dynamics and Their Impacts, Land Use Planning for BGI, Food Security, and Climate Resilience, Land Acquisition and Management for BGI and Food Production, and Case Studies of BGI and Land Use Integration, concluding with a summary of key insights.

9.2 Urban Land Use Dynamics and Their Impacts

Global patterns of urban growth and land use change reveal significant trends influenced by population dynamics and economic development. Urban land expansion globally increased from 2000 to 2020, with higher densities in urban areas where suburban and urban areas expanded, with notable growth in India and China, leading to increased human concentration in cities (Hanberry 2023), which indicate reduced stability and increased vulnerability to environmental challenges. In developing regions, particularly in Asia and Africa, urbanization is accelerating, with cities experiencing rapid population growth and unplanned sprawl, leading to fragmented land use and the emergence of new urban centers (Wang and Kintrea 2021). Interestingly, while urban growth is often highlighted, studies indicate that land use intensification—rather than mere expansion—constitutes a significant portion of urban land changes, particularly in megacities (Zheng et al. 2021). This multifaceted urbanization process underscores the need for sustainable development strategies to mitigate the adverse effects of rapid urban expansion and land use changes globally.

Urbanization significantly impacts agricultural land, food production, and ecosystem services (ESs), creating complex trade-offs. The impact of urbanization on agricultural land, food production, and ecosystem services is a complex and multifaceted issue. As urban areas expand, the competition for land intensifies, leading to the conversion and fragmentation of fertile agricultural land into urban spaces. This results in a decline in available agricultural land, making it less productive and harder to manage. The decline in agricultural land due to urbanization reduces food production capabilities, leading to lower crop yields and higher food prices. Urbanization also alters food consumption patterns, increasing demand for meat and processed foods while decreasing demand for grains and vegetables (Hovhannisyan and Devadoss 2020). Urbanization can cause land degradation, resource depletion, and environmental pollution, which weaken ecosystem functions (Cheng et al. 2023). This can result in areas with low ecosystem service supply but high demand, particularly at urban edges, leading to deficits in ecosystem services (Li et al. 2024). For instance, a study indicated that urbanization resulted in a 9.54% decrease in agricultural land, contributing to a loss of ecosystem service values by approximately 85.14 million from 1980 to 2022 (Nayak et al. 2024). The relationship between urbanization and ESs is nonlinear and varies spatially, with different impacts observed in old versus new urban areas (Li et al. 2023). Thus, effective land management and urban planning are essential to balance urban growth with the preservation of agricultural land and ecosystem services (Chen et al. 2024; Wu 2024). Overall, a comprehensive approach that integrates sustainable urban planning, agricultural support programs, and environmental conservation measures to ensure food security and ecological resilience in the face of rapid urban growth.

The relationship between urban land use change and disaster risk is increasingly evident, particularly in the context of flooding. As cities develop and expand, more people and infrastructure are concentrated in high-risk areas, heightening vulnerability to disasters and the absence of risk-based land use planning further exacerbates this danger (Su et al. 2021). Rapid urbanization leads to increased impermeable surfaces, which heightens the risk of rainstorm waterlogging disasters due to altered surface runoff dynamics (Xu et al. 2024a, b). Another study by Fitriyati et al. (2024) indicate that urban expansion exacerbates flood risks by increasing flood-prone areas, with significant implications for spatial planning and disaster mitigation strategies. For instance, in Guiyang, urban land use changes have resulted in diminished vegetation cover impacts infiltration capacity and increased flood vulnerability, particularly in central urban areas (Tang et al. 2024). Urban areas, due to conversion of natural landscapes into built environments, are typically warmer than rural areas due to the heat island effect, intensifying heatwaves and increasing related health risks (Su et al. 2021). Furthermore, integrated modeling frameworks demonstrate that effective land use management can significantly reduce flood risks, highlighting the importance of dynamic planning that accounts for ongoing urban changes (Ku 2024). Conversely, land use changes in disaster-prone areas, such as around Mount Semeru, can increase risks associated with volcanic eruptions, underscoring the multifaceted nature of land use impacts on disaster risk (Bachri et al. 2024). Thus, a comprehensive approach to land use planning is essential for mitigating disaster risks in urban environments across various contexts.

Urbanization and climate change significantly exacerbate vulnerabilities in urban populations and areas. Identifying vulnerable populations and areas within urban environments due to urbanization and climate change impacts is crucial for developing effective mitigation and adaptation strategies. Research shows that rapid urbanization in developing countries has heightened social vulnerability, particularly for marginalized groups like the urban poor and migrants, who often live in high-risk informal settlements prone to hazards like flooding and landslides due to inadequate living conditions and limited adaptive capacity (Xu et al. 2024a, b). Moreover, studies in post-industrial cities demonstrate that specific urban areas, particularly those with intensive development, are more susceptible to climate-related hazards, such as thermal risks and flooding (Bronder et al. 2024). High population density in urban areas heightens the risk of exposure to climate-related hazards like floods, storms, and heatwaves (Lulla et al. 2015). Additionally, vulnerable populations often consist of low-income households, elderly individuals, and marginalized communities who may lack the resources to effectively cope with climate impacts (Antal and Bhutani 2023). For instance, in cities like Wuhan, assessments by Xu et al. (2024a, b) revealed that other factors such as low education, poor health, and unstable employment contribute to a high vulnerability rate among certain populations. Climate change impacts including urban heat island effect further complicates this issue, as it disproportionately affects communities based on demographic and geographic factors, necessitating targeted adaptation strategies (Assaf and Assaad 2024). Effective urban policy must integrate vulnerability assessments to prioritize interventions for at-risk populations and areas, ensuring urban resilience to climate change.

9.3 Approaches for Integrating BGI to Address Food Security and Climate Change Challenges

To fully harness the benefits of BGI in urban areas, cities must implement comprehensive strategies for land use management, planning, and control. These strategies should aim to optimize land use, thereby enhancing the efficiency and effectiveness of BGI systems, ensuring food security, and bolstering resilience to climate-related disasters. By integrating such approaches, communities can better manage natural resources, mitigate the impacts of climate change, and foster sustainable environments that benefit both people and ecosystems. Globally, various approaches and strategies are being employed, including Risk-Informed Development, Risk-Sensitive Land Use Planning, Nature-Based Solutions (NbS), zoning regulations and incentives for BGI and urban agriculture, governance and stakeholder engagement, and innovative land use policies focused on food security and disaster resilience. Some of these approaches, considered cutting-edge and innovative, are outlined below.

9.3.1 Nature-Based Solutions (NbS)

NbS are increasingly recognized as essential strategies for addressing environmental and societal challenges, especially in urban areas. NbS is an umbrella concept that encompasses several related approaches, including Water Sensitive Urban Design (WSUD), Sustainable Urban Drainage Systems (SuDS), Ecosystem-based Adaptation (EbA), Ecosystem-based Disaster Risk Reduction (Eco-DRR), Blue-Green Infrastructure (BGI), Sponge City (SC), and Water Sensitive City (WSC). These solutions involve both natural and constructed systems that leverage ecological processes to provide benefits such as biodiversity enhancement, climate change mitigation, ecosystem restoration, amenity value, and resilience, all while being cost-effective and energy-efficient (O'Hogain and McCarton 2018). Various organizations, including The International Union for Conservation of Nature (IUCN) and the European Commission, have defined NbS in different ways to capture their diverse applications and benefits. The International Union for Conservation of Nature (IUCN) defines NbS as "actions to protect, sustainably manage, and restore natural or modified ecosystems, that address societal challenges effectively and adaptively, simultaneously providing human well-being and biodiversity benefits" (Cohen-Shacham et al. 2016). The European Commission has adopted NbS in initiatives like Horizon 2020, aiming to position Europe as a leader in sustainable urban practices (Moreau et al. 2022). European Commission defines NbS as solutions that are "inspired and supported by nature, which are cost-effective, simultaneously provide environmental, social and economic benefits and help build resilience. Such solutions bring more, and more diverse, nature and natural features and processes into cities, landscapes, and seascapes, through locally adapted, resource-efficient and systemic interventions" (European Commission 2023). Both definitions are similar in terms of their core essence; however, the latter definition provides more detailed insights on the scope of the NbS.

NbS encompass a range of interventions, from natural reserve conservation to engineered systems. In a study on the effectiveness of nature-based infrastructures for urban flood risk mitigation, Esraz-Ul-Zannat et al. (2024) identified 16 different NbS infrastructures. These include green roofs, rain gardens, swales, rainwater harvesting, permeable pavements, detention ponds, constructed wetlands, retention ponds, urban agriculture, infiltration trenches, riparian buffers, sand filters, urban forests, green walls, green streets, and blue roofs. Each of these infrastructures varies in scale, application, and geometric properties, particularly regarding their use in flood mitigation, as illustrated in Fig. 9.1. They are integral to urban sustainability, promoting resource recovery and minimizing environmental impacts (Kotsila et al.

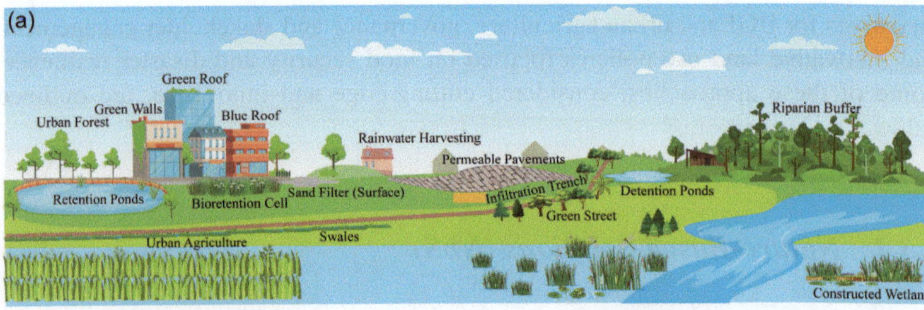

(b)

	Micro Scale	Meso Scale	Macro Scale	
Area	• Blue roof • Green roofs • Green walls • Permeable pavements		• Riparian buffer • Urban agriculture • Urban forest	Area
Linear		• Swales • Green street • Infiltration trench		Linear
Point	• Rainwater harvesting	• Rain gardens • Detention ponds • Retention ponds • Sand filters (surface)	• Constructed wetlands	Point
	Micro Scale	Meso Scale	Macro Scale	

Fig. 9.1 **a** Urban NbS Infrastructures across diverse scales in infographic form. **b** Scale of application and geometric properties of the identified flood mitigation measures. *Source* Esraz-Ul-Zannat et al. (2024)

2021; O'Hogain and McCarton 2018). Despite their potential, the concept of NBS remains vaguely defined, necessitating clearer frameworks for effective implementation (Moreau et al. 2022). While NbS offer promising solutions, their integration into existing urban frameworks can be contentious, often reflecting broader neoliberal trends in urban governance (Kotsila et al. 2021).

Effective land management and planning for BGI are crucial for enhancing food security and climate disaster resilience. NbS are increasingly recognized as vital strategies for enhancing food security and climate resilience through effective land management and planning. These solutions leverage natural processes to address environmental challenges while promoting sustainable agricultural practices. NbS can optimize the water-ecosystem-food nexus, for example in Mediterranean regions, by improving water management and enhancing food security and climate resilience while preserving ecosystems (Vanino et al. 2024). In agricultural landscapes, NbS help reduce trade-offs between food production and conservation, focusing on carbon sequestration and disaster-risk management (Miralles-Wilhelm 2023). NbS are essential for mitigating land degradation and disaster risks, for example, in vulnerable agroecosystems like vineyards of Oltrepò Pavese, Italy, where they can effectively reduce shallow landslide risks, enhance disaster risk reduction, and promote sustainable agriculture, aligning with food security and climate disaster resilience goals (Gambarani et al. 2024). Urban areas can also benefit from NbS, which enhance resilience including food security and climate disaster resilience against climate change impacts and promote public health through green strategies in land management and planning for the built environment (Sommese 2024; Vazin et al. 2024). While NbS present promising solutions, their implementation remains limited, necessitating greater awareness and collaboration among stakeholders to fully realize their potential in land management and climate resilience.

9.3.2 Risk-Informed Development Planning

Risk-informed development (RID) planning is a decision-making process that incorporates risk assessment to enhance the sustainability and resilience of development in the face of evolving threats (Opitz-Stapleton et al. 2019). RID employs risk-based frameworks to evaluate various threats, opportunities, uncertainties, and options, encouraging decision-makers to recognize the trade-offs involved in any development choice. Risk-based development decision-making encompasses three interconnected goals, as illustrated in Fig. 9.2. RID emphasizes understanding interconnected risks, particularly in the context of climate change and disasters, as highlighted by Belle et al. (2024). However, RID involves integrating disaster and climate change risks into everyday decision-making processes, making them a central part of development strategies, including policy, planning, budgeting, and monitoring at all levels (Selby and Perez-Dalena 2020). This approach ensures that disaster risk reduction (DRR) and climate change adaptation (CCA) are embedded in development activities, rather than being treated as separate add-ons, leading to more cohesive and effective actions.

This approach is increasingly recognized as essential for sustainable urban development, particularly in areas prone to disasters. RID and land use management are crucial for enhancing food security and climate disaster resilience. This approach

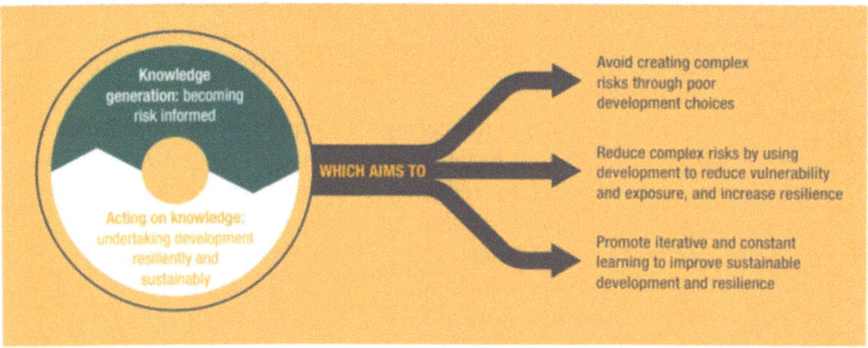

Fig. 9.2 Risk-informed development and its core aims. *Source* Opitz-Stapleton et al. (2019)

incorporates risk assessment into the planning process, ensuring that land use decisions are guided by an understanding of potential hazards. In the peri-urban areas, implementing climate-smart agricultural practices such as site-specific crop selection and efficient water management, can boost crop yields, enhance food security, improve soil health, and lower greenhouse gas emissions, fostering overall climate resilience (Karri and Nalluri 2023). Studies indicate that urban and peri-urban agriculture can mitigate risks associated with food deserts and unemployment, thereby enhancing food security for vulnerable populations (Santini 2022). While the integration of RID into land use planning shows promise, challenges remain in ensuring equitable access to resources and data, particularly for vulnerable populations. Addressing these disparities is vital for achieving comprehensive resilience and food security.

9.3.3 Risk-Sensitive Land Use Planning (RSLUP)

RSLUP focuses specifically on land use planning and management while RID provides a comprehensive framework for incorporating risk into all aspects of development. RSLUP approach is essential for enhancing urban resilience against climate disasters by integrating disaster risk management into land use planning by prioritizing safe areas for development based on vulnerability to hazards like earthquakes, landslides, and floods (Sharma et al. 2022). This approach employs risk assessment and mapping to evaluate disaster risk sensitivity, identifying high-risk zones and guiding policymakers on necessary interventions (Barua et al. 2023). Effective RSLUP incorporates community engagement and local knowledge, fostering adaptive capacity and disaster preparedness (Ghimire 2023), as seen in Nepal's decentralized governance structure. Multi-stakeholder dialogues are crucial for addressing trade-offs and ensuring equitable climate adaptation solutions (Gaborit 2022). However, the implementation of participatory policies remains limited, affecting the overall efficacy of RSLUP initiatives (Sharma et al. 2022). Along with RSLUP in city planning, regional development planning must incorporate disaster risk reduction principles, enhancing urban resilience and food security through informed spatial planning (Latue et al. 2023).

Risk-sensitive land use planning increasingly incorporates BGI to enhance urban resilience against climate-related hazards. BGI combines natural and engineered systems to manage stormwater by reducing combined sewer overflows (CSOs), mitigate flood risks, and improve ecological health by infiltrating and retaining stormwater, as demonstrated in Zurich, where various BGI combinations were evaluated for their CSO reduction potential (Cavadini et al. 2024). In addition, in flood-prone areas, land use planning can mitigate risks by integrating BGI, promoting a shift towards more sustainable water management practices (Dyca et al. 2024). BGI offers many benefits and co-benefits including various ecosystem services besides flood mitigation (Ferreira et al. 2020; Spahr et al. 2021). For example, BGI contributes to urban resilience by providing ecosystem services that reduce vulnerability to climate change, as seen in Bayraklı, İzmir, where existing BGI was assessed for its stormwater management capabilities (Coşkun Hepcan and Canguzel 2024). The integration of BGI in urban planning not only addresses environmental challenges but also fosters community engagement and social equity, essential for sustainable urban futures (Meemrot et al. 2024). While BGI presents significant benefits, challenges remain in its implementation, particularly regarding cost-effectiveness and the need for tailored strategies that consider local conditions and community needs.

9.4 Integrating BGI into Land Use Policy and Practices

(a) Land Use Policies

Sustainable urban land use policies that integrate BGI are crucial for ensuring food security and bolstering resilience against climate-related disasters. By incorporating BGI, cities can effectively address the challenges including managing water resources, reducing flood risks, and creating urban green spaces that contribute to local food production. Various land use policies around the world exemplify the successful integration of BGI, showcasing diverse approaches to harmonizing urban development with environmental sustainability. These policies highlight the importance of strategic planning in creating resilient, food-secure urban environments.

- **Comprehensive Land Use Planning**

Comprehensive land use planning is essential for integrating BGI to enhance food security and climate resilience in urban areas. BGI integration in urban planning connects agriculture and urban lands, promoting sustainable practices and enhancing food production and ecological stability (Pagac et al. 2023). This approach not only addresses urbanization challenges but also promotes sustainable development through effective resource management against other challenges including climate change impacts. It is crucial to develop integrated land use plans that prioritize BGI alongside traditional infrastructure. This should follow comprehensive climate risk assessments within the planning processes to identify vulnerable areas and determine the best locations for BGI. It is also essential to consider BGI early in urban development and strategically allocate land for BGI projects, including green spaces, parks, and water bodies. Implementing BGI enhances biodiversity, air and water quality, and mitigates urban heat effects, all of which are critical for climate resilience while offering economic benefits, such as higher property values and job creation (Adesina et al. 2024). In Estonia, comprehensive planning has identified critical blue and green assets, ensuring effective land-use regulation and ecological integrity (Kaskevich et al. 2022). The socio-economic benefits of BGI, such as increased access to green spaces, support food security by enhancing community engagement and resource availability (Yang et al. 2024). Moreover, Comprehensive planning can guide investments towards sustainable practices that bolster food security and climate resilience (Cooper 2024).

- **Zoning Regulations**

Zoning regulations, as a land use control tool, are vital for promoting sustainable urban land use policies, especially when it comes to integrating BGI to improve food security and resilience to climate disasters. Various land use zones and their associated regulations control urban growth and ensure that BGI elements, such as parks, green roofs, and permeable surfaces, are incorporated into urban planning (Bansal and Pandey 2024) while protecting agricultural land from urban encroachment. It is imperative to implement zoning regulations that promote BGI and restrict development in vulnerable areas along with protecting and restoring natural ecosystems to enhance their resilience. Zoning regulations can include risk assessments to restrict development in high-risk areas, thereby mitigating disaster impacts and enhancing community resilience (Domingo et al. 2021). Additionally, implementing mixed-use zoning is essential to encourage the coexistence of residential, commercial, and agricultural activities. For example, the integration of multifunctional green segments in industrial areas can revitalize urban spaces while adhering to sustainability principles (Nedyalkova 2024). However, BGI integration into zoning regulations is growing, but challenges remain in harmonizing policies and ensuring effective implementation.

- **Incentives for Green Development**

Incentives for green development through the integration of BGI in the forms of financial, policy, and community engagement are essential for sustainable urban land use policies. Offering financial support for BGI projects can reduce initial investment costs, encouraging municipalities to implement these solutions (Adesina et al. 2024). Moreover, tax incentives for property owners who integrate BGI elements can encourage private investment in sustainable practices (Yang et al. 2024). Additionally, green building certifications that acknowledge sustainable practices, such as the incorporation of BGI can be encouraged.

- **Land Conservation Policies**

Land conservation policies play a crucial role in integrating BGI into sustainable urban land use. These policies integrate blue and green elements, reducing costs and improving ecological outcomes (Perera et al. 2024). In addition to promoting new BGI initiatives, conservation policies safeguard natural ecosystems, wetlands, and agricultural land from conversion to non-agricultural uses (Feltynowski 2023; Gupta and De 2024). However, conservation policies support green spaces and biodiversity by using conservation easements to prevent soil erosion and maintain soil health.

- **Community Engagement and Education**

Community engagement and education are crucial for successfully integrating BGI into sustainable urban land use policies. Everett et al. (2023) showed that community involvement is crucial for effective BGI implementation, ensuring local needs are met and fostering active participation, which enhances accountability and sustainability in urban planning. Educational campaigns are crucial for raising awareness of the benefits of BGI, such as food security and resilience. In this regard, educational programs in schools can raise awareness and appreciation of BGI, encouraging sustainable practices, where schools can model BGI integration, demonstrating how green spaces enhance urban living and biodiversity (Sutkowska et al. 2024).

(b) Land Management Practices

Implementing sustainable land management practices that incorporate BGI is essential for dealing with the urban challenges including ensuring food security and enhancing resilience to climate disasters. Sustainable Land management practices in the cities can ensure the integration and functioning of BGI leading to generating multiple benefits including improving water management, reducing soil erosion, and enhancing biodiversity.

- **Integrated Water Management**

Integrated water management (IWM) that incorporates BGI contributes sustainable land management practices aimed at solving urban challenges including ensuring food security and enhancing climate disaster resilience. Ferreira et al. (2024) showed that when combined with IWM, BGI enhances urban water management, especially in informal settlements, by boosting community engagement and optimizing resource utilization. Incorporating water management into land use planning promotes environmental conservation and raises public awareness, encouraging stronger collaboration among stakeholders (Kalogiannidis et al. 2023). Innovative methods like sustainable drainage systems (SuDS) and rainwater harvesting can be used to manage stormwater and lower flood risks. For example, in Jakarta, the implementation of BGI such as retention ponds has successfully reduced flooding and improved water quality, highlighting its critical role in strengthening climate resilience (Agustine et al. 2023).

- **Sustainable Urban Agriculture**

Urban agriculture, which integrates BGI, is a key sustainable land management practice that supports food security and bolsters resilience to climate disasters. Urban agriculture is a vital strategy against food insecurity in rapidly urbanizing cities like Lahore, where community interest in farming is driven by self-reliance and health concerns, despite challenges such as land access and time constraints (Alam and Naeem 2024). Another study by Toromade et al. (2024) showed that urban farming can significantly reduce carbon footprints and enhance local food production through innovative practices like rooftop gardens and vertical farming. Encouraging the use of organic fertilizers and integrated pest management can improve soil health and crop resilience, leading to greater benefits in agriculture.

- **Ecosystem Restoration**

Ecosystem restoration using sustainable land management with BGI can solve many urban challenges including supplying fresh food and combating climate induced disasters. Restored BGI enhances urban ecosystems by integrating natural elements, which boost biodiversity, lower pollution levels, and improve overall environmental health (Yang et al. 2024). BGI systems like bioswales and green roofs manage stormwater and reduce urban flooding risks, thereby enhancing climate resilience (Coşkun Hepcan and Canguzel 2024; Czyża and Kowalczyk 2024). Ensuring equitable participation and resource allocation among diverse community groups is crucial for maximizing the benefits of these sustainable practices.

- **Disaster and Climate Adaptation Strategies**

Using BGI for disaster and climate adaptation transforms sustainable land management by strengthening urban resilience to climate-related hazards and delivering vital ecosystem services. BGI greatly diminishes urban vulnerability to climate change by managing stormwater and reducing flood risks. For example, in Bayraklı, İzmir, the current BGI was insufficient for handling extreme stormwater runoff, underscoring the need for enhanced capacity and planning (Coşkun Hepcan and Canguzel 2024). Ambrosi et al. (2024) found that Alpine cities benefit from BGI by balancing urban heat, flood risks, and water scarcity, which improves overall quality of life and ecological health. In India, incorporating BGI into urban planning boosts resilience and equity, tackling climate hazards and supporting economic transitions (Meemrot et al. 2024). Land management strategies should adapt to changing climate conditions and their effects on food security and infrastructure, with regular monitoring and adjustment of BGI initiatives to improve resilience.

- **Technological Integration**

Technological integration in BGI is pivotal for sustainable land management, enhancing environmental resilience, improving food security, and addressing climate change impacts. Technologies like IoT and AI enhance BGI functionality, optimizing urban planning and operations to boost biodiversity and reduce pollution (Yang et al. 2024). Moreover, innovative geospatial methods, like graph theory and gravity models, help identify and implement BGI networks for effective stormwater management (Kaur and Gupta 2022). Singh et al. (2023) found that integrated BGI systems reduce infrastructure costs, boost property values, and enhance residents' quality of life. However, ensuring equitable access and effective implementation, especially in informal urban areas, is a major challenge (Ferreira et al. 2024).

- **Governance and Stakeholder Engagement**

Effective governance and stakeholder collaboration encourage community involvement, ensuring that local knowledge and cultural values influence infrastructure development, including BGI. Yang et al. (2024) found that engaging local communities in BGI planning and maintenance fosters ownership and ensures projects address residents' needs. Engaging communities in BGI projects also improves public support and accountability, resulting in more sustainable outcomes highlighting co-production, shifting from passive to active involvement in decision-making (Everett et al. 2023). In addition, public education programs on BGI benefits can boost community support and involvement (Cavadini et al. 2024). Maher (2020) showed that multi-stakeholder governance is essential for tackling complex issues like urban drainage, where collaborative approaches facilitate the shift from grey to blue-green infrastructure, boosting urban resilience and sustainability.

9.5 Challenges to Land Acquisition and Strategic Allocation for BGI

Urban land acquisition and allocation for BGI face significant challenges, particularly in developing countries. Acquiring and allocating urban land for BGI involves several challenges including complex land tenure systems, weak governance, and socio-economic factors that hinder effective implementation.

9.5.1 High Land Cost and Competing Land Uses

Urban land is often expensive, making it challenging to acquire large areas for BGI projects (Keita and Kourouma 2024). The financial strain of land acquisition especially upfront cost and land maintenance for BGI, compounded by corruption and unofficial fees, further complicates access to urban land, particularly in developing countries (Ameyaw and De Vries 2021; Mushinge et al. 2022). Urban areas face multiple competing demands for land, including residential, commercial, and industrial purposes, which can reduce the availability of land for BGI projects (Keita and Kourouma 2024). It is essential to ensure fair compensation and resettlement of affected landowners and residents to avoid conflicts.

9.5.2 Land Tenure and Property Rights Issues

Unclear or complex land tenure systems, whether formal or customary, make it difficult to identify legitimate landowners and obtain consent for land acquisition. This is especially problematic in areas with a history of land conflicts or weak land governance institutions (Emiru et al. 2023). In Ghana, for instance, land acquisition is complicated by overlapping ownership claims and inadequate documentation, leading to disputes and delays in the approval process (Boafo et al. 2023). When multiple parties—such as individuals, families, communities, or the state—lay claim to the same piece of land, it leads to disputes and resistance to land acquisition. Resolving these competing claims necessitates thorough consultation, negotiation, and compensation. However, incomplete or inaccurate land records create difficulties in establishing land ownership, boundaries, and usage. This lack of clear information complicates the land acquisition process and heightens the risk of disputes (Emiru et al. 2023). Informal settlements often lack formal land rights, posing challenges for BGI implementation and potentially displacing vulnerable communities without adequate resettlement plans and compensation.

9.5.3 Weak Governance

Ineffective land governance poses significant obstacles to urban development and the implementation of BGI. Weak land governance, characterized by ineffective administration, unclear policies, and opacity, can undermine the public's trust in land acquisition processes (Emiru et al. 2023). Cumbersome public land allocation processes with bureaucratic inefficiencies and the involvement of unqualified middlemen contribute to inefficiencies, making it challenging to implement BGI effectively (Ameyaw and De Vries 2021). For example, accessing urban land for housing in Zambia faces challenges like high fees, limited supply, corruption, and speculation, leading to unplanned settlements and scarcity for legal uses like BGI (Mushinge et al. 2022). Weak administration along with other issues in land governance lead to unplanned settlements and reduce the availability of land for essential projects like BGI.

9.5.4 Community Engagement

Community engagement is vital for successful BGI implementation. Involving local communities in planning and decision-making ensures that BGI projects align with their needs and preferences. The expropriation of peri-urban land for housing development often displaces farming communities, leading to social unrest and inadequate compensation. A study in Bahir Dar, Ethiopia, illustrates the challenges of compensating farmers for land acquisition, highlighting the potential negative impacts on livelihoods (Adigeh and Abebe 2023). To ensure community involvement in BGI projects, conducting public consultations and workshops is essential and these platforms allow for gathering valuable input and fostering a sense of community ownership.

9.5.5 Public-Private Partnerships, and Financing Mechanisms for Land Acquisition and BGI Development

Public-private partnerships (PPPs) are essential for effective land management, especially in acquiring and allocating land for BGI. These partnerships can improve efficiency, transparency, and inclusivity in land administration processes. For example, Community-Based Public-Private Partnerships (CBP3) integrate community needs and BGI for stormwater management, improving water quality and well-being along with enhancing community engagement and flexibility, tailoring solutions to urban challenges (US EPA 2015). PPPs enable the distribution of risks and responsibilities to the party best suited to manage them, whether those are financial, legal, or technical (Zhao et al. 2023). However, while PPPs can improve project profitability, the low profitability of green projects remains a challenge, necessitating supportive legislation (Pagac et al. 2023). Additionally, despite the potential benefits, the implementation of PPPs in the green economy is still limited, highlighting the need for further incentives and legislative support to fully realize their advantages in land management for BGI. They support sustainable urban development by incorporating green technologies and practices into land management, leveraging public and private sector expertise, and reducing costs through shared resources and innovative financing mechanisms.

Financing mechanisms for land acquisition and the development of BGI are increasingly vital in urban planning and for the successful implementation of these projects, especially in the context of limited resources and growing infrastructure needs. PPPs can facilitate the financing of land acquisition and BGI development by leveraging the resources and expertise of both public and private sectors. Various innovative financing strategies including Land-Based Financing Mechanisms, Equity Financing Models, Fair Compensation Mechanisms, have emerged, each with unique advantages and challenges. Land-Based Financing Mechanisms are essential tools for funding urban infrastructure projects. In Ghana, a collaborative approach between state and non-state actors employs leasehold premiums and infrastructure levies to generate funds for urban development. This strategy underscores the importance of effective stakeholder management and understanding market conditions to optimize revenue collection (Biitir et al. 2022). Another valuable mechanism is Land Value Capture, which enables municipalities to recover the increased land value resulting from public investments. This approach provides a sustainable funding source for infrastructure projects by capturing a portion of the financial benefits generated through improved urban amenities and services (Tirumala and Tiwari 2023).

Equity Financing Models provide innovative approaches to overcoming the financial challenges of land acquisition for development projects. One such model is the "Flats for Land" system, which eliminates the need for upfront land purchase costs by allowing developers to finance land acquisition through equity-like arrangements, thereby reducing financial risk (Petris 2023). Another approach involves Auction Protocols, which simplify the land acquisition process by removing the requirement to set land reservation values. This method encourages competition among private entities, allowing them to finance infrastructure projects more effectively (Chowdhry 2022). Together, these models offer flexible solutions for financing land acquisition and development, making them valuable tools for urban planning and infrastructure expansion.

Fair compensation mechanisms are essential for equitable land acquisition processes, ensuring that affected communities are adequately compensated for their losses. A strong regulatory framework is key to achieving fairness in compensation. For example, Indonesian regulations require proper replacement guarantees for communities impacted by land acquisition. These regulations help protect the rights of those affected, ensuring they receive fair compensation and appropriate support during the relocation process (Adhisukmawati et al. 2023). While these financing mechanisms present promising solutions, their effectiveness often hinges on robust governance frameworks and stakeholder engagement, which can vary significantly across different regions and contexts.

9.6 Case Studies

Numerous cities around the world have demonstrated the successful integration of BGI into their land use planning strategies. These case studies offer valuable insights and lessons for other urban areas seeking to enhance food security and climate resilience. By learning from these case studies, cities can develop tailored strategies for BGI integration that address their specific needs and challenges. Here a few of them are discussed below.

9.6.1 The Active, Beautiful, Clean (ABC) Waters Program in Singapore

Singapore's leadership in BGI is underscored by ABC Waters Program, an innovative approach, to urban planning and sustainability. This program aligns with the vision of a "City of Gardens and Water" by integrating terrestrial and aquatic green spaces, resulting in lush greenery and abundant green areas that enhance aesthetics, recreation, and microclimate regulation (Liao 2019). This integration of BGI not only enhances resilience against climate change but also promotes efficient water management and ecological benefits. The city has adopted advanced stormwater management strategies, such as rainwater harvesting, retention ponds, buffer zones, and green roofs, with irrigation, achieving a zero-water balance in urban areas to enhance its resilience and adaptability to climate change (Keita and Kourouma 2024). This model allows for the reuse of stormwater, crucial for maintaining green spaces even during dry spells, thereby supporting food security through urban agriculture (Breulmann et al. 2024). The city's strategic use of bioretention basins and green roofs effectively manages increased rainfall intensity, which is crucial for maintaining urban ecosystems and enhancing climate resilience (Cavadini and Cook 2023). Although Singapore's initiatives are commendable, challenges persist in maintaining an adequate water supply for BGI during extended dry spells, requiring continuous innovation and strategic planning to effectively sustain these systems.

Singapore, a developed nation with limited land and resources, has successfully implemented BGI to boost food production. Around 56% of Singapore's land is covered by vegetation, with 27% actively managed (parks, gardens, lawns) and 29% spontaneous, while an additional 10% is allocated to public facilities and open spaces (Lim and Xenarios 2021). By leveraging innovative techniques like vertical farming and rooftop gardening, Singapore aims to meet a significant portion of its vegetable needs locally (Barthel and Isendahl 2013; Shah et al. 2021). For example, traditional farms and rooftop gardens together now meet approximately 35.5% of Singapore's vegetable needs, a significant increase from the previous 5% (Astee and Kishnani 2010). This strategic approach not only contributes to food security but also reduces reliance on imports and fosters sustainable urban development.

9.6.2 Sponge City in China

The Sponge City initiative, a multifaceted approach, in China offers a promising framework for sustainable urban development through integrating BGI into urban flood management and water scarcity. This framework also acts as a comprehensive climate change adaptation strategy, effectively managing water resources and promoting ecological balance. By incorporating natural water management systems such as BGI, sponge city manage stormwater, reduce flood risks and improve water quality, which is crucial for food production systems in urban areas (Balado and Solla 2024). BGI, through infrastructures like permeable materials and green spaces, absorbs and filters rainwater, replicating natural hydrological processes and significantly reducing the risk of urban flooding (Bian 2024). Three-Points Approach (3PA) into Sponge City concept, emphasizes a holistic approach, combining day-to-day water management with strategies for design and extreme weather events, thereby ensuring food systems remain functional during disasters (Van De Ven et al. 2024). In essence, enhancing stormwater resilience is crucial for maintaining agricultural productivity in urban areas like Jiangchuanlu Street in Shanghai, showing how ecological infrastructure can reduce flood risks while supporting food security (Daixin and Mingyang 2023). Such strategies can create multifunctional landscapes that promote urban agriculture and biodiversity, thereby enhancing community resilience (Chen and Song 2024). Despite its advantages, the Sponge City initiative faces challenges in implementation, including policy support and public engagement (Chan et al. 2024). Successful implementation of the Sponge City model requires overcoming challenges and ensuring comprehensive stakeholder engagement.

9.7 Conclusion

In conclusion, this chapter highlighted the critical importance of effective land use policies, planning, and management in successfully integrating BGI into urban environments. This integration is particularly crucial in addressing the challenges posed by changing urban dynamics and climate change. While various strategies exist for integrating BGI, ensuring food security, and enhancing climate disaster resilience, not all are suitable for addressing current and future challenges. By implementing appropriate approaches and targeted interventions, such policies can enhance food security, bolster resilience to climate-related disasters, and contribute significantly to sustainable urban development.

Current literature considers innovative solutions such as NbS, risk-informed development planning, risk-sensitive land use planning, and sustainable land use policies and management practices effective and applicable to tackle the urban challenges. Effective approach and target interventions align with the vision can optimize urban environments strategical allocation of blue and green infrastructures and their implementation for better food production, stormwater management, and climate risk mitigation as appeared in the case studies, Singapore in particular. Challenges to integrate BGI successfully, such as high land costs, competing land uses, land tenure issues, weak governance, community engagement, public-private partnerships, and financing for land acquisition and BGI development need to be addressed, particularly in securing and strategically allocating land within existing land management and planning frameworks. Analyzing successful BGI integration case studies with similar context including geographic, socio-political, economic, environmental aspects, allow cities to develop tailored strategies that address their unique needs and challenges. The growing recognition of BGI by stakeholders as a vital solution to urban challenges is a promising development, now increasingly reflected in practice. Significant advancements have already been made in this area, and further research aimed at fostering innovation in approaches and infrastructure has the potential to amplify these efforts, driving even greater impact.

Conflict of Interest Statement The author reports no conflict of interest for this publication.

References

Adem Esmail B, Geneletti D (2018) Multi-criteria decision analysis for nature conservation: a review of 20 years of applications. Methods Ecol Evol 9(1):42–53

Adesina AA, Okwandu AC, Sikhakhane Nwokediegwu ZQ (2024) Towards sustainable urban development: conceptualizing green infrastructure and its impact on urban planning. Int J Appl Res Soc Sci 6(7):1274–1296. https://doi.org/10.51594/ijarss.v6i7.1266

Adhisukmawati IGAKH, Handayani IGAKR, Karjoko L (2023) Aspek Keadilan Pengaturan Pemberian Ganti Rugi dalam Pengadaan Tanah untuk Pembangunan bagi Kepentingan Umum. In: Proceeding legal symposium, vol 1, pp 1–17. https://doi.org/10.18196/pls.v1i1.86

Adigeh DT, Abebe BG (2023) The practice of peri-urban land acquisition by expropriation for housing purposes and the implications: the case of Bahir Dar, Ethiopia. Urban Sci 7(2):41. https://doi.org/10.3390/urbansci7020041

Agustine V, Putra DD, Rilansari V (2023) Blue-green infrastructure for Jakarta's sustainable planning and resilience. IOP Conf Ser: Earth Environ Sci 1211(1):012018. https://doi.org/10.1088/1755-1315/1211/1/012018

Alam AN, Naeem S (2024) Geo-sustainable practices in urban agriculture: a study of sustainable land use and natural resource conservation at household level. Res J Soc Issues 6(2):412–438. https://doi.org/10.56976/rjsi.v6i2.234

Ambrosi L, Kleidorfer M, Einfalt T, Back Y, Jasper-Tönnies A, Fennig C, Hauser M, Funke F, Leitinger G (2024) Balancing Urban Heat, Flood and water scarcity: Blue-Green infrastructure in Alpine Cities. https://doi.org/10.5194/egusphere-egu24-15236

Ameyaw P, De Vries W (2021) Toward smart land management: land acquisition and the associated challenges in Ghana. a look into a blockchain digital land registry for prospects. Land 10(3):239. https://doi.org/10.3390/land10030239

Antal H, Bhutani S (2023) Identifying linkages between climate change, urbanisation, and population ageing for understanding vulnerability and risk to older people: a review. Ageing Int 48(3):816–839. https://doi.org/10.1007/s12126-022-09504-7

Assaf G, Assaad RH (2024) Assessing the vulnerability of communities exposed to climate change-related challenges caused by the Urban Heat Island effect using machine learning. Comput Civil Eng 2023:177–184. https://doi.org/10.1061/9780784485248.022

Astee LY, Kishnani NT (2010) Building integrated agriculture: utilising rooftops for sustainable food crop cultivation in Singapore. J Green Build 5(2):105–113. https://doi.org/10.3992/jgb.5.2.105

Bachri S, Shrestha RP, Sumarmi, Irawan LY, Masruroh H, Prastiwi MRH, Billah EN, Putri NRC, Hakiki ARR, Hidiyah TM (2024) Land use change simulation model using a land change modeler in anticipation of the impact of the Semeru volcano eruption disaster in Indonesia. Environ Chall 14:100862. https://doi.org/10.1016/j.envc.2024.100862

Balado J, Solla M (2024) Multi-criteria GIS for Sponge City planning with open data sources in Vigo (Spain). In: ISPRS annals of the photogrammetry, remote sensing and spatial information sciences, X-4/W5-2024, pp 41–48. https://doi.org/10.5194/isprs-annals-X-4-W5-2024-41-2024

Bansal S, Pandey S (2024) Legal frameworks for sustainable urban development: analysing the efficacy of zoning regulations in promoting environmental conservation. In: E3S web of conferences, vol 527, p 01022. https://doi.org/10.1051/e3sconf/202452701022

Barthel S, Isendahl C (2013) Urban gardens, agriculture, and water management: sources of resilience for long-term food security in cities. Ecol Econ 86:224–234. https://doi.org/10.1016/j.ecolecon.2012.06.018

Barua U, Ansary MA, Islam I, Munawar HS, Mojtahedi M (2023) Multi-criteria earthquake risk sensitivity mapping at the local level for sustainable risk-sensitive land use planning (RSLUP). Sustainability 15(9):7518. https://doi.org/10.3390/su15097518

Belle JA, Velásquez C, Oxley M, Getiashvili K (2024) Understanding and integrating systemic risk (SR) into disaster risk reduction (DRR) and risk informed development (RID) for long-term resilient development. In: New insights on disaster risk reduction [Working Title]. IntechOpen. https://doi.org/10.5772/intechopen.1004915

Bian Z (2024) The advantages of sponge cities in urban flood prevention and control. Appl Comput Eng 58(1):155–162. https://doi.org/10.54254/2755-2721/58/20240712

Biitir SB, Aziabah S, Ayitio J (2022) Providing urban infrastructure using land-based financing mechanisms. A collaboration between state and non-state actors in Ghana. Urban Forum 33(4):415–443. https://doi.org/10.1007/s12132-021-09445-5

Boafo HK, Twum-Bobie AY, Dinye RD (2023) Associated challenges of urban land acquisition and development procedures in Ghana: a case study of Kumasi. Int J Multidiscip Res Anal 06(09). https://doi.org/10.47191/ijmra/v6-i9-44

Breulmann M, Müller RA, Van Afferden M (2024) Modelling urban stormwater and irrigation management with coupled blue-green infrastructure in the context of climate change. Blue-Green Syst 6(1):100–113. https://doi.org/10.2166/bgs.2024.101

Bronder J, Kurianowicz A, Piasecka J, Krupanek J, Krzyżak J (2024) Delimitation and prioritization of climate-affected areas using the example of a post-industrial city. Sustainability 16(10):3957. https://doi.org/10.3390/su16103957

Cavadini GB, Cook L (2023) Blue Green Infrastructure in a future climate: can we reduce combined sewer overflows? https://doi.org/10.5194/egusphere-egu23-3148

Cavadini GB, Rodriguez M, Cook LM (2024) Connecting blue-green infrastructure elements to reduce combined sewer overflows. J Environ Manag 365:121465. https://doi.org/10.1016/j.jenvman.2024.121465

Chai K-C, Ma X-R, Yang Y, Lu Y-J, Chang K-C (2022) The impact of climate change on population urbanization: evidence from china. Front Environ Sci 10:945968. https://doi.org/10.3389/fenvs.2022.945968

Chan FKS, Zhu Y-G, Li G, Feng M, Xu M, Xie L, Lu X, Zhu Z, Wang J, Lu L, Liu S, Wang Z, Chen W-Q (2024) Urban water challenges in Chinese cities–Sponge City program: initiatives, progress, challenges and opportunities on NBS and urban agriculture. https://doi.org/10.5194/egusphere-egu24-5034

Chen H, Song F (2024) Study on the application of environmentally friendly materials in the construction of Sponge Cities in China and their impact on rainwater runoff. J Progr Eng Phys Sci 3(2):56–64. https://doi.org/10.56397/JPEPS.2024.06.07

Chen D, Wang Y, Bao H (2024) Sharing or sparing? The trade-offs among urban services, food production and ecosystem services. Habitat Int 147:103037. https://doi.org/10.1016/j.habitatint.2024.103037

Cheng Y, Kang Q, Liu K, Cui P, Zhao K, Li J, Ma X, Ni Q (2023) Impact of urbanization on ecosystem service value from the perspective of Spatio-temporal heterogeneity: a case study from the Yellow River Basin. Land 12(7):1301. https://doi.org/10.3390/land12071301

Chowdhry B (2022) Financing land acquisition for infrastructure projects. Financ Res Lett 47:102656

Cohen-Shacham E, Walters G, Janzen C, Maginnis S (2016) Nature-based solutions to address global societal challenges. IUCN: Gland, Switzerland, vol 97, pp 2016–2036

Cooper DG (2024) Participatory urban planning. In: Christens BD (ed) The Cambridge Handbook of community empowerment, 1st edn. Cambridge University Press, pp. 241–261. https://doi.org/10.1017/9781009153720.012

Coşkun Hepcan Ç, Canguzel A (2024) The impact of blue-green infrastructure on climate resilience for hydrometerological hazards: the case of Bayraklı, İzmir. Düzce Üniversitesi Orman Fakültesi Ormancılık Dergisi. https://doi.org/10.58816/duzceod.1413255

Czyża S, Kowalczyk AM (2024) Applying GIS in Blue-Green infrastructure design in urban areas for better life quality and climate resilience. Sustainability 16(12):5187. https://doi.org/10.3390/su16125187

Daixin D, Mingyang B (2023) A resilience enhancement approach to the Sponge City based on ecosystem-based disaster risk reduction—taking the urban design of Jiangchuanlu Street in Shanghai, China as an example. J Resour Ecol 14(6). https://doi.org/10.5814/j.issn.1674-764x.2023.06.001

Debangshi U, Ghosh P, Tiwari H, Maurya DK, Kumar M (2022) Urban resiliency towards climate change. Int J Environ Climate Change 2037–2055. https://doi.org/10.9734/ijecc/2022/v12i1131194

Domingo D, Palka G, Hersperger AM (2021) Effect of zoning plans on urban land-use change: a multi-scenario simulation for supporting sustainable urban growth. Sustain Cities Soc 69:102833. https://doi.org/10.1016/j.scs.2021.102833

Dyca B, Hartmann T, Tempels B (2024) Land use planning and management for flood risk management. In: Lamond J, Proverbs D, Bhattacharya N (eds) Research Handbook on flood risk

management. Edward Elgar Publishing, pp 71–85. https://doi.org/10.4337/9781839102981.00013
EEA (2021) Nature-based solutions in Europe: policy, knowledge and practice for climate change adaptation and disaster risk reduction. European Environment Agency
Emiru MA, Adam AG, Dadi TT (2023) Exploring institutional capacity of urban land delivery and administration for housing development in Bahir Dar, Ethiopia: institutional analysis. Cogent Soc Sci 9(2):2278208. https://doi.org/10.1080/23311886.2023.2278208
Esraz-Ul-Zannat Md., Dedekorkut-Howes A, Morgan EA (2024) A review of nature-based infrastructures and their effectiveness for urban flood risk mitigation. WIREs Climate Change e889. https://doi.org/10.1002/wcc.889
European Commission (2019) EU guidance on integrating ecosystems and their services into decision-making. European Commission Brussels, Belgium
European Commission (2023) Nature-based solutions. https://research-and-innovation.ec.europa.eu/research-area/environment/nature-based-solutions_en
Everett G, Adekola O, Lamond J (2023) Developing a blue-green infrastructure (BGI) community engagement framework template. URBAN DESIGN Int 28(3):172–188. https://doi.org/10.1057/s41289-021-00167-5
Feltynowski M (2023) Urban green spaces in land-use policy–types of data, sources of data and staff–the case of Poland. Land Use Policy 127:106570
Ferreira C, Mourato S, Kasanin-Grubin M, Ferreira AJD, Destouni G, Kalantari Z (2020) Effectiveness of nature-based solutions in mitigating flood hazard in a Mediterranean Peri-Urban catchment. Water 12(10):2893. https://doi.org/10.3390/w12102893
Ferreira J, Costa Dos Santos D, Campos LC (2024) Blue-green infrastructure in view of Integrated Urban Water Management: a novel assessment of an effectiveness index. Water Res 257:121658. https://doi.org/10.1016/j.watres.2024.121658
Fitriyati N, Arifin HS, Kaswanto RL, Marimin (2024) Enhancing land use planning through integrating landscape analysis and flood inundation prediction Bekasi City's in 2030. Geomat Nat Hazards Risk 15(1):2360623. https://doi.org/10.1080/19475705.2024.2360623
Gaborit P (2022) Resilience and climate disaster management in cities: transformative change and conflicts. J Peacebuild Dev 17(3):257–273. https://doi.org/10.1177/15423166221128793
Gambarani A, Vercesi A, Bordoni M, Giarola A, Vivaldi V, Meisina C (2024) Nature-based Solutions in vineyards of Oltrepò Pavese (Northern Italy): state of the art and analysis of their effectiveness in shallow landslide mitigation. https://doi.org/10.5194/egusphere-egu24-17275
Ghimire M (2023) Approach to multi-disaster risk assessment and policy recommendation for disaster resilient urban development: a study of Birgunj-Simara Corridor. Geograph J Nepal 1–34. https://doi.org/10.3126/gjn.v16i01.53476
Gupta A, De B (2024) A systematic review on urban blue-green infrastructure in the south Asian region: recent advancements, applications, and challenges. Water Sci Technol wst2024014. https://doi.org/10.2166/wst.2024.014
Hanberry BB (2023) Urban land expansion and decreased Urban Sprawl at Global, National, and City Scales during 2000 to 2020. Ecosyst Health Sustain 9:0074. https://doi.org/10.34133/ehs.0074
Hovhannisyan V, Devadoss S (2020) Effects of urbanization on food demand in China. Empir Econ 58(2):699–721. https://doi.org/10.1007/s00181-018-1526-4
Huang C-C, Wang C-L (2024) Enhancing urban flood resilience: interdisciplinary integration of climate adaptation, flood control, and land-use planning from 3PA to 4PA. J Water Climate Change 15(4):1961–1968. https://doi.org/10.2166/wcc.2024.125
Humbal A, Chaudhary N, Pathak B (2023) Urbanization trends, climate change, and environmental sustainability. In: Pathak B, Dubey RS (eds) Climate change and urban environment sustainability. Springer Nature Singapore, pp 151–166. https://doi.org/10.1007/978-981-19-7618-6_9

Kalogiannidis S, Kalfas D, Giannarakis G, Paschalidou M (2023) Integration of water resources management strategies in land use planning towards environmental conservation. Sustainability 15(21):15242. https://doi.org/10.3390/su152115242

Karri V, Nalluri N (2023) Enhancing resilience to climate change through prospective strategies for climate-resilient agriculture to improve crop yield and food security. Plant Sci Today. https://doi.org/10.14719/pst.2140

Kaskevich V, Villoslada Peciña M, Ward R, Sepp K (2022) Mapping Blue-Green infrastructure to evaluate conditions in the Estonian coastal zone. https://doi.org/10.5194/egusphere-egu22-1739

Kaur R, Gupta K (2022) Blue-Green Infrastructure (BGI) network in urban areas for sustainable storm water management: a geospatial approach. City Environ Interact 16:100087. https://doi.org/10.1016/j.cacint.2022.100087

Keita K, Kourouma S (2024) Challenges and opportunities in adopting Blue-Green Infrastructure in Cities of the developing world. In: Joshi PK, Rao KS, Bhadouria R, Tripathi S, Singh R (eds) Blue-Green infrastructure for sustainable urban settlements. Springer Nature, Switzerland, pp 139–165. https://doi.org/10.1007/978-3-031-62293-9_6

Kim S, Andrew SA, Ramirez De La Cruz E, Kim W-J, Feiock RC (2024) Impacts of local government perceptions of disaster risks on land resilience planning implementation. Land 13(7):1085. https://doi.org/10.3390/land13071085

Kotsila P, Anguelovski I, Baró F, Langemeyer J, Sekulova F, Jt Connolly J (2021) Nature-based solutions as discursive tools and contested practices in urban nature's neoliberalisation processes. Environ Plan E: Nat Space 4(2):252–274. https://doi.org/10.1177/2514848620901437

Koukoula M, Torelló-Sentelles H, Peleg N (2024) Urbanization and climate change impacts on future heavy summer rainfall in Milan, Italy. https://doi.org/10.5194/egusphere-egu24-4612

Ku C-A (2024) Evaluating the effects of land-use strategies on future flood risk reduction in urban areas. Cities 150:104989. https://doi.org/10.1016/j.cities.2024.104989

Latue T, Latue PC, Sihasale DA, Rakuasa H (2023) Regional development planning and policy in the aspects of vulnerability and disaster resilient cities: a review. Nusantara J Behav Soc Sci 2(3):67–74. https://doi.org/10.47679/202333

Li J, Xie B, Dong H, Zhou K, Zhang X (2023) The impact of urbanization on ecosystem services: Both time and space are important to identify driving forces. J Environ Manag 347:119161. https://doi.org/10.1016/j.jenvman.2023.119161

Li J, Wang J, Zhou W (2024) Different impacts of urbanization on ecosystem services supply and demand across old, new and non-urban areas in the ChangZhuTan urban agglomeration, China. Landsc Ecol 39(6):107. https://doi.org/10.1007/s10980-024-01900-5

Liao K-H (2019) The socio-ecological practice of building blue-green infrastructure in high-density cities: what does the ABC Waters Program in Singapore tell us? Socio-Ecol Pract Res 1(1):67–81. https://doi.org/10.1007/s42532-019-00009-3

Lim M, Xenarios S (2021) Economic assessment of urban space and blue-green infrastructure in Singapore. J Urban Ecol 7(1):juab020. https://doi.org/10.1093/jue/juab020

Liu Y, Li Y (2017) Revitalize the world's countryside. Nature 548(7667):275–277. https://doi.org/10.1038/548275a

Lulla V, Stanforth A, Prudent N, Johnson D, Luber G (2015) Modeling vulnerable urban populations in the global context of a changing climate. In: Ahn R, Burke TF, McGahan AM (eds) Innovating for healthy urbanization. Springer US, pp 193–208. https://doi.org/10.1007/978-1-4899-7597-3_9

Maher J (2020) Better Urban Drainage for Liveable Cities—Creating blue green infrastructure through science translation, collaboration and multi-stakeholder governance. IOP Conf Ser: Earth Environ Sci 588(5):052005. https://doi.org/10.1088/1755-1315/588/5/052005

Malmström A, Klein J, Käyhkö J, Juhola SK (2024) Land use planning. In: Fiorino DJ, Eisenstadt TA, Ahluwalia MK (eds) Elgar encyclopedia of climate policy. Edward Elgar Publishing, pp 338–342. https://doi.org/10.4337/9781802209204.ch64

Meemrot R, Meena RL, Derwal N, Meena P, Balkh S, Jangid JB (2024) Greening the concrete jungle: enhancing urban resilience through Blue-Green infrastructure. Int J Res Appl Sci Eng Technol 12(4):2931–2935. https://doi.org/10.22214/ijraset.2024.60578

Miralles-Wilhelm F (2023) Nature based solutions in agricultural landscapes for reducing tradeoffs between food production, climate change, and conservation objectives. Front Water 5:1247322. https://doi.org/10.3389/frwa.2023.1247322

Moreau C, Cottet M, Rivière-Honegger A, François A, Evette A (2022) Nature-based solutions (NbS): a management paradigm shift in practitioners' perspectives on riverbank soil bioengineering. J Environ Manag 308:114638. https://doi.org/10.1016/j.jenvman.2022.114638

Mushinge A, Munshifwa EK, Chileshe RA, Jain N (2022) Urban societal needs and sustainability in Zambia. In: Chatterjee U, Biswas A, Mukherjee J, Mahata D (eds) Sustainable urbanism in developing countries, 1st edn. CRC Press, pp 365–374. https://doi.org/10.1201/9781003131922-24

Nayak D, Shukla AK, Devi NR (2024) Decadal changes in land use and land cover: impacts and their influence on urban ecosystem services. AQUA-Water Infrastruct Ecosyst Soc 73(1):57–72. https://doi.org/10.2166/aqua.2024.211

Nedyalkova I (2024) Integrating blue-green infrastructure in modern production complexes. Zbornik Radova Sa Nacionalne Konferencije Sa Međunarodnim Učešćem – Zelena Gradnja 2024 - Zbornik Radova, 233–238. https://doi.org/10.5937/greenb24032N

Nolon JR (2022) Local resilience, land use law, and disaster planning. In: Kuo SS, Marshall JT, Rowberry R (eds) The Cambridge Handbook of disaster law and policy, 1st edn. Cambridge University Press, pp 151–180. https://doi.org/10.1017/9781108770903.010

O'Hogain S, McCarton L (2018) Nature-based solutions. In: O'Hogain S, McCarton L (eds) A technology portfolio of nature based solutions. Springer International Publishing, pp 1–9. https://doi.org/10.1007/978-3-319-73281-7_1

Opitz-Stapleton S, Nadin R, Kellett J, Calderone M, Quevedo A, Peters K, Mayhew L (2019) Risk-informed development: from crisis to resilience. Overseas Development Institute

Pagac J, Adamec S, Kiss V (2023) Blue-green infrastructure as a tool for connecting agricultural and urban land. 609–616. https://doi.org/10.5593/sgem2023V/6.2/s27.74

Parvin GA, Esraz-Ul-Zannat Md., Ahsan R, Islam I (2022) Disaster risks and resilience of urban Bangladesh: role of Blue-Green infrastructure. In: Dhyani S, Basu M, Santhanam H, Dasgupta R (eds) Blue-Green infrastructure across Asian Countries. Springer Singapore, pp 419–440. https://doi.org/10.1007/978-981-16-7128-9_19

Perera ACS, Davies PJ, Graham PL (2024) A global review of urban blue-green planning tools. Land Use Policy 140:107093. https://doi.org/10.1016/j.landusepol.2024.107093

Petris P (2023) A real option analysis of the "Flats for Land" system—An idiosyncratic equity financing mechanism on real estate investments. Int Real Estate Rev 26(4):551–568. https://doi.org/10.53383/100373

Pinto LV, Inácio M, Pereira P (2023) Green and blue infrastructure (GBI) and urban nature-based solutions (NbS) contribution to human and ecological well-being and health. Oxford Open Infrastruct Health 1:ouad004. https://doi.org/10.1093/ooih/ouad004

Santiago B (2024) Sustainability in action: green infrastructure as a marker of sustainable urban development. FIU Undergrad Res J 2(1). https://doi.org/10.25148/URJ.020109

Santini G (2022) The role of urban and Periurban agriculture to build food secure cities and resilient food systems. Acta Horticult 1356:1–8. https://doi.org/10.17660/ActaHortic.2022.1356.1

Selby S, Perez-Dalena MG (2020) Risk-informed development: a strategy tool for integrating disaster risk reduction and climate change adaptation into development. UNDP

Shah AM, Liu G, Meng F, Yang Q, Xue J, Dumontet S, Passaro R, Casazza M (2021) A review of urban green and blue infrastructure from the perspective of food-energy-water nexus. Energies 14(15):4583. https://doi.org/10.3390/en14154583

Sharma A, Donovan K, Krishnamurthy S, Creed M (2022) Exploring the scope of public participation for risk sensitive land use planning in Nepal: a policy review. Sustainability 14(21):14137

Singh YV, Banerjee S, Sagar D (2023) Integrated blue green infrastructure approach for storm water management. EPRA Int J Climate Resour Econ Rev 1–5. https://doi.org/10.36713/epra13053

Sommese F (2024) Nature-based solutions to enhance urban resilience in the climate change and post-pandemic era: a taxonomy for the built environment. Buildings 14(7):2190. https://doi.org/10.3390/buildings14072190

Spahr KM, Smith JM, McCray JE, Hogue TS (2021) Reading the green landscape: public attitudes toward green stormwater infrastructure and the perceived nonmonetary value of its co-benefits in three US cities. J Sustain Water Built Environ 7(4):04021017. https://doi.org/10.1061/JSWBAY.0000963

Su Q, Chen K, Liao L (2021) The impact of land use change on disaster risk from the perspective of efficiency. Sustainability 13(6):3151. https://doi.org/10.3390/su13063151

Sukhwani V, Shaw R (2022) An integrated governance approach towards a water-energy-food nexus and climate change. Edward Elgar Publishing. https://doi.org/10.4337/9781800371613.00053

Sutkowska M, Stefańska A, Vaverkova MD, Matsui Y (2024) Fostering sustainable Urban Development: integrating school areas for supporting the City's Blue-Green infrastructure. Acta Scientiarum Polonorum. Architectura 23:12–31. https://doi.org/10.22630/ASPA.2024.23.2

Tang J, Liu D, Shang C, Niu J (2024) Impacts of land use change on surface infiltration capacity and urban flood risk in a representative karst mountain city over the last two decades. J Clean Prod 454:142196. https://doi.org/10.1016/j.jclepro.2024.142196

Tirumala RD, Tiwari P (2023) The appeal of land-based financing instruments. In: Tirumala RD, Tiwari P (eds) Advances in infrastructure finance. Springer Nature Singapore, pp 135–158. https://doi.org/10.1007/978-981-99-0440-2_6

Toromade AS, Soyombo DA, Kupa E, Ijomah TI (2024) Urban farming and food supply: a comparative review of USA and African cities. Int J Adv Econ 6(7):275–287. https://doi.org/10.51594/ijae.v6i7.1304

US EPA (2015) Financing green infrastructure—Is a community-based public-private partnerships (CBP3) right for you? [Overviews and Factsheets]. U.S. Environmental Protection Agency. https://www.epa.gov/G3/financing-green-infrastructure-community-based-public-private-partnerships-cbp3-right-you

Van De Ven FHM, Zevenbergen C, Avellar Montezuma M, Ding Z, Veerbeek W, Chen S (2024) The three-points sponge policy approach; toward an enhanced multi-level resilience strategy. Front Water 6:1361058. https://doi.org/10.3389/frwa.2024.1361058

Vanino S, Baratella V, Pirelli T, Ferrari D, Di Fonzo A, Pucci F, Nikolaidis NP, Lilli MA, Doğan ZA, Topdemir T, Awabdeh S, Al-Hadidi L, Bani Hani N, Panagopoulos A, Pisinaras V, Chatzi A, López E, Papadaskalopoulou C, Tassopoulos D et al (2024) Nature-based solutions for optimizing the water–ecosystem–food nexus in Mediterranean countries. Sustainability 16(10):4064. https://doi.org/10.3390/su16104064

Vazin F, Chan DWM, Hanaee T, Sarvari H (2024) Nature-based solutions and climate resilience: a bibliographic perspective through science mapping analysis. Buildings 14(6):1492. https://doi.org/10.3390/buildings14061492

Wang YP, Kintrea K (2021) Urban expansion and land use changes in Asia and Africa. Environ Urbaniz ASIA 12(1_suppl):S13–S17. https://doi.org/10.1177/0975425321999081

Wang Q, Zhao G, Zhao R (2024) Resilient urban expansion: Identifying critical conflict patches by integrating flood risk and land use predictions: a case study of Min Delta Urban Agglomerations in China. Int J Disaster Risk Reduct 100:104192. https://doi.org/10.1016/j.ijdrr.2023.104192

Wu H (2024) The impact of urbanization on ecosystem services and ecological security patterns. https://doi.org/10.5194/egusphere-egu24-20661

Wu X, Moustakas S, Bezak N, Radinja M, Alivio MB, Mikoš M, Dohnal M, Bares V, Willems P (2024) Evaluating the performance of Blue and Green infrastructures in an urban area through a fine-scale water balance model. https://doi.org/10.5194/egusphere-egu24-1677

Xu H, Gao J, Yu X, Qin Q, Du S, Wen J (2024a) Assessment of rainstorm waterlogging disaster risk in rapidly urbanizing areas based on land use scenario simulation: a case study of Jiangqiao Town in Shanghai, China. Land 13(7):1088. https://doi.org/10.3390/land13071088

Xu J, Takahashi M, Li W (2024b) Identifying vulnerable populations in urban society: a case study in a flood-prone district of Wuhan, China. Nat Hazard 24(1):179–197. https://doi.org/10.5194/nhess-24-179-2024

Yang Y, Li X, Wang D, Liu Z (2024) Optimizing urban ecosystems through green and blue infrastructure: strategies for sustainability, resilience, and justice. Appl Comput Eng 72(1):78–83. https://doi.org/10.54254/2755-2721/72/20240985

Zhao H, Cheng X, Gao J, Yu G (2023) A study on risk sharing of smart city PPP projects based on EW-G1 and TOPSIS-UT: a case study in China. Kybernetes. https://doi.org/10.1108/K-05-2023-0824

Zheng Q, Weng Q, Wang K (2021) Characterizing urban land changes of 30 global megacities using nighttime light time series stacks. ISPRS J Photogramm Remote Sens 173:10–23. https://doi.org/10.1016/j.isprsjprs.2021.01.002

Chapter 10
Assessing Urban Resilience for Climate Disasters in Phnom Penh, Cambodia

Yeonji Lee, Indrajit Pal, Seak Sopath, and Ayush Baskota

Abstract Cambodia has achieved a remarkable milestone of sustaining robust and inclusive economic growth over about twenty years and is suitable to set up the concept of building resilience with long- and short-term development achievement. The frequency and impacts of climate hazards are increasing regularly, particularly in certain developing nations and their cities. It is critical to better understand how natural disasters affect economies and people. However, urban resilience has weakened due to the high demands of developing countries pursuing rapid economic growth. The rationale of this study is to comprehend the traits of the city, evaluate the urban resilience (UR) to hydrological catastrophes, and indicate the apt schemes to remove the gaps and reduce the restrictions. The study evaluates the socioeconomic, natural and institutional capacity of Phnom Penh for flood resilience. This research utilized qualitative and quantitative methods. In the initial step, group discussions were conducted with public and private sectors using structured questionnaire to identify UR parameters, followed by interviews with key informants to score each UR parameter factor. Quantitative data is then collected through questionnaires, spatial analysis for statistical analysis and disaster frequencies. The results of the study will be useful in bridging current gaps in multi-stakeholder collaboration and data insufficiency in Phnom Penh for resilient urban policy and planning.

Keywords Urban resilience · Flood · Climate hazards · Urban planning · Stakeholder

Y. Lee · I. Pal (✉) · S. Sopath · A. Baskota
Department of Civil and Infrastructure Engineering (CIE), Asian Institute of Technology (AIT), Pathum Thani, Thailand
e-mail: indrajit-pal@ait.ac.th

M. G. Kibria et al. (eds.), *Urban Blue-Green Infrastructure Approach for Food Security and Climate Disaster Resilience*, Disaster Risk Reduction,
https://doi.org/10.1007/978-981-96-8746-6_10

10.1 Introduction and Background

Cambodia has experienced a remarkable achievement of sustaining robust and inclusive economic growth over about twenty years and is suitable to set up the concept of building resilience with long- and short-term development achievement. The country covers 181,035 sq.km. of which 45% are tropical forests and Tonlé Sap, the largest freshwater lake in South East Asia.

Following a fast-transforming market with a high dependence on agriculture, Cambodia's economy has rapidly increased towards lower-middle income status. Its average growth rate of more than 7% in the past decade, and the forecasted increase in growth in the short term place it among the rapidly growing economies in the world, and Cambodia aspires to reach an upper middle-income status by 2030 by developing tourism, textile exports, and construction industry (Ly et al. 2024).

More than 75% of the country's population lives in rural areas, higher than the global average among lower-middle-income countries. Subsequently, agriculture plays an important role in the economy, contributing around 22% to the GDP and employing a third of the workforce (ADB 2021). Agriculture and fishing are also key sources of revenue and are critical to domestic food security. However, roughly 88% of Cambodians are still employed in the informal economy (Giulio 2024), which indicates a higher level of socioeconomic vulnerability. Many of the individuals who work outside of the social safety nets are vulnerable to external shocks that may disrupt their lives and livelihoods.

Despite being one of the world's least urbanized regions, cities in the Greater Mekong Subregion (GMS) are expanding and have a profound global economic influence (ADB 2016). While the current levels of urbanization in the region remain low, many researchers predict that the GMS will be pushed to urbanize as Southeast Asia's urbanization rate grows from 42% in 2010 to 65% in 2050, leading to profound changes in its economy and global transportation links. This growth is also expected to have tremendous social, environmental and economic implications, including exacerbation of disaster and climate risks (Barkham, R et al. 2014).

One of the key challenges faced by urbanizing cities in Asia Pacific is to mitigate risks to natural hazards and climate change. Due to the expanding urban functions, high population density, and complexity of land use, the extent of disaster and climate-induced damages is very likely to spread to secondary and tertiary levels beyond the main catastrophe range. Although it is difficult to eliminate the impacts of climate change and natural hazards, building an urban system can limit losses and damages while also ensuring the long-term growth of cities by enhancing urban disaster prevention and resilience.

The UN Sustainable Development Goals (SDGs) recommend that all nations improve their ability to plan and manage collaborative, sustainable, and integrated human settlement planning and management by 2030. Furthermore, integrated disaster risk management was proposed to be established and implemented at all levels. Cities are both parties to the production of greenhouse gases, which are the principal drivers of climate change and are the most impacted by its impacts.

The accepted "New Urban Agenda" at UN-Habitat III in 2016 also emphasized the roles of cities and the execution of initiatives by local governments for sustainable development. While most of the earliest climate change adaptation research and programming focused on ecosystems and agriculture (Monteiro et al. 2022), there is an increasing acknowledgement of the need for interventions to aiding cities to adapt to climate change (Tyler et al. 2016). Urban resilience is also available via an adaptive circulation model (Pickett et al. 2011). As a result, there is a significant push towards approaching urban resilience through broader scale of stressors and cascading impacts, and as an operational tool for sustainability (Chelleri et al. 2015).

The rationale of this study is to comprehend the traits of the city and evaluate the urban resilience to hydrological catastrophes, indicating the apt schemes to remove the gaps and reduce the restrictions. This research utilized qualitative and quantitative methods. In the initial step, group discussions were conducted with public and private sectors using structured questionnaire to identify UR parameters, followed by interviews with key informants to score each UR parameter factor. Quantitative data is then collected through questionnaires, spatial analysis for statistical analysis and disaster frequencies. Municipal plans, policies, government and international agency reports were reviewed and synthesized using secondary data review process to aid analysis and reporting of findings (ACAPS 2014).

This chapter aims to investigate the connections between climate change adaptation, catastrophe risk management, governance and urban resilience. The study uses the Rockefeller Foundation's City Resilience Index (Rockefeller Foundation & Arup 2014) to assess Cambodia's urban resilience to hydrological hazards through multiple dimensions. The indicators and approaches in this study might be utilized to develop adaptation plans and risk management strategies for the two metropolitan regions to enhance urban resilience in response to the increasing challenges posed by climate change and natural hazards. The conceptual framework for gauging urban resilience is depicted in Fig. 10.1.

10.2 Climate Change and Urban Resilience

Cities worldwide are encountering ecological, economic and political shocks and stresses exacerbated by climate change impacts. To boost urban resilience, cities must become more robust to a more extensive range of overlapping and interrelated stresses. Adaptation, flexibility, diversity and the capacity to accept and innovate are essential traits of resilient individuals, communities, cities, and systems (Tyler & Moench 2012). More specifically, these traits must be present in the public and private sectors at the forefront of technological development to build sustainable urban infrastructure. Initiatives to increase urban resilience to climate change, including measures for adaptation and mitigation must be included in development strategies and plans. Policies already in place to address other vulnerability drivers in cities, such as housing in high-risk regions and social safety nets may commonly be altered at little or no expense to increase resilience.

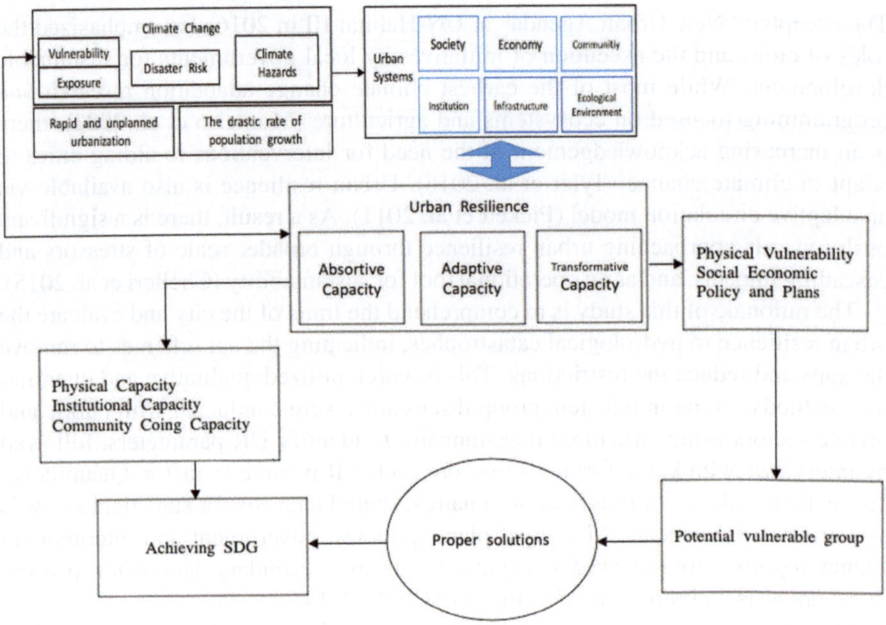

Fig. 10.1 Conceptual framework on resilience cities

The resilience theory has proved effective in managing climate change risks and unforeseen occurrences and boosting efforts to live and prosper in the face of climate change (ADB 2014). Urban climate change resilience (UCCR) recognizes the complexity of rapidly expanding urban settings and the unpredictability of climate change, including adaptation, mitigation and climate resilient development (ADB 2014). This viewpoint emphasizes how important it is to think of cities as complex systems that may change and adapt in the face of unanticipated shocks or pressures, enabling them to survive and prosper.

Long-term issues exist for cities worldwide, especially those with urban disadvantaged populations, in maintaining the welfare of their residents. Pre-existing susceptibility exacerbates these problems, partly brought on by both direct and indirect effects of climate change. Urban resilience also allows people to survive and prosper despite the pressures or shocks they experience, especially the poor and powerless (ADB 2014).

A city cannot become climate resilient with a single action. Instead, resilience is developed over time through a series of actions that depend on one another. As individuals and organizations learned from their previous failures and applied what they had learned to future actions, these acts would become more sophisticated and advanced (Folke 2006; Leichenko 2011). The notion of urban resilience to climate change is shown in Fig. 10.2, along with how efforts to reduce catastrophe risk and urban poverty are required but insufficient to maintain urban functioning in the face of direct and indirect effects of climate change. As a result, one of the most crucial

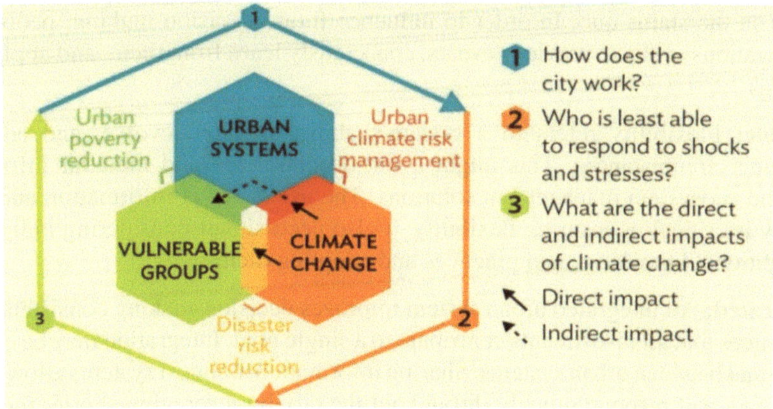

Fig. 10.2 Conceptualizing urban resilience to climate change (ADB 2014)

methods for boosting urban climate resilience is cross-sectoral collaboration (Tyler & Moench 2012).

10.3 Urban Development and Modernization

Cambodia has various potentials to build competitive, sustainable, and inclusive cities. Phnom Penh is both the nation's largest and fastest-growing metropolis and the gateway to the international economy. The city's turnaround into a hub of quick urbanization and economic growth from a city torn apart by conflict and instability is remarkable. However, sustainable development has been hindered due to unexpected and rapid growth without adequate consideration and investment in infrastructure development, potentially increasing vulnerabilities of both the urban environment and its population (Saltsman et al. 2021).

10.4 Urban Resilience Index Framework

This section will discuss the framework in multiple dimensions from a comprehensive point of view. First, the qualities of resilient systems are divided into seven parts: reflective, flexible, integrated, robust, redundant, inclusive, and resourceful (Tyler & Moench 2012). Each dimension has its own characteristics and significance in urban resilience.

Reflective: Reflective systems acknowledge the innate instability and constant change of the modern environment. They have systems to constantly adapt their standards or norms in response to new data, rather than seeking permanent solutions based on the status quo. In order to influence future decision-making, people and organizations evaluate previous events, consciously learn from them, and apply their findings.

Flexible: Flexibility refers to a system's ability to alter, evolve, and adapt to changing circumstances. This might assist decentralized and modular infrastructure and ecosystem management solutions. The use of novel information and technology is critical in ensuring flexibility. It also necessitates considering indigenous or traditional knowledge and practices and adopting them.

Integrated: An integrated urban system improves decision-making consistency and guarantees that all investments contribute to a single goal. Integration may be evident inside and between urban systems. Sharing information amongst systems allows them to interact and respond quickly throughout the city by leveraging shorter feedback loops.

Robust: Robust systems are formed of well-planned, well-built, and well-managed physical assets that can resist the effects of hazard occurrences without significant damage or loss of function. Robust design foresees probable system failures and incorporates protections to guarantee that failure is predictable, safe, and not disproportionate to the cause. Over-reliance on a single asset, cascade failure, and design thresholds that, if surpassed, might lead to catastrophic failure are all avoided.

Resourceful: During a shock or under stress, resourceful people and institutions may quickly devise new tactics to achieve their goals or meet their requirements. Investing in the ability to predict future challenges, determine priorities, and respond, for example, by mobilizing and coordinating larger human, financial, and physical resources, might be a component. The ability of a city to restore critical system operation, even under severely limited conditions, is based on its resourcefulness.

Redundant: Redundancy is the intentional establishment of extra capacity inside systems to absorb interruptions, extreme pressures, or surges in demand. It has variety: there are various ways to fulfil a particular need or accomplish a specific job. Dispersed infrastructure networks and resource reserves are two examples. Instead of being an unexpected result of poor design, redundancies should be planned for, cost-effectively managed, and prioritized on a city-wide scale.

Inclusive: Inclusion emphasizes the need for broad community dialogue and involvement, particularly in the most disadvantaged populations. Addressing one sector's, location's, or community's shocks or pressures in isolation from others is antithetical to the concept of resilience. An inclusive strategy fosters a sense of shared ownership or a common vision for municipal resilience.

10.5 The Four Dimensions of Urban Resilience Index

10.5.1 The Dimension of Health and Wellbeing

This component concerns people, namely the health and happiness of all city residents and employees. This section evaluates how properly the city provides for all the people's fundamental demands for water, housing, and food, particularly in times of need. It assesses how effectively the city accommodates various kinds of revenue, such as a straightforward approach to corporate investment and assistance for society. Eventually, it explains whether routine and emergency medical procedures can help a city safeguard its residents' health (Fig. 10.3).

This dimension focuses on how cities are organized and how social and economic systems function to ensure social well-being and productivity. Systems for preserving law and order, as well as financial management are included in this dimension. Open spaces and cultural resources are essential in creating a municipal environment that fosters collective identity and mutual support. Only if their physiological demands

Fig. 10.3 The dimension of health and wellbeing of CRI (Rockefeller Foundation and Arup 2014)

are satisfied by providing a minimal amount of food, water, cleanliness, energy, and shelter is this conceivable.

10.5.2 *The Dimension of Economy and Society*

Socioeconomic conditions are important indicators to assess a society's development status. Previous assessment results show that poor urban communities' low socioeconomic status significantly threatens people's well-being, especially that of vulnerable groups. Most families lack sufficient resources to respond to social needs such as health care or providing education to their children (Fig. 10.4).

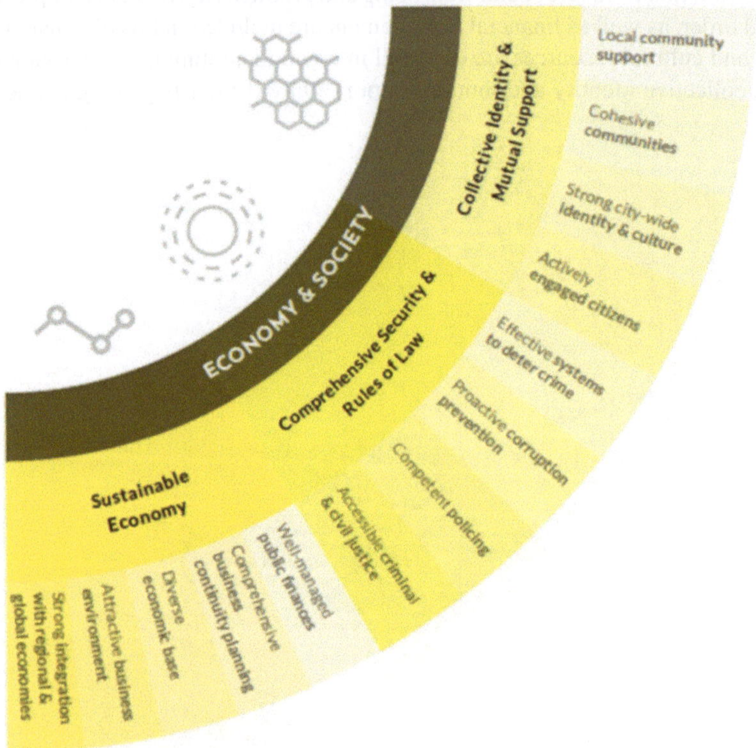

Fig. 10.4 The Dimension of Economy & Society of CRI (Rockefeller Foundation and Arup, 2014)

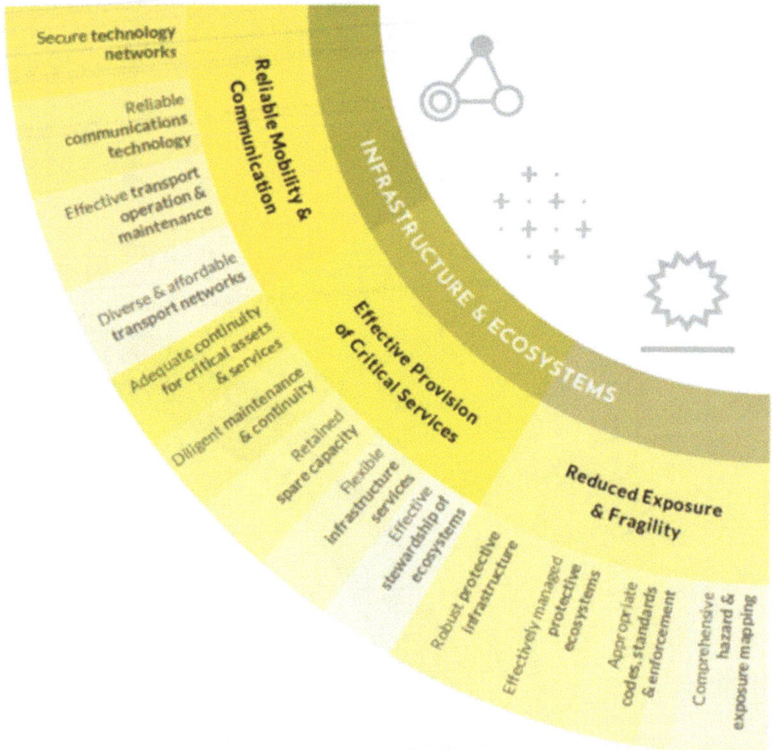

Fig. 10.5 The Dimension of infrastructure & ecosystems of CRI (Rockefeller Foundation and Arup 2014)

10.5.3 The Dimension of Infrastructure and Ecosystems

This dimension focuses on geographical structures and ecosystems that support, sustain, and link humans. The CRI considers the ecosystem's and infrastructure's resilience to natural hazards. Maintaining essential services in the face of shock or stress is equally important. For example, robust transportation systems make it easier for people, commodities, services, and information to move around, particularly those that handle essential services such as solid waste management, energy distribution, and water delivery are vital in ensuring urban resilience Fig. 10.5.

10.5.4 The Dimension of Leadership and Strategy

A resilient city corrects its mistakes and reacts appropriately to situations in line with ground situations. Strong urban management and leadership are essential for a

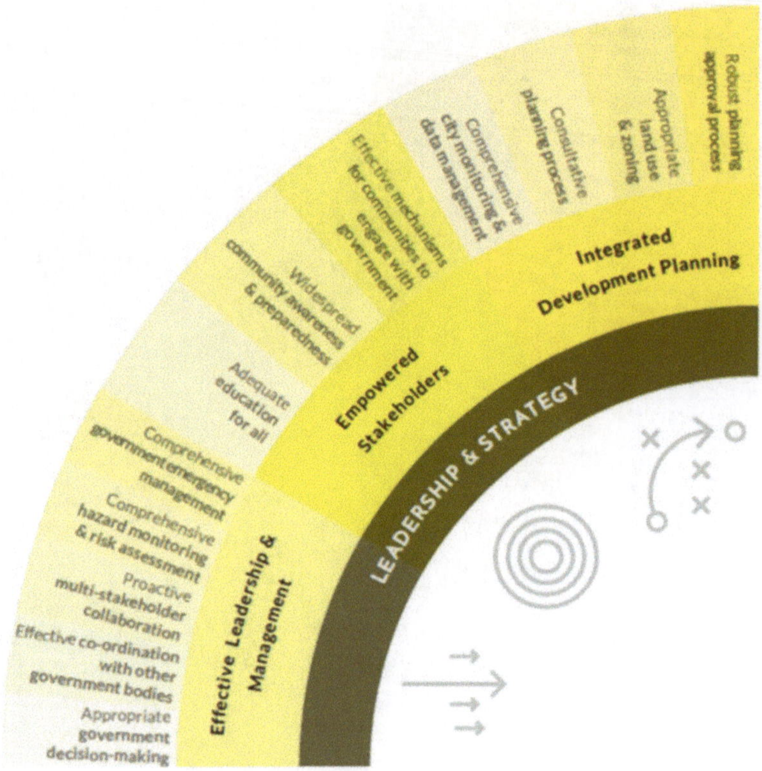

Fig. 10.6 The Dimension of Leadership & Strategy of CRI (Rockefeller Foundation and Arup 2014)

city, and inclusive governance is essential for realistic decision-making. A city has to educate and inform its constituents to enable people and organizations to make the right decisions. Additionally, it is crucial to see that the city develops holistically, with the vision of the city aligned with sectoral goals and efforts Fig. 10.6.

10.6 Study Area and General Information of Phnom Penh City

10.6.1 Study Area

Phnom Penh City (PPC) is located in Cambodia's south-central area, near the junction of Tonle Sap, Mekong and Basic rivers (Fig. 10.7). The Mekong River runs 486 km from north to south across Cambodia, passing through Phnom Penh, providing a vital riverine ecosystem that sustains the city's social, economic and cultural fabric.

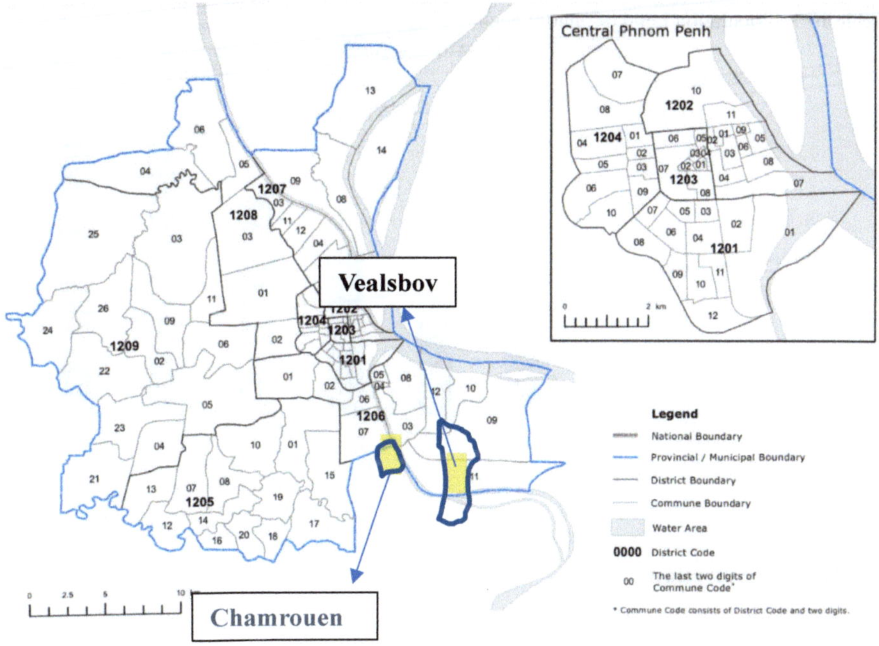

Fig. 10.7 Map of administrative areas in phnom penh municipality, ministry of planning of Cambodia (2013)

The city has a total area of 692.46 sq.km. and an estimated population of 2,078,000 people in 2020. The city is divided into 14 Khans, 105 Sankats, and 953 villages.

As shown in Table 10.1, the population of Phnom Penh more than doubled between 1995 (0.92 M) and 2019 (2.1 M); the city's growth and the formation of three additional districts aided in its population growth. According to JICA's research on the PP Transport Master Plan, the population inside the city's expanded administrative borders is estimated to reach roughly 2.86 million in 2035.

Chamroeun and Vealsbov in Phnom Penh are the research regions. Over 95% of the Mekong's flow has entered the river since it entered Cambodia. The Tonle Sap flows into the Mekong during the dry season when the two rivers meet at Phnom Penh. This reversal in Tonle Sap flow and associated floods benefits wildlife, agriculture, and even the integrity of the Angkor Temple foundation. However, due to the increased frequency and instability in Cambodia's climate, intense rains have driven the Mekong River over its banks more severely each year. As a result, to assess urban resilience to climatic risks, particularly hydrological disasters, (1) climate hazards impact urban systems and urban vulnerable groups, and (2) the role of urban systems on the urban poor has been assessed and analyzed. Furthermore, data on the current UR of Cambodia through the Arup's UR Index in four dimensions has been collected.

Table 10.1 Population trend of phnom penh

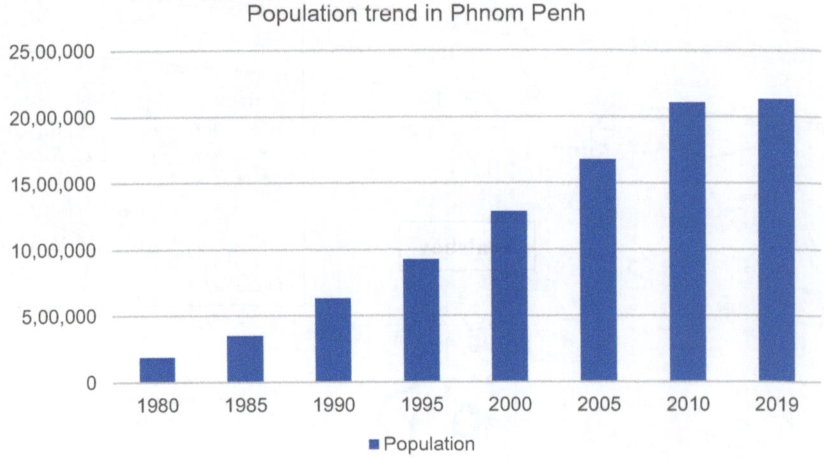

Study Area Selection

The targeted study area was chosen based on the criteria below.

- High vulnerability to hydrological disasters, which have affected the well-being of residents of a local area.
- Area with a high rate of urban dwellers in informal settlements
- Rapid population shift from rural to urban areas
- Highly densely populated urban areas.
- An area with limitations of building UR to hydrological hazards.

Over the last twenty years, rapid urbanization has created new sorts of catastrophe hazards in Cambodia. The urban poor tend to live in places most vulnerable to catastrophes, yet little is known about their susceptibility or resistance to disasters (Flower et al 2018).

10.6.2 Urban Environmental Challenges and Climate Change

PPC faces various issues due to growing urbanization, including uncontrolled development, traffic, congestion, unregulated parking, waste management, riverbank erosion, pollution, and floods (Sasin & Sokha 2015). These difficulties are heightened by the prospect of climate change repercussions. Cambodia has been designated as among the most susceptible Lower Mekong Basin countries to climate change impacts (Notre Dame Global Adaptation Initiative 2022), partly because of its relatively low height, closeness to the coast and the Mekong River, and insufficient capacity to manage climate-related threats. Flooding and altering precipitation patterns may also significantly exacerbate the impacts. Other dangers include the effect of tropical cyclones, which have grown in frequency and intensity over the previous several decades and are expected to increase further. Rising sea levels will also impact the supply of fresh water in PPC by increasing saltwater intrusion into low-lying regions and adding to coastal erosion.

10.6.3 Problem Tree Analysis

Like coastal cities worldwide, PPC has immense potential but is also exposed to hydrological hazards and climate risks. A problem tree analysis was conducted to understand better the root causes of the city's vulnerability to hydrological hazards. Climate change, a lack of institutional capacity, fast and unplanned urbanization, and poverty can all be underlying causes of vulnerability (Fig. 10.8). The problem tree provides a qualitative synthesis of the classification and risk characteristics of the PPC to determine some of the most susceptible locations affected by CC impacts and DRM by establishing the underlying causes of the vulnerabilities. By utilizing the problem tree, we define the vulnerabilities of PPC across five major sectors:

1. Climate change impacts
2. Built environment
3. Systems to respond to hazards and disasters
4. Governance structure, financial resources, and city management
5. Political and economic impacts

10.7 Vulnerabilities of Phnom Penh

This section outlines the causes of vulnerability in the PPC metropolitan area to hydrological catastrophes, including physical, institutional, and socioeconomic vulnerabilities, based on the self-vulnerability assessment in the submitted data-gathering process.

Main Interviewees for the Assessment

(1) Civil Servants (Key stakeholders)
(2) Public Sectors (NGOs and International Agencies)

10.7.1 Physical Vulnerability

PPC has seen seasonal flooding and flash floods, although flooding research in Cambodia has concentrated primarily in the rural regions. During the 2011 floods,

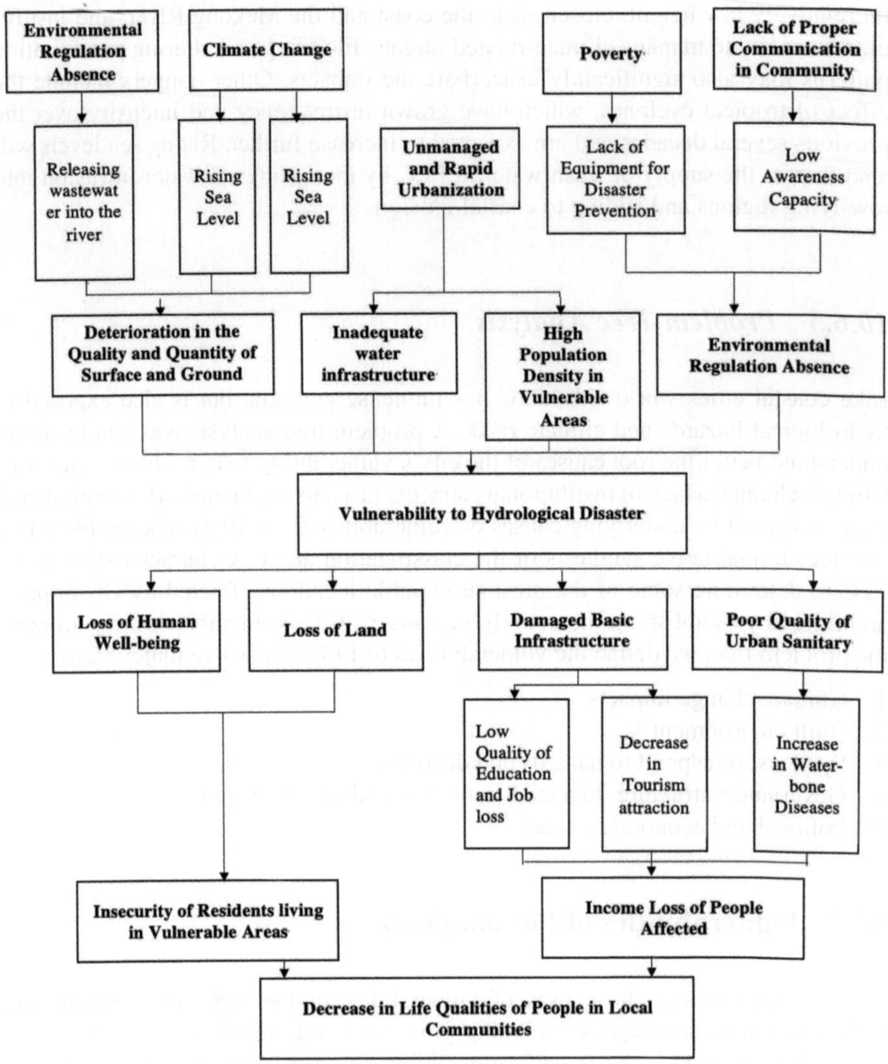

Fig. 10.8 Problems tree analysis

precipitation was 24% greater than the September average, and by October, the Mekong and Tonle rivers had reached 100-year flood levels, implying that without any consideration of upcoming changes, the potential for more significant future flooding existed. Climate change is anticipated to increase the severity and duration of these flood episodes, hence exacerbating the impact of floods on Cambodian cities in the future.

Most notably, flood duration in peri-urban areas far surpasses that in urban areas, with more than a third of poor communities suffering flooding for more than three months each year. In comparison, 80% of flooding in metropolitan areas lasts less than a day. In addition, according to a poll conducted by the NGO Think-Hazards on the climate change vulnerability index, Cambodia ranks ninth out of 127 nations. A predictive flood model suggests that potentially disastrous and life-threatening river floods are likely at least once in the next ten years. Despite the country's abundant water resources, groundwater levels are currently insufficient to allow large-scale agriculture, and in certain locations, such as Southern Cambodia, the supply is extensively contaminated with arsenic. Floods also result in contamination of water sources due to inadequate sanitation infrastructure, leading to water-borne diseases and negative health outcomes (Davies et al. 2014).

10.7.2 Institutional Vulnerability

The urban system of PPC is expanding rapidly in terms of quantity but slowly in terms of quality. Furthermore, Phnom Penh has seen a remarkable metamorphosis from a metropolis plagued by violence and instability to a hub of fast urbanization and economic growth, supported by increased attention and investment by government and development partners (World Bank 2017). However, like with other fast-growing Asian cities in the 1970s and 1980s, the city's expansion and development has been mainly unplanned, putting increasing demand on the city's infrastructure. Rapid in-migration, a lack of urban planning and enforcement capacity, along with real estate investment and growth offer obstacles to Phnom Penh's sustainable urban development.

Cambodia has steadily implemented its decentralization policy for urban planning and development since 2005, when the Council of Ministries formally endorsed the National Strategic Framework for De-concentration and Decentralization (D&D) (Eng 2016). While the centralized system remains at the heart of day-to-day management and administration, crucial duties and responsibilities are being delegated to lower levels of government systematically. A fundamental goal of the D&D policies was to improve the effectiveness and efficiency of public service delivery through sub-national councils such as Municipal, Provincial, Khan, and communal councils.

For many years, Phnom Penh City Hall has played an essential role in repairing and growing the city. Nevertheless, relations between ministries, institutions, and departments at various levels remain unaddressed, owing to a delay in the transfer of authority and function from national ministries to sub-national organizations, and an absence of explicit national or municipal policy committed to urban resilience (Beckwith 2022).

10.7.3 Socioeconomic Vulnerability

Similar to many other urbanizing cities in Asia Pacific, Phnom Penh also suffers from socioeconomic disparity resulting in vulnerable and marginalized urban poor. The city also hosts a substantial number of informal settlers, comprising nearly 15% of total population (Phnom Penh Municipality 2012) who live in risk-prone areas and are exposed to high levels of hydrological hazards. A lack of flood risk data and mapping, particularly for the urban and peri-urban poor, is a key impediment to adequately comprehending vulnerability at the city levels. Flood risk varies among metropolitan areas, according to secondary sources, community survey data, and interviews with public authorities.

Officials say that peri-urban regions such as Charmroeun and Vealsbov and the impoverished are the most vulnerable to floods. According to an STT assessment of nearly 400 poor urban communities in Phnom Penh, 66 percent of communities in peri-urban regions suffer floods, compared to 27% in urban areas (STT 2018). In such communities, disaster risks are exacerbated by the interaction of multiple socio-economic vulnerability contexts: poor quality dwellings, recurrent diseases and health issues, tenure insecurity and lack of access to formal markets and urban services (Flower et al. 2018; Khemro & Payne 2004).

10.7.4 Institutional Capacity

There is just one city plan and report linked to urban disasters based on data collecting. In addition, Cambodia now has various policies and plans in place to deal with climate change and the natural or manmade calamities that accompany it. However, finding the necessary documentation to address the concerns at the city level is challenging. However, via an interview with officials of the local governments, it was possible to get a city-level plan comprising of emergency scenarios by catastrophes and how to regulate and deal with them in detail on the plan.

10.7.5 Policy and Planning Capacities

An interview with Phnom Penh city hall personnel revealed that the city had no municipal-level climate change adaptation and mitigation policies. Furthermore, the city lacks any precise city-level climate change scenarios and strategies based on them. The city has a coping plan for hydrological catastrophes, which is backed by several international organizations and nations. However, the scope of the project for the strategies is limited. Therefore, no formal paper exists. Furthermore, it was discovered that there are currently no social supporting mechanisms for the urban poor in informal settlements to adapt and cope with calamities. However, inter-governmental and international organizations play a crucial role in supporting urban development.

10.7.6 Physical and Natural Capacities of the City

When considering Phnom Penh's physical capabilities, one of the most important indicators is the availability of water resources. It is situated at the confluence of three rivers: the Mekong, Tonle Sap, and Bassac, all of which supply fresh water to its inhabitants. This distinguishes Phnom Penh Capital as a one-of-a-kind clean water metropolis, having been honored by ADB in Manila, the Philippines in 2004 for "Clean Water Supply, Water for All" and won the Stockholm Industry Water Award in 2010. Phnom Penh Capital has four Water Treatment Plants that can create 136,020,000 m3 of water per year and 430,000 m3 of water per day, corresponding to 90% of the Phnom Penh area and supplying 250,070 consumers. It has a total water supply network length of 2156 km and an average water pressure of 10–25 m. The three rivers are vital industrial, agricultural, and domestic water sources. Despite this, water shortage is a serious concern throughout the country, especially during El Nino periods. Groundwater sources support more than half of the population, especially when there is insufficient surface water.

Three elements have influenced PPC's vulnerability and capacity to hydrological disasters. The foremost are the natural and physical dimensions, such as land usage, water supply, and geography. According to data research, PPC is one of the most vulnerable cities to climate change consequences, including river level increase and sand water intrusion. These variables have several harmful implications, including physical vulnerability. The second component is related to the city's institutional capabilities. Through data collection and analysis, it has become clear that PPC lacks adequate policies, strategies, and plans, as well as active coordination between municipal agencies and private sectors, to mitigate and adapt to climate change and its effects and to protect the environment. The municipal administration still lacks a dedicated department to guide all stakeholders and manage work consistency.

Chamrouen, one of the two designated regions, is located on the outskirts of the municipal district of Chbar Ampov. The majority of the communities in this region began settlement in 1979. The bulk of inhabitants in the restricted area work as informal garbage collectors. Furthermore, Chamroeun and Vealsbov are particularly vulnerable to flooding. Because of poor urban planning and unregulated settlement growth, the threats have shifted from flash river floods due to low elevation to development-induced flooding, especially in wetlands where landfilling is ongoing for redevelopment (Handley 2020). Chamroeun exemplifies the ever-changing threat of flood threats. Furthermore, neither area currently has flood protection.

10.8 Urban Resilience Assessment in Phnom Penh City

10.8.1 Overview of Urban Resilience

The preceding portion of this chapter recognized PPC's risk and capacity to deal with hydrological catastrophes. However, compared to other hazardous towns in Cambodia, PPC has a far greater potential to become a resilient metropolis to natural hazards in terms of adaptation and mitigation at the domestic and minimal levels. It is necessary to identify its resilience based on an evaluation using long-term indicators for development in order to comprehend the city's potential and highlight the gaps and weaknesses in capability impacting the urban development process.

The resilience of complex urban systems often influences the resilience of urban regions. Disruptions to the essential services they offer can have long-term consequences well beyond municipal limits. The complication of cities makes resilience building difficult, especially when the political economy of urbanization is not adequately interpreted and incorporated (Béné et al. 2017). Focusing on an objective policy, such as climate protection, without considering others might result in unfavorable effects. Thus, stakeholders' actions may result in definite arrangements, unforeseen effects, or an integration of the two. Constructing and improving a flexible city thus necessitates a multi-sectoral, comprehensive perspective. It will also require a different approach to governance, one that incorporates the local and regional contexts and specific attributes of the system that determine their resilience and adaptability (Walker et al. 2004).

The following attributes should be considered when evaluating UR.

(1) **Robustness:** Systems contain well-designed, built, and managed resources.
(2) **Reflectiveness:** Acceptance of today's fundamental and ever-increasing unpredictability and change.
(3) **Flexibility:** In reaction to changing conditions, systems can alter, evolve, and adapt.
(4) **Integration:** It encourages consistency in decision-making and ensures all investments contribute to a shared goal.
(5) **Resourcefulness:** During a shock or stress, people and institutions can quickly create new ways to achieve their aims or satisfy their demands.

To establish UR in PPC, four aspects are analyzed: financial and technological capacity, physical and social abilities and an ability to build strategies and policies. Furthermore, there are four resilience classifications, comprised of 12 parameters used to assess the UR and will be utilized to evaluate the resilience of PPC (Rockefeller Foundation and Arup, 2014).

- **Health and Well-being:** Guaranteeing the health and well-being of all city residents and workers.
- **Economy and society:** The social and economic mechanisms permitting metropolitan people to coexist happily and collaboratively.

- **Infrastructure and environment:** Anthropogenic and natural systems providing essential services, conserving, and linking city dwellers; and
- **Leadership and strategy:** Needs of a city for educated, inclusive, integrated, and participatory decision-making.

These factors are used to evaluate UR not just for climate change impacts but also for recovery from emergency shocks such as terrorism and economic meltdown. They are not simply the inverse of vulnerability but also the potential to shift an urban system toward sustained development.

10.8.2 Resilience Capacities

This section examines four resilience indicators, also noted as four capacities of PPC mentioned in the preceding section Table 10.2.

10.8.2.1 Policy and Planning Capacity

Social stability, security, low human vulnerability, physical exposure and combined plans for development in the city are some of the indicators used to assess Phnom

Table 10.2 Resilience factors and core capacities for resilience (*Source* Self-assessment data collection

Resilience factor	Core capacities for resilience (Parameters)			
	Policy and planning capacity	Physical and natural capacity	Social capacity	Technical and financial capacity
1. Health and well-being (People)	1–1. Minimal human vulnerability	1–2. Safeguards to human life and health		1–3. Diverse livelihoods and employment
2. Infrastructure and envi-ronment (place)	2–1. Reduced exposure and fragility	2–2. Continuity of critical services		2–3. Reliable mobility and communications
3. Economy and society (organization)	3–1. Comprehensive Security and Rules of Law		3–2. Collective Identify & Mutual Support	3–3. Sustainable economy
4. Leadership and strategy (knowledge)	4–1. Integrated development planning		4–2. Empowered stakeholders	4.3 Effective leadership and management

Penh's policy and planning capacities. The Phnom Penh Sustainable City Plan 2018–2030 (Phnom Penh Capital Administration et al 2019) and Phnom Penh Green City Strategic Plan 2017–2030 (National Council for Sustainable Development et al. 2016) pave way for institutionalization and policy arrangement for the implementation of priority actions and city projects for sustainable and resilience PPC by tackling climate change, while pursuing economic growth, poverty alleviation, strengthening development, and social inclusion.

To better understand the stakeholders who participate in the city's performance, institutional mapping was used to analyze the roles. This analysis is presented in Table 10.3.

10.9 Potential Solutions and Recommendations to Improve Urban Resilience

The process of developing measures to improve UR and close gaps in the social economy category is described in this section. According to the preceding section, the study examined and identified the city of Phnom Penh's UR performance. Figure 10.9 depicts the relationship between Phnom Penh's capacity and its vulnerability to natural disasters. This section emphasizes the two-way links between city capacity and vulnerability for implementing action plans to address gaps and satisfy urban people's aspirations to adapt to natural disasters and improve their quality of life.

10.9.1 Promoting Stakeholder Engagements

This study emphasized the relevance of multi-stakeholder participation in creating resilience in the prior section by examining the UR categories in the city.

- **Key Stakeholders at national level**:

  1. Strengthening institutions, governance, urban planning (building regulations), and execution
  2. Investing in long-term urban infrastructure
  3. Promoting inclusive urbanization
  4. Ensuring regulation and policies at multiple levels to deter crime and tackle political corruption.
  5. Securing financial support for the local.
  6. Clearly define the roles and responsibilities of each stakeholder.
  7. Dispersing the actual meaning of (urban)resilience at multiple levels
  8. Efficiently improving capacity constraints in planning, implementation, and operation and maintenance of infrastructure

Table 10.3 Institutional capacity mapping across resilience and vulnerability indicators of urban resilience

Resilience indicators	Nature of vulnerability	institution	Institutional capacity
Comprehensive security and rules of law	Systems to deter crime	Transparency International Cambodia	NGO
	Corruption prevention	Anti-Corruption Unit (A.C.U), and Legal Affairs & Human Rights Dept	City Government
	Policing & civil justice	Legal Affairs & Human Rights Dept	City Government
Reduced exposure and vulnerability	Environmental Policy	Department of Economy and Finance	City Government
	Safeguards for critical infrastructure	Urbanization dept	City Government
	Building codes and standards	Urbanization dept	City Government
Minimal human vulnerability" basic needs	Food supply	The intl. Organizations, Association of Cambodian Local Economic Development Agencies (ACLEDA) and Cambodia people's party	City Government, Foreign Sectors, and domestic public and private sectors
	Effective sanitation	Cambodia people's party, international organizations, and foreign sectors	City Government, Foreign Sectors, and domestic public and private sectors
	Energy and water supply	Ministry of Industry Mines and Energy, Phnom Penh Water Security Authority (PPWSA, Intl. and Domestic NGOs	National Government, Municipal agencies, and international and Domestic sectors
	Safe Housing	*Phnom Penh Ministry* of Economic, Finance Urban Development and Labor, General Department of Housing (GDH), Ministry of Land Management	National Governments, Municipal government, and international governments
Integrated development planning	City Monitoring & Data Management	Department of Land management, Urban planning, construction and cadastral (DLMUPCC)	City Government

(continued)

Table 10.3 (continued)

Resilience indicators	Nature of vulnerability	institution	Institutional capacity
	Land Use & Zoning	Department of Land management, Urban planning, construction and cadastral (DLMUPCC)	City Government
	Planning approval process	Urbanization dept	City Government

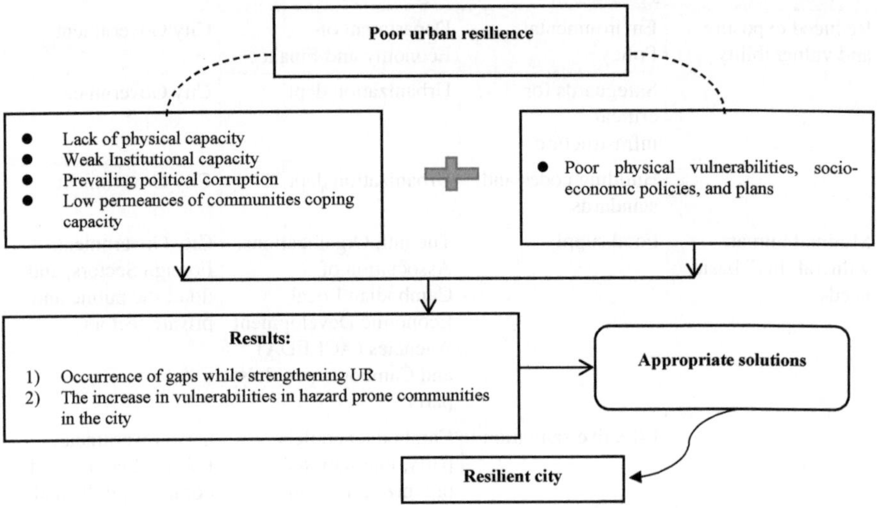

Fig. 10.9 The process in decision making to make solutions for resilient city

- **Primary stakeholders at sub-national municipal and regional levels:**

1. Encouraging stakeholders and consensus building processes
2. Specifying and identifying strategic plans and detailed land use plans
3. Encouraging the locals to participate in discussion for residents regularly.
4. Securing financial support for the local.
5. Improving transparency in rural areas.
6. Strengthening the capacities of local authorities (Mainly in rural areas) to implement conduct feasibility studies, risk evolution, and prepare Public–Private-Partnership projects (PPPs).

- **Secondary and tertiary stakeholders:**
    1. Suggesting resilience education from children to adults
    2. Encourage learning and sharing ideas from others through posters, newsletters, social media, mobile applications, etc.
    3. Continually creating some environments for learning.
    4. Making partnerships with key suppliers and customers.

10.9.2 Combating Political Corruption for a Better Institutional Capacity

This study identified the existing UR of PPC by category in the preceding section. Several challenges to the city's resiliency were discovered throughout the investigation. The Climate Change Strategic Plan and the Rectangular Strategy identify institutional capacity as hindering successful adaptation, urban resilience, and development strategy implementation. The issues with inter-institutional ministries are explained in the proposed plan.

The primary policy goals for UR will determine whether PPC becomes a sustainable and inclusive city in the future. According to the findings of the Phnom Penh UR study, the most significant impediment to improving UR was "political corruption." According to the United Nations Office on Drugs and Crime, corruption has far-reaching consequences. There are several problems caused by political corruption.

Three representative examples.

(1) Comprehending the Sustainable Development Goals (SDG).
(2) Economic damages and inefficiency
(3) Inequality & poverty

Corruption impedes the fulfilment of the United Nations Sustainable Development Goals in the previous scenario (SDGs). The SDGs are wide, and their propensity to corruption is understandable: "a greater and more continuous future for everyone" frequently runs against the interests of a few and can be failed by various forms of corruption. Nations fail to alleviate poverty, eliminate hunger, provide adequate health care and high-quality education for their populations, other fundamental rights, secure gender equality, improve inequality, and so on in the face of dwindling state capacity. When public discontent with corruption and cronyism reaches a critical mass, the political benefits of tackling corruption may exceed the dangers of upsetting interests. In the absence of broader reform attempts, enhanced and more open procedures, professional accountability systems, and the use of cutting-edge technical equipment to collect, analyze, and share data to avoid, identify, and deter corrupt behaviors are considered to make progress.

10.10 Conclusion

This study elaborates on a technique for identifying the resilient ability of urban systems and generating countermeasures to address differentials and raise the ability to manage hydrological hazards by utilizing parameters of long-range development achievements. Institutional Mechanisms, Self-vulnerability evaluation, benchmarking, and urban resilience evaluation are all part of the technique. According to the assessment, empowered stakeholders and increased relationships between local communities and the municipal government will be able to value the plan to construct PPC's Urban Resilience. Identifying the UR is necessary to respond to climate change and induced natural hazards. Furthermore, analyzing the effectiveness of UR using indicators from long-term development goals is an efficient technique to detect gaps and shortcomings.

Furthermore, using indicators from the long-term development goals to assess UR performances would be an effective strategy to identify gaps and weaknesses in current capabilities while minimizing potentially vulnerable populations. The study found that engaging stakeholders in UR development can help the city's long-term growth. Consequently, the city can adapt to natural disasters and increase its citizens' quality of life. Moreover, long-term combined policy and planning and a monitoring and evaluation (M&E) system are essential, as are efficient actions to help in implementation, which can contribute to the achievement of the proposed solution in this research.

The study's findings may be extended to cities with comparable difficulties or geographical situations. This research reveals the intricate nuances of Phnom Penh. However, this study mainly focuses on UR to hydrological hazards and does not provide a comprehensive assessment. It is suggested that more studies be conducted focused on particular solutions to improve city institutional capacities, such as how to combat corruption and how to encourage locals to engage in initiatives and discussions for their communities. Furthermore, for more particular future research results, specific components closely associated with natural catastrophes should be addressed.

References

ACAPS (2014) Secondary data review—Cambodia. The Assessment Capacities Project, Geneva

ARUP (2021) City resilience index-understanding and measuring City resilience. Retrieved from https://www.arup.com/-/media/arup/files/publications/c/170223_cri-brochure.pdf

ADB (2016) Urban development in the greater mekong subregion Retrieved from https://www.adb.org/sites/default/files/publication/185008/urban-development-gms.pdf

ADB (2014) Urban climate change resilience—a synopsis. Retrived from https://www.adb.org/sites/default/files/publication/149164/urban-climate-change-resilience-synopsis.pdf

ADB (2021) Cambodia agriculture, natural resources, and rural development sector assessment, strategy, and road map. https://doi.org/10.22617/tcs210256-2

Barkham R, Brown K, Parpa C, Breen C, Carver S, Hooton C (2014). Resilient cities: A Grosvenor research report. Grosvenor

Béné C, Mehta L, McGranahan G, Cannon T, Gupte J, Tanner T (2017) Resilience as a policy narrative: potentials and limits in the context of urban planning. Climate Dev 10(2):116–133. https://doi.org/10.1080/17565529.2017.1301868

Beckwith L (2022) Cambodia's resilience agenda: understanding how local institutions and actors accept, contest and accommodate an externally driven approach. Geoforum 128:125–134. https://doi.org/10.1016/j.geoforum.2021.12.015

Chelleri L, Waters JJ, Olazabal M, Minucci G (2015) Resilience trade-offs: Addressing multiple scales and temporal aspects of urban resilience. Environ Urban 27:181–198

Davies GI, McIver L, Kim Y, Hashizume M, Iddings S, Chan V (2014) Water-Borne diseases and extreme weather events in cambodia: review of impacts and implications of climate change. Int J Environ Res Public Health 12:191–213

Eng N (2016) Decentralization in cambodia: new wine in old bottles. Publ Admin Devel 36:250–262. https://doi.org/10.1002/pad

Flower B, Fortnam M, Kol L, Sasin P, Wood RG (2018) Using participatory methods to uncover interacting urban risks: a case study of three informal settlements in Phnom Penh, Cambodia. Environ Urban 30:301–316. https:// doi.org/https://doi.org/10.1177/0956247817735481

Folke C (2006) Resilience: the emergence. of a perspective for social-ecological systems analyses. Glob Environ Chang 16(3): 253–267

Giulio B (2024) Workers in the Cambodian informal economy insights into labour statistics from the 2019 LFS. In ILO eBooks. https://doi.org/10.54394/dpoy3021

Handley E (2020) The largest lake filling in our history: Cambodian satellite city puts 1 million at risk of flooding. ABC News. Available online at: https://www.abc.net.au/news/2020-07-28/cambodia-satellite-city-sand-fill-lake-megamall-flooding-risk/12495180

Khemro BHS, Payne G (2004) Improving tenure security for the urban poor in phnom penh, cambodia: an analytical case study. Habitat Int 28(2):181–201

Leichenko R (2011) Climate change and urban resilience. Curr Opin Environ Sustain 3(3):164–168. https://doi.org/10.1016/j.cosust.2010.12.014

Ly Sodeth, Pynnonen Lauri Veli Johannes, Sondergaard Lars M, Katwal Saurav, Hill Samuel Christopher, Tong Kimsun, Hayati Fayavar, Fukao Tsuyoshi, Fata No, Phim Runsinarith. Cambodia economic update: export revival and trade shifts—strengthening cambodia's education system for future growth (English). Washington, D.C.: World Bank Group. http://documents.worldbank.org/curated/en/099060324061538453/P1773401c2325e0f81a05f159ec748e81f0

Monteiro A, Ankrah J, Madureira H, Pacheco MO (2022) Climate risk mitigation and adaptation concerns in urban areas: A Systematic review of the impact of IPCC Assessment reports. Climate 10(8):115. https://doi.org/10.3390/cli10080115

National Council for Sustainable Development, Phnom Penh City Hall & Global Green Growth Institute (2019), Phnom penh green city strategic plan 2017–2026, https://ncsd.moe.gov.kh/sites/default/files/2019-05/Phnom%20Penh%20Green%20Strategic%20Plan.pdf

Notre Dame Global Adaptation Initiative (2022) Ranking—Notre dame global adaptation initiative—university of notre dame. Notre Dame Glob Adapt Initiat. Available online at: https://gain.nd.edu/our-work/country-index/rankings/

Phnom Penh Municipality (2012) The phnom penh urban poor assessment: a baseline survey on the social and economic situations and capacity of existing services in urban poor communities, Phnom Penh

Pickett S, Cadenasso M, Grove J, Boone CG, Groffman PM, Irwin E, Kaushal SS, Marshall V, McGrath BP, Nilon C, Pouyat R, Szlavecz K, Troy A, Warren P (2011) Urban ecological systems: Scientific foundations and a decade of progress. J Environ Manage 92(3):331–362. https://doi.org/10.1016/j.jenvman.2010.08.022

Phnom Penh Capital Administration, National Council for Sustainable Development & Global Green Growth Institute (2019), Phnom Penh Sustainable City Plan 2018–2030, https://gggi.org/wp-content/uploads/2019/06/SUBSTAINABLE-CITY-REPORT_EN_FA3.pdf

Rockefeller Foundation & Arup (2014) "City Resilience Framework" https://www.rockefellerfoundation.org/app/uploads/City-Resilience-Framework1.pdf

Saltsman A, Bertrais D, Fauveaud G, Ngo N, Pak K, Thuon T (2021). Limits to growth: A vulnerability approach to understanding urbanization in Cambodia (p. 34). Cent Khmer Stud (CKS)

Sasin P, Sokha S (2015) City at risk? phnom penh: hazard, vulnerability and capacity assessment. people in need—cambodia. Available online at: https://www.peopleinneed.net/city-at-risk-phnom-penh-hazard-vulnerability-and-capacity-assessment-567pub

STT (2018) The Phnom Penh Survey: A. study on Urban Poor Settlement in Phnom Penh, 52–53

Tyler S, Nugraha E, Nguyen HK, Van Nguyen N, Sari AD, Thinpanga P, Tran TT, Verma SS (2016) Indicators of urban climate resilience: A contextual approach. Environ Sci Policy 66:420–426. https://doi.org/10.1016/j.envsci.2016.08.004

Tyler S, Moench M (2012) A framework for urban climate resilience. Climate Dev 4(4):311–326. https://doi.org/10.1080/17565529.2012.745389

Walker B, Holling CS, Carpenter SR, Kinzig A (2004) Resilience, adaptability and transformability in social–ecological systems. Ecol Soc 9(2): 5. URL: http://www.ecologyandsociety.org/vol9/iss2/art5/

World Bank (2017) Urban Development in Phnom Penh (Working Paper No. 121692). World Bank Group. Available online at: https://documents.worldbank.org/en/publication/documents-reports/documentdetail/286991511862455372/Urban-development-in-Phnom-Penh

Chapter 11
Urban Energy and Waste Management for BGI, Food Security and Climate Disaster Resilience

Md. Anwarul Abedin, Sadiya Arefin Juthee, Md. Hosenuzzaman, Nina Takashino, and Rajib Shaw

Abstract Food insecurity has become a burning issue in many regions of the world and climate change is aggravating it severity. Urban areas can lessen the effects of climate change by implementing Blue-Green Infrastructure (BGI), which is made up of natural and semi-natural systems. The integration of urban energy systems and waste management plays a pivotal role in enhancing blue-green infrastructure (BGI), ensuring food security, and building climate disaster resilience. The synergies between urban energy and waste management practices are imperative to support sustainable urban environments in the face of climate change and increasing urbanization. By examining innovative approaches to energy generation, waste-to-energy conversion, and circular resource management, the chapter highlights strategies that reduce environmental footprints while enhancing urban resilience. BGI, integrated with sustainable waste and energy systems, can mitigate climate impacts by improving stormwater management, reducing urban heat, and providing green spaces. Moreover, efficient waste management systems, such as organic waste composting and bioenergy production, contribute to soil fertility and local food production, enhancing food security. This chapter will focus on how urban energy and waste management contribute to BGI, food security and disaster resilience in growing cities.

Md. A. Abedin (✉) · Md. Hosenuzzaman
Department of Soil Science, Bangladesh Agricultural University, Mymensingh, Bangladesh
e-mail: m.a.abedin@bau.edu.bd

Md. A. Abedin · N. Takashino
College of Policy Science, Osaka Ibaraki Campus, Ritsumeikan University, Osaka, Japan

S. A. Juthee
Department of Crop Botany, Bangladesh Agricultural University, Mymensingh, Bangladesh

Md. Hosenuzzaman
UWA School of Agriculture and Environment, The University of Western Australia, Perth, Australia

R. Shaw
Graduate School of Media and Governance, Keio University, Minato, Japan

M. G. Kibria et al. (eds.), *Urban Blue-Green Infrastructure Approach for Food Security and Climate Disaster Resilience*, Disaster Risk Reduction,
https://doi.org/10.1007/978-981-96-8746-6_11

Keywords Green energy · Waste management · BGI · Food security · Disaster resilience

11.1 Introduction

Currently, 54% of the world's population resides in cities, and this proportion is expected to rise in the coming decades (United Nations 2014). By 2030, urban land cover is projected to nearly triple, expanding by approximately 1.2 million km^2 (Seto et al. 2012). Urbanization will continue across both developed and developing regions, with urban populations expected to reach 86% in developed areas and 64% in less developed areas by 2050 (Koop & van Leeuwen 2017). This rapid urban expansion generates vast amounts of solid waste. Global solid waste production varies by income level, ranging from 0.1–0.8 tons per capita per year in low-income countries, 0.2–0.5 t/cap/yr in middle-income countries, and 0.3–0.8 t/cap/yr in high-income countries—closely linked to per capita energy consumption, GDP, and private consumption (Gichamo & Gökçekuş 2019). Inefficient waste management systems contribute to water, air, and soil pollution, while open and unsanitary landfills increase the risk of disease transmission through contaminated drinking water. Addressing solid waste management remains a critical challenge for urban areas of all sizes, from megacities to small towns and large villages (UN-Habitat 2010).

The urbanisation process continues to have an increasingly strong impact on climate change as it progresses worldwide. Methane, nitrous oxide, and carbon dioxide from solid waste account for roughly 5% of greenhouse gas emissions, making it a significant contributor to the emissions of greenhouse gases. Urbanization causes regional warming and urban heat island effect, extreme events such as high temperature, heat wave and heavy rainfall increase in frequency, and also leads to increased urban flood risk (Xinyue & Mingxing 2019). Consequently, there is a multi-scale superposition effect at both the global and regional levels as a result of urbanization's influence on climate change. Consequently, there is a multi-scale superposition effect at both the global and regional levels as a result of urbanization on climate change. Most of the world's population will be exposed to anthropogenic climate change in urban areas (IPCC 2014). Therefore understanding the changing risk cities (and their urban populations) are facing is a critical global change issue (Argüeso et al. 2015).

The local and regional effects of urbanization, solid waste and global climate change will undoubtedly have significant impacts on blue green infrastructure, food security and climate disaster resilience. Our current lack of understanding of how urban energy and waste management will develop in urban areas makes it more difficult for us to respond to the growing risk of disaster and food security. The world grapples with rapid urbanization, climate change, and the imperative for sustainable development, the integration of urban energy and waste management becomes critical. This chapter explores the interconnected strategies that leverage these systems to enhance blue-green infrastructure (BGI), promote food and nutrition security, and fortify urban resilience against climate disasters.

11.2 Global Scenario of Urban Energy, Waste Production, Disasters, and Food Security

11.2.1 Urban Energy Systems

Urban energy systems depict "the combined processes of acquiring and using energy" to fulfill the energy service demands of an urban population (Jaccard 2006). In the past, these needs were quite straightforward and primarily consisted of cooking and space heating needs. Nonetheless, a broad range of energy services such as communications, mobility services, interior and outdoor lighting, electric power for appliances, building heating and cooling, and so forth support modern city activities. According to recent studies, urban areas are responsible for more than 66% of energy consumption globally, which is predicted to increase to 73% by the end of 2030. Consequently, these demands are responsible for more than 70% of the CO_2 emissions in the world (IEA 2022). Therefore, cities are essential components of the modern energy system and have taken charge in attempts to transition away from fossil fuels and towards a more sustainable foundation.

11.2.2 Generation of Municipal Solid Waste

The migration of people from rural to the city and the growing economy have greatly accelerated the municipal solid waste (MSW) generation rate in urban areas (Table 11.1) (Hassan 2000; Minghua et al. 2009; Singh et al. 2011). Currently, the waste generated from urban centers of the world is around 2.01 billion tonnes per year (0.11–4.54 kg/capita/day) that is projected to rise to 3.40 billion tonnes per year by 2025 (Kaza et al. 2018). According to UNEP (2015), an amount of 7–10 billion tonnes of urban waste produced each year which is fueled by population growth, urbanization and rising consumption, the volumes of waste are likely to even double in lower-income African and Asian cities by 2030. The level of urbanization (or percentage of urban population) and waste generation is the lowest in low income countries, generating about 219 kg/capita/year in 2010 which is projected to reach to 343 kg/capita/year by 2025 (World Bank 2012). The total quantity of waste generated in low-income countries is expected to increase significantly (more than three times) by 2050 (Kaza et al. 2018).

Table 11.1 Municipal waste generation of selected countries

Countries	Year	Waste generation (Million metric tons)	Reference
United States (USA)	2018	265.2	Alves (2023)
China	2017	215.2	
Germany	n/a	53.7	
Japan	2020	41.7	
France	n/a	38.0	
Israel	n/a	6.2	
United Kingdom (UK)	2020	31.0	
Costa Rica	n/a	1.6	
New Zealand	2018	3.7	
Spain	n/a	22.4	
Poland	n/a	13.7	
Netherlands	n/a	9.0	
Colombia	2018	12.1	
Australia	2019	12.6	
Chile	2018	8.2	
Turkey	2020	34.6	
South Korea	2020	22.5	
Italy	2020	28.9	

n/a means data not available

11.2.3 Climate-Induced Disasters in Cities

Predictions associated with climate change indicate a rise in the frequency and severity of natural catastrophes that will impact both the environment and the global community (EEA 2017; IPCC 2018). Rapid urbanization under climate change is exposing cities around the globe to enormous risks, particularly cities built along waterways or near coasts (Kumar & Saroj 2014; Re Swiss 2020). Gu (2019) studied the exposure and vulnerability of 1,860 cities and urban areas to natural disasters (floods, cyclones, earthquakes, droughts, landslides, and volcanic eruptions). Nearly 58% of the metropolitan regions under study have high levels of exposure to at least one of the six natural hazardsi.e., floods, cyclones, earthquakes, droughts, landslides, and volcanic eruptions. In general, the most of the riskiest cities of the world are located in East Asia, Taiwan, China, Japan and the Philippines on the basis of their degree of exposure to disasters (Gu 2019). Since many of these cities are coastal, they frequently experience earthquakes, floods, hurricanes, and other natural calamities.. Overall, 43.9, 32.8, and 18.2% of people experienced damage by floods, earthquakes, and wind storms, respectively, while an additional 5.2% of people were at risk of being completely destroyed by storm surges and tsunamis. The effects of a natural hazard in some densely occupied cities can be catastrophic (Debele et al. 2019). To combat the growing effects of disasters, urban planners and decision-makers must therefore strengthen resilience and adaptation techniques.

11.2.4 Urbanization and Food Security

Agricultural land is converted to residential and industrial areas due to expanding cities and rapid urbanization. Hence, the availability, stability, safety and access (four dimensions of food security) to foods in the urban areas become limited. For instance, 2600 hectares of agricultural land are converted to urban areas annually (Maxwell et al. 2000). The consequence of land use changes is the shifting to peri-urban and rural agriculture, that play an important role for foods to the cities.

11.3 Urban Energy Management

Based on three design-driven principles, it aims to: reduce waste and pollution; recycle materials and products to the fullest extent possible; and restore the environment.

11.3.1 Renewable Energy Integration

Adopting sustainable and renewable energy sources such as wind, solar, and hydropower in urban spaces including waste-generated biogas from anaerobic digestion and thermal electricity during pyrolysis and incineration contributes to a sustainable energy mix (Fig. 11.1). Microgrids and decentralized energy systems enhance the resilience of urban areas by reducing dependence on centralized grids vulnerable to climate-related disruptions.

11.3.2 Energy-Efficient Buildings

An energy-efficient structure maximizes resource efficiency by ensuring proper living conditions within the home using the least energy. Implementing energy-efficient building designs, utilizing green technologies, and promoting energy conservation measures reduce the overall energy demand of urban areas. Smart infrastructure, including energy-efficient lighting and appliances, and renewable energy integration such as solar energy, further optimize energy efficiency. According to energy statistics data from the Ministry of Statistics and Program Implementation in India, buildings consume the most energy after industries (Gupta & Chakraborty 2021). In

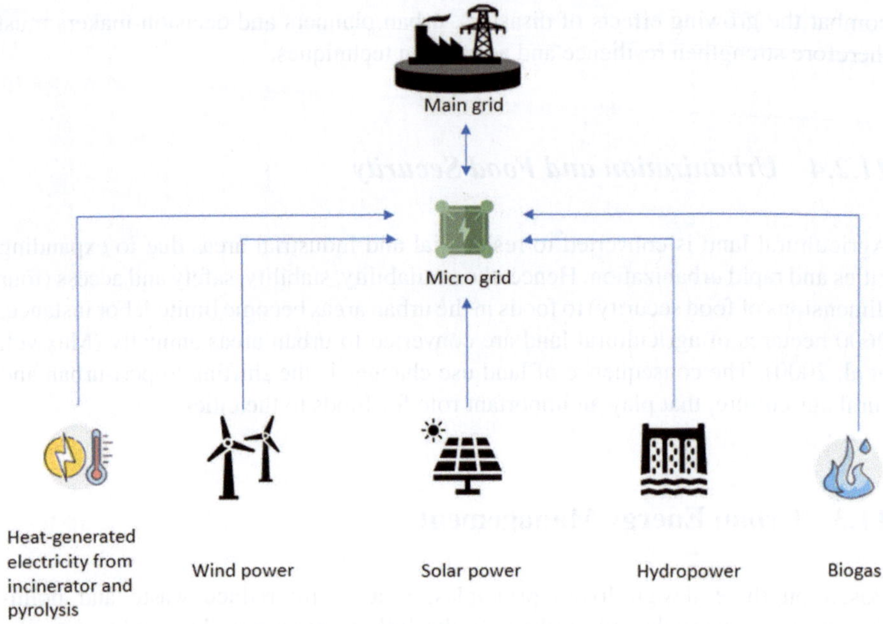

Fig. 11.1 Sustainable urban energy sources

order to balance all aspects of energy use in a structure, an energy-efficient building offers the best possible combination of renewable energy sources, energy-efficient equipment, and passive solar design principles. For example, permeable pavements with 80% grass coverage reduced the mean average air temperature, reduced cooling consumption (25%), CO_2 emissions (14.7%), and electricity consumption (53.4%) and integration of green roofs (such as rooftop garden) can offer optimized energy efficiency and environmental sustainability (Habibi & Kahe 2024).

11.3.3 Waste-To-Energy Solutions

Innovative waste-to-energy techniques, such as incineration and anaerobic digestion, can convert organic waste into bioenergy (Fig. 11.3). These solutions not only address waste management challenges but also contribute to the generation of renewable energy.

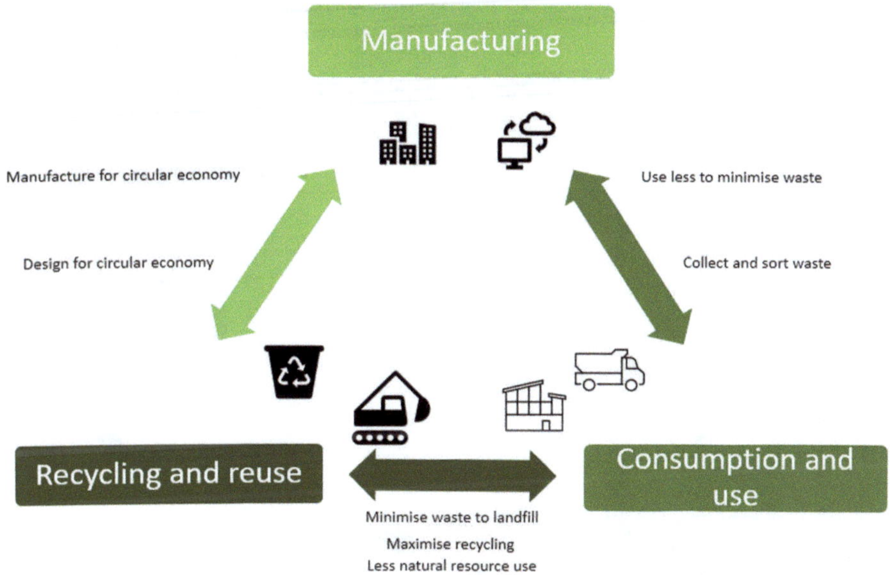

Fig. 11.2 Circular economy principles

11.4 Waste Management Strategies

11.4.1 Circular Economy Principles

A circular economy model reduces waste production and encourages material reuse and recycling. Blue-green infrastructure incorporating circular principles aims to reduce the environmental impact of waste and enhance resource efficiency (Fig. 11.2).

11.4.2 Smart Waste Management Systems

Proper and efficient waste management is the main challenge for highly populous towns and cities. Smart technology-based waste management systems, such as sensor-equipped waste bins and data analytics, provide data on waste generation patterns and behaviour. Hence, it optimizes waste collection routes and schedules. Efficient waste management decreases the environmental footprint and improves the overall resilience of urban areas. Technology-based innovative solutions are imperative to handle enormous amounts of waste in densely populated urban and peri-urban areas. An automated lid system, a display system, a communication system, and an identification system make up a smart IoT-based integrated system that ensures a clean environment and sustainable urban living. The sensor in the waste bins identifies and measures the level of garbage. The device keeps track of the rubbish within

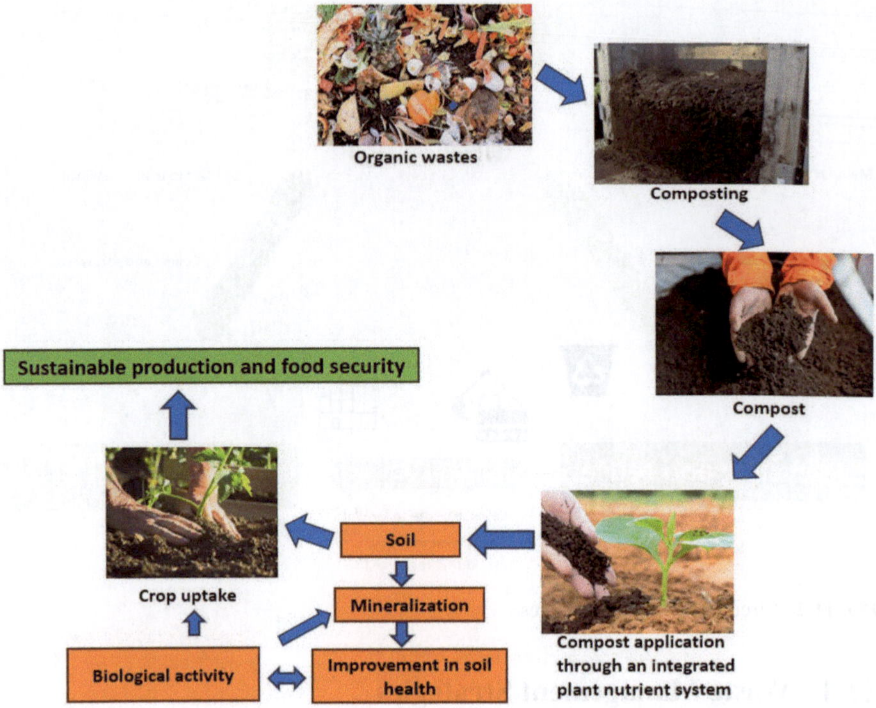

Fig. 11.3 Organic wastes for crop production and food security

the trash bin on a continual basis and displays the proportion filled up on a liquid crystal display. When the trash bin is filled, the communication system makes use of a worldwide system for mobile communications module to notify the appropriate authority to pick up the rubbish. This IoT-based waste management system is much more efficient than any other conventional waste management system as it uses less manpower, avoids spillover of waste, reduces carbon emission through vehicle use optimization, saves time and more economical (Sohag & Poddar 2020).

11.4.3 Organic Waste for Agricultural Use

Organic waste can be repurposed through composting to enrich soil quality for urban agriculture and contribute to sustainable food security (Fig. 11.3). This circular approach not only diverts waste from landfills but also contributes to local food production.

11.4.4 Blue-Green Infrastructure Integration

Waste recycling results in a considerable decrease in greenhouse gas (GHG) emissions. The greenhouse gas emissions resulting from open dump landfilling are around 1000 kg CO_2-eq. tonne^{-1}, but conventional landfilling can significantly lower these emissions to 300 kg CO_2-eq. tonne^{-1}. Actually, it can even be a net sink of carbon when most material is recycled or the energy is recovered (Manfredi et al. 2009).

Recently, a circular economy proposal was unveiled by the European Commission. One of the causes is the 600 million tonnes of materials that are lost annually in Europe's waste that could be recovered or used again. Turning waste into a resource is an essential part of increasing resource efficiency and part of this circular economy package (European Commission 2015). The organic waste converted to compost can be used in the blue-green infrastructure as source of nutrients for permeable pavements and rooftop gardening.

11.5 Sustainable Urban Planning

Integrating BGI principles into urban planning ensures the creation of green spaces, water management systems, and climate-resilient landscapes. BGI elements, such as green roofs and permeable surfaces, enhance energy efficiency and mitigate the urban heat island effect (Fig. 11.5).

- **Stormwater management**

When precipitation hits the surface of the Earth in a natural setting, it either seeps into the ground or runs as runoff until it reaches a body of water like a lake or river. Rapid urbanisation is causing impermeable surfaces to replace natural landcover, drastically changing the patterns and processes of the urban environment and leading to improper storm water management and subsequent flooding risks. In addition to its many environmental advantages, integrated Blue-Green Infrastructure (BGI), a landscape conservation strategy with nested networks of blue and green spaces (permeable pavements, rain gardens, bioswales, urban tree cover, wetlands, and small ponds), can offer an alternative to traditional storm water management. For suitability study, BGI incorporates five essential factors: slope, land cover, drainage density, hydrologic soil properties, and proximity to roadways. A network of corridors created by the least cost path model and gravity model connects the node, which is made up of a variety of blue and green open areas found throughout the city, like parks, gardens, and lakes. BGI principles contribute to effective stormwater management by absorbing and slowing down water runoff. Sustainable drainage systems, wetlands, and green corridors help prevent flooding and improve overall water quality (Kaur & Gupta 2022).

11.6 Food Security Enhancement

11.6.1 Urban Agriculture and Green Spaces

Allocating spaces for urban agriculture within the city promotes local food production. Green spaces also contribute to the well-being of urban residents and enhance biodiversity. Urban agriculture areas are increasingly perceived as a recreation space for urban dwellers to enjoy nature (Van Leeuwen et al. 2010). Urban agriculture is practiced by citizens for a number of purposes, such as recreational, urban greening, neighbourhood enhancement and food production. (Rosol 2010; Thibert 2012; Cohen & Reynolds 2014). Grassroots organizations are practicing urban agriculture to transform the city using underused spaces (Thibert 2012). They are also working to city beautification by converting underutilized areas into valuable green spaces (Krasny & Tidball 2017) and providing 'green areas' where parks are lacking in the city (Schukoske 2000). Some cities have adopted urban agriculture as an urban green strategy. In USA and Canada community gardens are officially recognized within the city's definition of 'Parks and Open Areas' (Schukoske 2000). The 'Urban Landscape Strategy' develops green spaces under three themes: 'Beautiful City', 'Productive Landscape' and 'Urban Nature' in Berlin. The 'Productive Landscape' theme connects allotments, agricultural and urban gardening spaces with interim users and space pioneers (Thierfelder & Kabisch 2016). Opoku et al. (2024) reported that implementing sustainable urban farming systems with green urban space design can lessen human-caused environmental problems such as climate change. Moreover, urban agriculture provides a potential contribution to food security by supporting local food supply and reducing reliance on the rural–urban food value chain (Armanda et al. 2019).

11.6.2 Resource Recovery from Waste

Recycling organic waste into compost enriches the soil for urban farming, creating a closed-loop system. By minimizing food waste and repurposing organic matter, cities can contribute to food security and sustainability (Fig. 11.4). Industrial, urban/suburban, and agricultural operations produce a significant amount of solid and liquid waste. Nowadays, managing and processing the garbage generated is a major concern for local authorities everywhere in the world, not just in metropolitan areas (i.e., municipalities). However, considering that the name "waste" is imprecise and actually refers to a resource, resource recovery from wastes has been a focus of management and study since it has enormous potential for social and environmental sustainability. Research worldwide is currently concentrated on recovering different resources from waste/wastewater produced by human activities, including energy, nutrients, bioproducts, and metals.

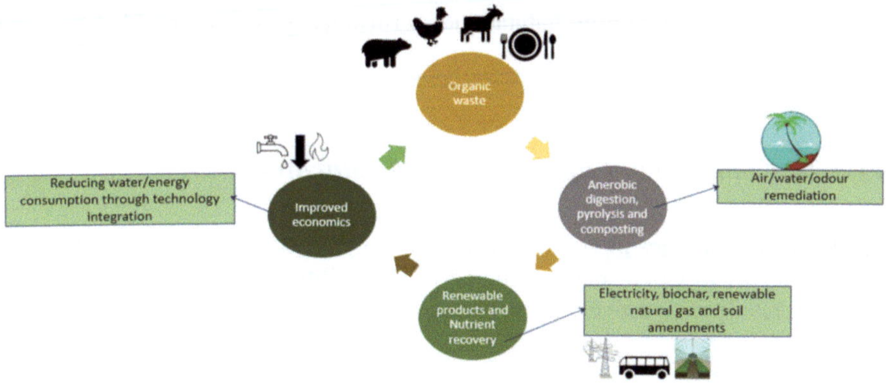

Fig. 11.4 Resources from waste

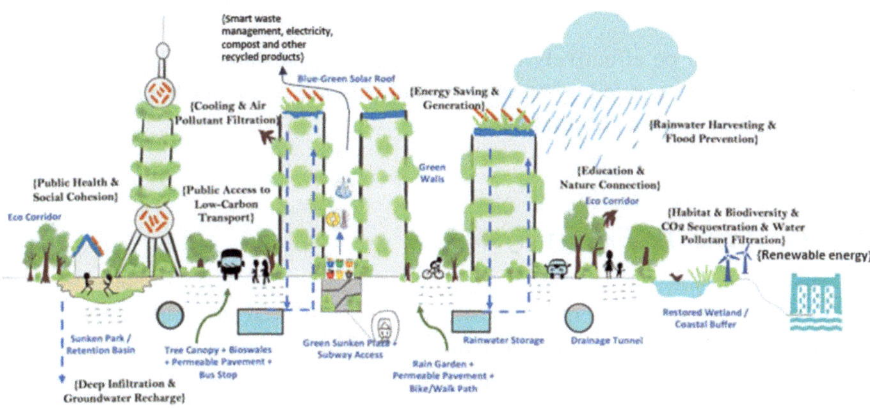

Fig. 11.5 Blue-green infrastructure for climate resilience (adopted and modified from Seihr et al., 2022)

11.7 Climate Disaster Resilience

11.7.1 Reducing Vulnerability to Climate Disasters

Integrating renewable energy and waste management systems reduces vulnerabilities to climate-related disruptions in energy supply and waste infrastructure. Sustainable urban planning and BGI enhance overall climate disaster resilience by addressing multiple facets of urban vulnerability (Fig. 11.5). Blue-green infrastructure restores natural ecosystems, which benefits cities and their natural surroundings. Permeable pavements, healthy catchments, and wetlands, for instance, can lessen the effects of stormwater runoff and flooding. Urban heat island effect is lessened by trees and green areas, and wildlife habitat and stormwater reuse are provided by green corridors.

11.7.2 Community Engagement and Education

Involving communities in waste reduction, energy conservation, food security and BGI initiatives fosters a sense of shared responsibility. Educational programs can empower residents to actively contribute to climate disaster resilience.

11.8 BGI, Urban Energy and Waste Management Nexus for Food Security and Climate Disaster Resilience

Urban energy can be effectively utilized by adopting sustainable practices such as BGI that support integrated urban systems and enhance resilience to climate impacts (Fig. 11.6). Key approaches include:

11.8.1 Waste-To-Energy Systems

Urban waste, particularly organic waste, can be converted into bioenergy through anaerobic digestion, gasification, or incineration. This process not only reduces landfill waste but also generates electricity and heat that can power urban systems. By repurposing waste, cities can decrease dependency on non-renewable energy sources and improve waste management, aligning with circular economy principles.

11.8.2 Renewable Energy Integration

Using solar, wind, and geothermal energy sources within urban settings helps reduce greenhouse gas emissions and decreases reliance on fossil fuels. Installing solar panels on buildings and integrating small-scale wind turbines, especially within BGI, can create sustainable microgrids that provide backup power during climate-related disasters, strengthening resilience and energy independence.

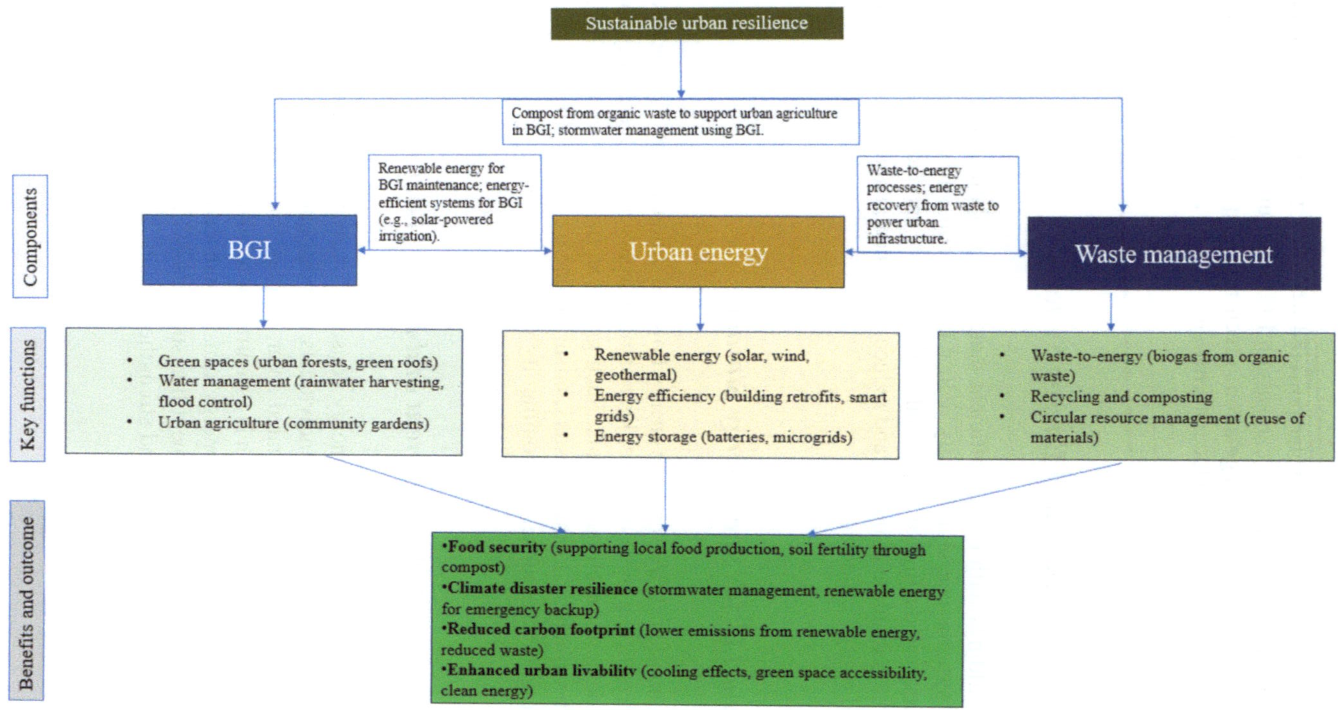

Fig. 11.6 BGI, urban energy and waste management nexus to ensure food security and disaster resilience

11.8.3 Energy-Efficient BGI Maintenance

Energy-efficient systems, such as solar-powered irrigation for urban agriculture or rainwater-harvesting infrastructure, can sustain BGI without a high energy burden. This is crucial for maintaining vegetation, green roofs, and community gardens, which provide cooling effects, reduce stormwater runoff, and contribute to urban food security.

11.8.4 Building-Integrated Energy Systems

Retrofitting buildings with energy-efficient materials, insulation, and smart energy management systems can significantly reduce energy consumption in heating and cooling, particularly during extreme weather. Green roofs and facades, which can be part of BGI, help insulate buildings and reduce urban heat, enhancing both climate resilience and energy efficiency.

11.8.5 Smart Grids and Energy Storage

Establishing smart grids that manage energy flow and utilize energy storage systems can improve resilience by allowing cities to store surplus renewable energy for use during peak demand or climate events. Such grids can prioritize energy allocation to critical infrastructure during disasters, ensuring energy availability for essential services, emergency response, and community shelters.

11.8.6 Community Engagement and Energy Literacy

Educating urban communities on energy conservation, renewable energy options, and waste reduction can empower individuals to make energy-conscious choices. Community energy projects, like shared solar, increase local access to renewable energy and foster engagement, which is essential for resilient urban systems.

Through these methods, urban energy can be managed effectively to create a resilient, low-carbon urban environment that supports BGI, enhances food security through sustainable resource use, and strengthens cities' abilities to withstand and recover from climate disasters.

11.9 Conclusion

BGI identifies the opportunity to manage water, enhance climate resilience reducing environmental carbon footprint through integrated infrastructure and urban design approaches. The integration of urban energy and waste management systems is a powerful strategy for creating sustainable, resilient, and climate-smart cities. By adopting a holistic approach that incorporates BGI, enhances food security, and fortifies urban resilience, cities can navigate the challenges of the twenty-first century while fostering a healthier and more sustainable urban environment.

References

Alves B (2023) Global municipal waste generation 2021, by select country. Retrieved from: https://www.statista.com/statistics/916749/global-generation-of-municipal-solid-waste-by-country/ (Accessed on 12 June 2024)

Argüeso D, Evans JP, Pitman AJ, Di Luca A (2015) Effects of city expansion on heat stress under climate change conditions. PLoS ONE 10(2):e0117066

Armanda DT, Guinée JB, Tukker A (2019) The second green revolution: innovative urban agriculture's contribution to food security and sustainability–a review. Glob Food Sec 22:13–24

Cohen N, Reynolds K (2014) Urban agriculture policy making in New York's "New Political Spaces" strategizing for a participatory and representative system. J Plan Educ Res 34(2):221–234

Debele SE, Kumar P, Sahani J, Marti-Cardona B, Mickovski SB, Leo LS, Di Sabatino S (2019) Nature-based solutions for hydro-meteorological hazards: Revised concepts, classification schemes and databases. Environ Res 179:108799

EEA (2017) Climate change impacts and vulnerability in europe—an indicator-based report. EEA Report European Environment Agency. Available online at: https://www.eea.europa.eu/publications/climate-change-impacts-and-vulnerability-2016 (accessed on 12 June 2024).

European Commission (2015) Closing the loop—an EU action plan for the circular economy (communication from the commission to the european parliament, the council, the european economic and social committee and the committee of the regions no. COM(2015) 614/2). European Commission, Brussels

Gichamo T, Gökçekuş H (2019) Interrelation between climate change and solid waste. J Environ Pollut Control 2(1):104

Gu D (2019) Exposure and vulnerability tonatural disasters for world'scities. DESA 4, 1–43. Available online at: https://www.un.org/en/development/desa/population/publications/pdf/technical/TP2019-4.pdf (accessed on 12 June 2024)

Gupta J, Chakraborty M (2021) Energy efficiency in buildings. In Sustain Fuel Technol Handb (pp 457–480). Academic Press

Habibi A, Kahe N (2024) Evaluating the role of green infrastructure in microclimate and building energy efficiency. Buildings 14(3):825

Hassan MN (2000) Policies to improve solid waste management in developing countries: some insights in Southeast Asian coutries. In proceedings of the 2nd international conference on solid waste management, Taipei, Taiwan. Environ Prot Adm, Repub China, 2000 (pp 191–207)

IEA P (2022) World energy outlook 2022. Paris, France: International Energy Agency (IEA)

IPCC (2014) Climate change 2014: impacts, adaptation, and vulnerability. Part A: global and sectoral aspects. Contribution of working group II to the fifth assessment report of the intergovernmental panel on climate change [Field CB, Barros VR, Dokken DJ, Mach KJ, Mastrandrea MD, Bilir TE, Chatterjee M, Ebi KL, Estrada YO, Genova RC, Girma B, Kissel ES, Levy AN, MacCracken

S, Mastrandrea PR, White LL (eds)]. Cambridge University Press, Cambridge, United Kingdom and New York, NY, USA

IPCC (2018) Managing the risks of extreme events and disasters to advance climate change adaptation: special report of the intergovernmental panel on climate change. Cambridge: Cambridge University Press. Available online at: https://www.ipcc.ch/report/managing-the-risks-of-extreme-events-and-disasters-to-advance-climate-change-adaptation/ (accessed on 12 June 2024)

Jaccard M (2006) Sustainable fossil fuels: the unusual suspect in the quest for clean and enduring energy. Cambridge University Press

Kaur R, Gupta K (2022) Blue-Green Infrastructure (BGI) network in urban areas for sustainable storm water management: a geospatial approach. City Environ Interact 16:100087

Kaza S, Yao L, Bhada-Tata P, van Woerden F (2018) What a waste 2.0: a global snapshot of solid waste management to 2050. World Bank Publications

Koop SHA, van Leeuwen CJ (2017) The challenges of water, waste and climate change in cities. Environ Dev Sustain 19:385–418. https://doi.org/10.1007/s10668-016-9760-4

Krasny ME, Tidball KG (2017) Community gardens as contexts for science, stewardship, and civic action learning. Urban horticulture: ecology, landscape, and agriculture, 267

Kumar P, Saroj DP (2014) Water–energy–pollution nexus for growing cities. Urban Climate 10:846–853

Manfredi S, Tonini D, Christensen TH, Scharff H (2009) Landfilling of waste: accounting of greenhouse gases and global warming contributions. Waste Manage Res 27(8):825–836

Maxwell D, Levin C, Armar-Klemesu M, Ruel M, Morris S, Ahiadeke C (2000) Urban livelihoods and food and nutrition security in Greater Accra, Ghana. IFRI Research Report 112. Washington DC: International Food Policy Research Institute

Minghua Z, Xiumin F, Rovetta A, Qichang H, Vicentini F, Bingkai L, Yi L (2009) Municipal solid waste management in Pudong new area. China. Waste Manag 29(3):1227–1233

Opoku A, Duff A, Yahia MW, Ekung S (2024) Utilisation of green urban space for food sufficiency and the realisation of the sustainable development goals–UK stakeholders perspective. Geogr Sustain 5(1):13–18

Re Swiss (2020) Re Databases 9980–2018. Available online at: https://swissre.com (accessed on 12 June 2024)

Rosol M (2010) Public participation in post-Fordist urban green space governance: the case of community gardens in Berlin. Int J Urban Reg Res 34(3):548–563

Schukoske JE (2000) Community development through gardening: state and local policies transforming urban open space. NYUJ Legis. & Pub Pol'y 3:351

Seto KC, Güneralp B, Hutyra LR (2012) Global forecasts of urban expansion to 2030 and direct impacts on biodiversity and carbon pools. Proc Natl Acad Sci 109(40):16083–16088

Siehr SA, Sun M, Aranda Nucamendi JL (2022) Blue-green infrastructure for climate resilience and urban multifunctionality in Chinese cities. Wiley Interdiscip Rev: Energy Environ 11(5):e447

Singh RP, Agrawal M (2010) Variations in heavy metal accumulation, growth and yield of rice plants grown at different sewage sludge amendment rates. Ecotoxicol Environ Saf 73(4):632–641

Singh RP, Tyagi VV, Allen T, Ibrahim MH, Kothari R (2011) An overview for exploring the possibilities of energy generation from municipal solid waste (MSW) in Indian scenario. Renew Sustain Energy Rev 15(9):4797–4808

Sohag MU, Podder AK (2020) Smart garbage management system for a sustainable urban life: an IoT based application. Internet of Things 11:100255

Thibert J (2012) Making local planning work for urban agriculture in the North American context: a view from the ground. J Plan Educ Res 32(3):349–357

Thierfelder H, Kabisch N (2016) Viewpoint berlin: strategic urban development in berlin-challenges for future urban green space development. Environ Sci Policy 62:120–122

Un-Habitat (2010) Solid waste management in the world's cities: water and sanitation in the world's cities 2010. Routledge

UNEP (2015) The mounting problem: world's cities produce up to 10 Billion tonnes of waste each year, UN study estimates. Retrieved from: https://unep.org/news-and-stories/press-release/mounting-problem-worlds-cities-produce-10-billion-tonnes-waste-each (accessed on 12 June 2024)
Van Leeuwen E, Nijkamp P, de Noronha Vaz T (2010) The multifunctional use of urban greenspace. Int J Agric Sustain 8(1–2):20–25
World Bank (2012) What a waste: a global review of solid waste management. Urban development series knowledge papers
Xinyue L, Mingxing C (2019) Research progress on the impact of urbanization on climate change. Adv Earth Sci 34(9):984

Chapter 12
Smart Agriculture in Smart Cities Across the Globe

Kenichiro Onitsuka, Gulsan Ara Parvin, and Reazul Ahsan

Abstract The evolving realms of smart agriculture and smart cities are gaining global prominence as strategic solutions to the challenges posed by rapid urbanization and the need for food security. Historically, agriculture has been an intrinsic component of urban environments, but it is only in recent times that urban agriculture has been acknowledged in global policy frameworks. In an increasingly urbanized world, the integration of agriculture into cityscapes becomes imperative. However, the scarcity of essential resources such as land, water, and energy in urban settings presents significant challenges. Moreover, urban agriculture could potentially exacerbate pollution and strain the already limited urban resources and infrastructure. Therefore, there is a pressing need for intelligent and sustainable agricultural practices within urban areas, particularly in smart cities where resource management and operations are optimized for efficiency. This chapter provides a comprehensive review of smart agricultural practices in various cities, exploring how smart cities are integrating smart agriculture to bolster their blue-green infrastructure.

Keywords Smart agriculture · Smart cities · Issues · Challenges · Technologies

K. Onitsuka
Graduate School of Global Environmental Studies (GSGES), Kyoto University, Kyoto, Japan

G. A. Parvin (✉)
College of Policy Science, Ritsumeikan University, Osaka-Ibaraki Campus, Osaka, Japan
e-mail: niruurp@yahoo.com

R. Ahsan
Department of City and Metropolitan Planning, University of Utah, Salt Lake City, USA

M. G. Kibria et al. (eds.), *Urban Blue-Green Infrastructure Approach for Food Security and Climate Disaster Resilience*, Disaster Risk Reduction,
https://doi.org/10.1007/978-981-96-8746-6_12

12.1 Agriculture in Cities

Urban agriculture (UA) refers to food and fuel production within cities or peri-urban areas, intended for market sale or home consumption, often marketed directly by farmers or their associates (Smit et al. 1996). Though the term gained prominence in the past three decades, the practice of cultivating food in urban areas dates back to the earliest cities. As ancient settlements developed around agricultural lands, inhabitants transferred their farming knowledge to sustain themselves and their communities (Steel 2013). In the twenty-first century, urban agriculture is increasingly linked to modern urban planning, sustainable development, and technological advancements, experiencing its most significant growth in history (Dobele and Zvirbule 2020). However, urban-agricultural interactions have a much deeper history. In Asia and Europe, city farming has long been practiced, primarily due to underdeveloped transportation networks that necessitated local food production. During the Industrial Revolution in Europe, urban allotment gardens helped impoverished city dwellers supplement their food supply in the absence of social security systems. Similarly, urban agriculture became essential during World War I and II, as supply chain disruptions forced cities to enhance local food production (Engel et al. 2019). Over time, agriculture has become an integral part of urban landscapes, from megacities to small towns, in both developing and developed countries. Various forms of urban agriculture have emerged, including community gardens, institutional gardens, vertical farming, rooftop gardens, veranda gardens, plant factories with artificial lighting, zero-acreage farming, agro-parks, and agro-tourism. These diverse approaches highlight the evolving role of urban agriculture in fostering food security and sustainability across different urban settings (Fig. 12.1).

a. Urban Agriculture in Kyoto City, Japan

b. Urban Agriculture in Khulna city, Bangladesh

Fig. 12.1 Urban agriculture in developed and developing countries. *Source* Authors

According to The UN Food and Agriculture Organization (UNFAO) over one-tenth of the global population, approximately 800 million people, practicing urban agriculture worldwide (Guttupalli 2024). In the age of rapid and uncontrolled urbanization, urban agriculture (UA) has garnered attention for its potential contribution to the food supply (of specific, mostly horticultural food products), to income generation, and to make the cities green. Generally, residents of cities had little understanding of where or how food is produced and distributed. However, in recent years with the increase of importance and attention, urban farming is allowing citizens to be closer to the production of food, and even active participants. Today, urban agriculture is revitalizing the connection between city dwellers and agricultural products (Guttupalli 2024). During Covid-19 pandemic the necessity of urban agriculture is realized once again, since due to the agricultural production in cities the proximity of producers and consumers can minimize transport costs and the possible losses incurred. Agriculture in cities offers numerous benefits including poverty alleviation of the urban poor, enhancing urban resilience by minimizing climate change impacts and environmental challenges (Zeeuw et al. 2011). However, there are different challenges associated with urban agriculture. Potential health risk by using contaminated land and water for urban agriculture, environmental pollution by chemical fertilizer, pesticide and insecticide used in urban agriculture and use of scarce land water and energy of urban areas for agriculture are the main concerns of urban agriculture. At the same time, in this urbanized world, food security and food safety is a key concerning issue. Therefore, all the challenges associated with the urban agriculture are required to be minimized by smart technologies, knowledge and innovation. In this era of diversified challenges and information and communication, people, space, environment and technologies are needed to be smarter. Therefore, cities and agriculture are transforming into smarter aspects. Scholars argue that smart cities should be smarter not only from a technological point of view, but also for its economic, social and environmental sustainability. In order to enhance socio-economic progress, smart cities need to mobilize all knowledge and information and communication technologies (ICT) into innovation hubs (Dos Santos 2016; Halegoua 2020). Since urban agriculture is one of the principal approaches to ensure food security of rapidly growing urban population, smart cities innovations can lead to flourish smart urban agriculture.

12.2 Smart Agriculture: Concept and Need

12.2.1 Definition and Overview of Smart Agriculture

Smart agriculture, also known as precision agriculture, smart farming or digital farming, refers to the application of modern information and communication technologies to optimize agricultural production systems. Smart agriculture is also synchronized with Agriculture 4.0 concept. It is a farming management concept that leverages data-driven insights and automation to increase the quantity and quality of agricultural output while improving sustainability and resource efficiency (IBM 2023). In Japan, smart agriculture is conceptualized as solving problems at production sites with advanced technologies, realizing Society 5.0, a vision of Japan Government for future societies where technologies have been advanced (Cabinet Office of Japan n.d.), in the field of agriculture (MAFF 2023). At its core, smart agriculture aims to observe, measure, and respond to inter- and intra-field variability in crops. This approach allows farmers to manage their fields more precisely, applying inputs (such as water, fertilizers, and pesticides) exactly where and when they are needed, rather than treating entire fields uniformly (MIT D-Lab n.d.).

12.2.2 Key Technologies of Smart Agriculture

Smart agriculture leverages a wide array of cutting-edge technologies to optimize farming practices, increase productivity, and enhance sustainability. The following key technologies are at the forefront of this agricultural revolution:

12.2.2.1 Internet of Things (IoT) Sensors and Devices

IoT sensors and devices are fundamental to smart agriculture, providing real-time data on various parameters such as soil moisture, temperature, humidity, and crop health. These sensors enable continuous monitoring of field conditions, allowing for timely interventions and data-driven decision-making. For example, soil moisture sensors help optimize irrigation schedules, reducing water waste and ensuring crops receive the right amount of water.

12.2.2.2 Artificial Intelligence (AI) and Machine Learning

AI and machine learning algorithms analyze vast amounts of agricultural data to provide predictive insights and prescriptive recommendations. These technologies can anticipate crop diseases, optimize planting schedules, and recommend optimal input levels, helping farmers make informed decisions to maximize productivity and profitability. AI-powered systems can also automate tasks such as irrigation and pest control, further enhancing efficiency.

12.2.2.3 Big Data Analytics

Big data analytics involves processing and interpreting large volumes of agricultural data to extract actionable insights. This enables farmers to identify patterns, trends, and correlations that can inform management strategies and improve overall farm performance. For instance, big data can help predict crop yields, optimize resource use, and enhance supply chain management.

12.2.2.4 Robotics and Automation

Robotics and automation technologies, including autonomous tractors, drones, and robotic weeders, perform tasks with minimal human intervention. These technologies streamline farm operations, reduce manual labor requirements, and enhance operational efficiency across various stages of the farming process. For example, autonomous tractors can plow, plant, and harvest crops with high precision, while robotic weeders can identify and eliminate weeds without harming crops.

12.2.2.5 Drones and Satellite Imaging

Drones and satellites equipped with multispectral cameras and other sensors capture high-resolution imagery of fields. This data provides valuable insights into crop health, growth patterns, and pest infestations, enabling early detection of issues and proactive management. Drones can also be used for tasks such as spraying pesticides and monitoring livestock.

12.2.2.6 GPS and Geospatial Technologies

GPS and geospatial technologies enable precise mapping and navigation for targeted field operations. These technologies are fundamental to precision agriculture practices, allowing for variable rate application of inputs and accurate planting patterns. GPS-guided equipment ensures that seeds, fertilizers, and pesticides are applied exactly where needed, reducing waste and improving efficiency.

12.2.2.7 Cloud Computing and Mobile Applications

Cloud computing platforms enable data storage, processing, and sharing across devices and stakeholders. These technologies facilitate real-time access to farm data and analytics, supporting on-the-go decision-making. Mobile applications allow farmers to monitor field conditions, receive alerts, and manage farm operations from anywhere, enhancing flexibility and responsiveness.

The integration of these technologies creates a more connected and intelligent farming ecosystem. For example, IoT sensors in the field might collect data on soil moisture, which is then analyzed by AI algorithms to determine optimal irrigation schedules. This information can be sent to automated irrigation systems or to farmers' mobile devices, allowing for precise and timely water application (Rajak et al. 2023; Morchid et al. 2023). As these technologies continue to evolve and become more accessible, they have the potential to transform agriculture on a global scale, addressing challenges such as food security, resource efficiency, and environmental sustainability. However, successful implementation will require ongoing innovation, collaboration between technology providers and agricultural stakeholders, and efforts to overcome barriers such as high initial costs and the need for specialized training (MIT D-Lab n.d.).

12.2.3 Issues Driving the Need for Smart Agriculture

The adoption of smart agriculture is driven by several pressing global challenges that impact food production and sustainability. These challenges highlight the necessity for innovative agricultural practices to ensure food security and environmental resilience.

12.2.3.1 Growing World Population and Food Demand

The global population is projected to reach nearly 10 billion by 2050, necessitating a significant increase in food production. Estimates suggest that food production must increase by 60% to 110% to meet the demands of this growing population (Van Dijk et al. 2021; FAO 2009). This surge in food demand is driven not only by population growth but also by rising incomes and changing dietary preferences, particularly in developing countries (United Nations 2021). By leveraging advanced technologies and data-driven approaches, smart agriculture can address these challenges, ensuring a sustainable and resilient food system for the future.

12.2.3.2 Climate Change Impacts on Agriculture

Climate change poses a substantial threat to agricultural productivity. Changes in temperature, precipitation patterns, and the frequency of extreme weather events can adversely affect crop yields and livestock health (EPA n.d.). For instance, increased temperatures can lead to heat stress in plants and animals, while altered rainfall patterns can cause droughts or floods, impacting soil health and water availability. Additionally, climate change can exacerbate pest and disease pressures, further challenging food production systems (EPA n.d.).

12.2.3.3 Resource Constraints (Land, Water, etc.)

Agriculture is a major consumer of natural resources, accounting for approximately 70% of global freshwater withdrawals and occupying about 50% of the Earth's habitable land. However, these resources are becoming increasingly scarce due to overuse, pollution, and competition from other sectors. Soil degradation, water scarcity, and loss of arable land are critical issues that threaten the sustainability of agricultural production.

12.2.3.4 Environmental Concerns and Sustainability

Conventional agricultural practices contribute significantly to environmental degradation, including deforestation, soil erosion, water pollution, and greenhouse gas emissions (WWF n.d.). Sustainable agriculture aims to mitigate these impacts by promoting practices that conserve resources, enhance biodiversity, and reduce emissions. However, achieving sustainability in agriculture requires a shift towards more knowledge-intensive and environmentally friendly farming systems (OECD 2023).

12.2.3.5 Labor Shortages in Agriculture

Many regions face labor shortages in agriculture due to urbanization, aging rural populations, and the physically demanding nature of farm work (FAO 2009). These labor constraints can limit the capacity to manage farms effectively and reduce overall productivity. Smart agriculture technologies, such as automation and robotics, offer potential solutions by reducing the reliance on manual labor and increasing operational efficiency.

12.2.3.6 Food Security and Safety Issues

Despite advances in agricultural productivity, food insecurity remains a significant global challenge. Nearly 690 million people were undernourished in 2019, and this number has been rising due to economic slowdowns, conflicts, and climate-related disasters. Ensuring food security involves not only increasing food production but also improving food distribution, access, and safety. Smart agriculture can enhance food traceability, reduce post-harvest losses, and ensure the production of safe and nutritious food (Union of Concerned Scientists 2019).

12.2.4 Key Challenges and Barriers to Adopt Smart Agriculture

While smart agriculture offers numerous benefits, several challenges and barriers hinder its widespread adoption, particularly among smallholder farmers in developing countries. The main challenges include: High Upfront Costs; Lack of Rural Connectivity; Need for Farmer Training and Education; Data Privacy and Security

Concerns; Limited knowledge and awareness of smart agriculture practices among farmers; Market challenges in finding buyers for climate-smart agricultural products; Policy and regulatory barriers that may impede the adoption of certain technologies; Social and cultural barriers, especially in traditional farming communities; Difficulties in demonstrating the long-term benefits of smart agriculture to farmers.

Addressing these challenges requires a multi-faceted approach involving governments, private sector actors, and international organizations. Strategies may include developing tailored financial products for smart agriculture investments, improving rural infrastructure, enhancing agricultural extension services, and creating supportive policy frameworks to encourage adoption.

12.3 Smart Cities and Smart Urbanism

According to the estimation of United Nations in 2018, more than half of the population is living in urban areas and it is projected to increase into 68% by 2050 (UN 2018). Due to rapid increase of population in cities, cities started to face diversified challenges and scarcity of resources, services and facilities for growing population. Therefore, at the beginning of 1990s, attempts were undertaken to form a unified approach on the way of establishing and advanced urban environment reflected in the so-called "Smart City" concept (Vishnivetskaya and Alexandrova 2019).

At present, in an increasingly digital and technology-driven world, cities must adopt new techniques, creative technologies, and smart infrastructure to have smart management with smart technologies and approaches and foster environments conducive to community and economic progress. Smart technology refers to a set of advanced technologies used in cities, such as wireless sensor networks, IOT, big data, cloud computing, artificial intelligence, applications, systems, mobile devices, mobile applications, wireless smartphones. The smart technology refers to each of the smart city elements, including people, economy, environment, infrastructure, governance, safety and security (Winkowska et al. 2019). "Smart city" is normally understood as safe and efficient urban environment based on up-to-date technologies, stimulating consistent economic growth and improving the population living standards (Hall et al. 2000). Smart cities are characterized by their use of data-driven approaches, integration of infrastructure and resources, and emphasis on environmental protection, resource efficiency, social equity, and economic growth (Siam Commercial Bank 2024). By employing data-driven methodologies, smart cities can make informed decisions that optimize their systems, improve efficiency, respond effectively to residents' needs, reduce costs, and enhance overall sustainability (McDaniel 2022).

Smart cities can ensure better urban living in several ways:

1. Improved Quality of Life: By optimizing city services and infrastructure, smart cities can reduce commute times, enhance public safety, and provide better access to healthcare and education.

2. Environmental Sustainability: Smart cities often incorporate green technologies and efficient resource management, reducing carbon emissions and improving environmental quality.
3. Economic Growth: By fostering innovation and attracting businesses, smart cities can create new job opportunities and stimulate economic development.
4. Enhanced Public Services: Digital platforms and data analytics enable cities to deliver more efficient and responsive public services.
5. Increased Citizen Engagement: Smart city technologies can facilitate greater participation in local decision-making and community initiatives (Shea 2020).

Despite having numerous opportunities and potentialities, critical concern in smart urbanism is the technocratic governance of cities through data and code, which can reduce complex urban characteristics and processes to simplified visualizations (Batty 2017). This approach, while efficient, must be balanced with an understanding of the qualitative aspects of urban life that are not easily quantifiable.

12.4 Smart Agriculture in Smart Cities

Modern urban management using technical tools that offer state-of-the-art technologies, considering the applicable ecological standards while saving resources and achieving the expected results in the principal goal of the smart city concept (Winkowska et al. 2019). Therefore, in smart cities agricultural practices need to adopt such approaches that can ensure sustainable management of ecology and use of resources.

Smart agriculture can be an integral part of smart city planning, contributing to urban environments' overall sustainability and efficiency (McDaniel 2022). Smart farming technologies, such as IoT devices, data analytics, and precision agriculture, can seamlessly integrate into smart city initiatives. These technologies enable urban farmers to optimize resource use, reduce waste, and enhance productivity, aligning with the goals of sustainable smart cities (Adamson 2023).

For example, two soilless cultivation methods, hydroponics, and aeroponics, can thrive in urban settings where space is limited, as they do not require fertile soil. These innovative farming techniques can significantly improve urban food production's efficiency and sustainability when coupled with automated nutrient delivery systems and environmental control technologies. Additionally, smart city initiatives can promote the integration of vertical farming, rooftop gardens, and community-based urban agriculture, further expanding the possibilities for local food production and enhancing food security within urban areas. By incorporating smart agriculture into smart city planning, urban areas can achieve greater self-sufficiency, reduce the environmental impact of food production and distribution, and provide their residents with access to fresh, nutritious produce (Adamson 2023).

12.5 Smart Agriculture in Smart Cities: Approaches Around the World

As urban populations continue to grow, integrating food production into the fabric of smart cities will be essential for creating truly sustainable urban systems. Smart urban agriculture offers a promising pathway for feeding cities while regenerating urban ecosystems and strengthening local communities. The integration of smart agriculture into smart cities represents a powerful approach for addressing interconnected challenges of urbanization, food security, and sustainability. By leveraging advanced technologies and data-driven approaches, cities can optimize food production within urban environments, enhance resource efficiency, and create more resilient local food systems. Key benefits of smart urban agriculture include improved access to fresh food, reduced transportation emissions, productive use of urban spaces, enhanced community engagement, and circular economy opportunities. However, realizing these benefits at scale will require overcoming challenges such as limited space, resource constraints, and regulatory barriers. Different innovative urban farming practices, technologies and government initiatives are emerging all over the world. Some examples of smart urban agriculture in different cities of Asia, Europe and America would provide future directions to minimize the challenges and thus to enhance the benefits from smart agriculture.

12.5.1 Smart Agriculture in Asian Cities

In Asia, among different countries, Singapore, Japan, Korea, China and India are trying to manage cities using different smart technologies and making cities smarter. Due to rapid urbanization in Asia and high population density, it is hard to incorporate agriculture in cities in Asia. However, despite having numerous limitations and challenges, Asian cities are also trying to incorporate smart agriculture in their smart cities' initiatives. Smart agriculture initiatives of Singapore, Tokyo and Bangalore are mentioned worthy.

Smart agricultural initiatives in Singapore

Singapore, a city-state with over 90% of its food imported, has embraced smart agriculture to boost local production and reduce external dependencies. Supported by government funding, research grants, and policies like the "30 by 30" goal to produce 30% of its nutritional needs by 2030, Singapore is advancing towards agricultural self-sufficiency. In this regard, its key strategy are high-tech farming techniques within urban settings, using robotics, IoT sensors, and data analytics for precision farming. Notable initiatives include:

- Sky Greens: The world's first low-carbon, hydraulic-driven vertical farm, using rotating towers and gravity-fed irrigation to optimize sunlight and water use, thereby maximizing efficiency and space (Fig. 12.2a).

a. A Sky Green farm tower in Singapore.
Photo: Kalinga Seneviratne/IPS.

b. Indoor rice paddy and bittergourd in Tokyo City

Fig. 12.2 Smart agricultural practices in Singapore and Tokyo. *Source* **a** https://ourworld.unu.edu/en/farming-in-the-sky-in-singapore, accessed on July 25, 2024; **b** https://hackerfarm.jp/2017/05/pasona-indoor-experimental-growing-exhibit-in-tokyo/an, accessed on July 25, 2025

- Comcrop: A 600-square-meter rooftop farm employing aquaponics to cycle nutrients between fish and plants, saving water and space atop a city mall.
- Floating Farms: Innovative farms on city reservoirs using hydroponics to utilize water surfaces and reduce evaporation.

Despite high operational costs and a demand for skilled labor, Singapore plans to expand these technologies, enhancing efficiency and maintaining its lead in urban agricultural innovation.

Smart agricultural initiatives in Tokyo

Tokyo city is the capital of Japan, where urban density meets technological advancement, has innovatively integrated agriculture into its urban landscape. Given the city's limited space and high population density, Tokyo has adopted vertical farming and rooftop gardens to integrate greenery into urban life. The city utilizes advanced technologies in its urban farms, including automated systems for climate, lighting, and watering, alongside IoT sensors that monitor moisture, nutrients, and growth rates. These technologies enable precise farming, ensuring high yields and quality produce.

Innovative Urban Farming Models in Tokyo:

- Pasona O2: Situated in the Otemachi district, Pasona O2 transforms an underground office space into a vibrant urban farm using hydroponic and soil-based farming techniques to grow crops like rice, tomatoes, lettuce, and herbs. This project not only supplies fresh produce to the local community but also serves as an educational and therapeutic space for city dwellers and employees (Fig. 12.2b).

- Odaiba Farm: Leveraging unused city spaces, particularly on Odaiba's artificial island, this initiative employs innovative hydroponic systems on commercial building rooftops, demonstrating how urban farming can be seamlessly integrated into existing structures.
- The Tokyo Metropolitan Government actively supports urban farming through subsidies and legislation, with projects like the Edible Green Curtain encouraging the growth of climbing plants and vegetables on walls and windows to combat urban heat islands. Furthermore, Tokyo advocates for transforming unused commercial and public rooftops into productive green spaces.

Despite the promise of Tokyo's urban farming initiatives, challenges remain, including high initial setup costs, ongoing technological maintenance, and cultural integration. Future plans focus on expanding these initiatives, improving integration with renewable energy sources, and enhancing public engagement through educational programs.

Smart agricultural initiatives in Bangalore

Bangalore, often referred to as the Silicon Valley of India, merges its technological prowess with deep agricultural traditions, despite challenges such as rapid urbanization, shrinking arable land, and water scarcity. This makes the adoption of sustainable agricultural practices essential for its growth.

Innovative Urban Farming Models in Bangalore:

- Hydroponics and Aeroponics: The city has witnessed a surge in hydroponics and aeroponics, which significantly reduce water usage compared to traditional farming and eliminate the need for soil. Set up on rooftops and other underutilized urban spaces, these farms supply fresh produce directly to urban consumers. Startups like "UrbanKisaan" are pioneering vertical farming techniques to maximize the use of space and resources.
- IoT Integration: Numerous Bangalore startups have integrated IoT into agriculture, enhancing precision farming. Sensors monitor soil moisture, nutrient levels, and weather conditions, facilitating automated systems that optimize irrigation, nutrient delivery, and overall plant health.

Government Initiatives:

- Krishi Marata Vahini: This online marketplace, initiated by the Karnataka state government, connects farmers directly with consumers and retailers, sidestepping the conventional supply chain, thereby reducing costs and boosting farmer profits.
- Subsidies and Training: The government offers subsidies for residents to establish rooftop gardens and conducts training sessions on sustainable farming techniques.

Community Engagement and Education:

- Workshops and Seminars: Frequent events are organized to raise awareness about the benefits of urban farming and sustainable practices.

- School Programs: Programs like "Seeds of Change" are integrated into school curriculums to spark interest in agriculture among young students, promoting it as a viable career path.

Challenges and Future Outlook:

Despite progress, Bangalore's urban agriculture still faces hurdles such as high initial costs and operations. Future directions include expanding technology-driven farms, implementing more robust government policies to support urban agriculture, and fostering greater collaboration between the technology and agricultural sectors.

12.5.2 Smart Agriculture in European Cities

In smart city development initiatives Europe is focusing on the aspects of transportation, energy, public services and infrastructure. Among these initiatives, urban blue-green infrastructure is one of the priorities, where smart urban agriculture is getting popular in different cities in Europe.

Smart Agriculture in Amsterdam City

This city renowned for its commitment to urban sustainability and innovative planning, has strategically incorporated city farming into its ecological strategy to enhance food security, minimize environmental impact, and strengthen community bonds. In the farming practices Amsterdam city has incorporated different smart technologies. IoT sensors and automated systems fare used for greenhouses optimize growth and conserve energy by regulating climate, lighting, and nutrients. Further, different urban farms are equipped with solar panels to sustainably power their operations. Table 12.1 denotes different urban farming practices, government and community initiatives and challenges in smart urban agriculture in Amsterdam.

12.5.3 Smart Agriculture in Berlin, Germany

Berlin, celebrated for its rich history and cultural diversity, is also at the forefront of urban agriculture, enhancing food security, community engagement, and sustainability.

Innovative Urban Farming Practices:

- Community Gardens: Notable are the "Prinzessinnengärten" and similar projects transforming vacant plots into lush community-managed spaces that facilitate social interaction and sustainable urban living (Fig. 12.3a).

Table 12.1 Smart urban farming practices of Amsterdam city

Innovative urban farming practices	Government and community initiatives	Challenges and future outlook
Rooftop Farms: Sprawling across the city, Amsterdam hosts some of Europe's largest rooftop farms, including a notable 4,000-square-meter farm atop a former industrial building. These farms employ lightweight substrates that enhance water retention without the soil weight ***Greenhouses and Vertical Farming***: Amsterdam maximizes its vertical spaces with advanced greenhouses using hydroponics and aquaponics systems. These systems efficiently cycle nutrient-rich water, reducing water use and soil dependency. A prime example is "De Schilde", which vertically integrates the production of vegetables and fish	***Municipal Support***: The city government fosters urban agriculture through grants and subsidies, driving local food production and fostering innovation ***Community Programs***: Urban farms double as community hubs where locals engage in sustainable farming workshops and directly purchase fresh produce ***Educational Impact and Social Cohesion*** : Urban farms in Amsterdam serve as dynamic educational platforms offering internships and volunteer opportunities, connecting residents and students with sustainable food sources and promoting healthy lifestyles	Despite their successes, these urban farms confront issues like scalability and high initial costs, necessitating ongoing technological advancements. Future plans are geared towards expanding these initiatives, diversifying crop production, and further integrating urban agriculture into Amsterdam's circular economy goals

a. Prinzessinnengärten in Kreuzberg, Berlin organic urban farm

b. Volunteers work in the fields at the Oakland Avenue Urban Farm in Detroit, Michigan on July 17, 2023. Daniel Mears

Fig. 12.3 Smart agriculture in Berlin and Detroit cities. *Source* **a** Posivakova et al. (2019), **b** Aguilar (2023)

- Vertical and Rooftop Farming: Berlin embraces vertical and rooftop gardens using hydroponics and aeroponics to efficiently utilize small spaces, reducing the urban heat island effect and the city's carbon footprint.

Technology Integration:

- Water Recycling Systems: Advanced systems in urban farms support rainwater harvesting and gray water reuse, aligning with Berlin's sustainable resource management goals.
- Solar Power: Many urban farms integrate solar panels to reduce dependence on non-renewable energy and enhance sustainability.

Government and Community Initiatives:

- Zoning and Land Use Policies: The city's zoning laws favor urban agriculture, enabling the use of vacant land for farming.
- Subsidies and Grants: Financial incentives support community groups and entrepreneurs in initiating urban farming projects.

Educational and Social Impact:

- Educational Programs: Urban farms partner with local educational institutions to provide practical learning experiences in sustainable agriculture.
- Community Cohesion: These spaces foster community bonds by uniting diverse groups in collaborative food production and conservation efforts.

Challenges and Future Outlook:

Berlin's urban agriculture encounters issues like climatic variability and space constraints, necessitating ongoing innovation. Future plans include expanding smart agriculture technologies and enhancing community engagement initiatives.

12.5.4 Smart Agriculture in North American Cities

North America first attempted smart city approach with assistance from IBM, using the IoT system to connect various public resources that can intelligently responsible to the residents. Along with the different smart technologies in city development and management, smart agricultural practices and initiatives are gradually embraced by urban farmers and city developers in different cities of Canada and USA. Urban farming initiatives and innovations of Toronto and Detroit city are briefly highlighted hereunder.

Toronto's Urban Agriculture Strategy

Toronto, Canada's largest city, has robustly integrated urban agriculture into its strategy to enhance food security, reduce environmental impact, and strengthen community bonds. The city's approach to urban farming is comprehensive and community-focused.

Key Urban Farming Initiatives:

- Community Gardens: Toronto features over 100 community gardens in public parks, residential areas, and institutional lands, facilitated by the city's Community Gardens Program. These spaces allow residents to cultivate their own food, enhancing social cohesion and nature connectivity.
- Rooftop Farming: Notably, projects like the Ryerson University Urban Farm, a quarter-acre rooftop space, produce over 10,000 pounds of fresh produce annually using hydroponic systems. This contributes to local food supplies and serves educational purposes.

Technology and Innovation:

- Greenhouse Technology: Urban farms in Toronto employ advanced greenhouses with automated climate controls and energy-efficient designs to boost productivity and extend growing seasons.
- Water Conservation Techniques: Techniques like rainwater harvesting and drip irrigation are prevalent, supporting Toronto's sustainability goals.

Government and Policy Support:

- Urban Agriculture Strategy: The city is developing a strategy to further blend food production into Toronto's urban fabric, aiming to increase public land access for farming, support new farmers, and improve local food distribution.
- Grants and Financial Assistance: Toronto offers grants to foster community gardening and farming, reducing entry barriers and promoting inclusive participation.

Community Engagement and Educational Programs:

- Youth Programs: The Toronto Youth Food Policy Council, among others, engages young people in urban agriculture, nurturing future farmers.
- Educational Workshops and Training: The city offers extensive training in urban farming techniques and sustainable food production to all residents.

Challenges and Future Directions:

Despite its achievements, urban agriculture in Toronto encounters challenges such as limited space in a dense urban setting, climate restrictions, and the need for continuous funding and policy development. Future plans focus on leveraging vertical farming technologies, expanding educational efforts, and enhancing food distribution networks to allow all communities access to locally grown produce.

Smart Agriculture in Detroit, USA

Detroit, once a thriving industrial hub, now leverages urban agriculture for city revitalization, addressing vacant land issues, food insecurity, unemployment, and community fragmentation.

Key Urban Farming Initiatives:

- Community and Market Gardens: Detroit supports numerous community and market gardens, such as "Keep Growing Detroit," which bolster food sovereignty and educate on self-sustenance, serving both as food sources and community hubs (Fig. 12.3b).
- Large-Scale Urban Farms: Significant projects like the Michigan Urban Farming Initiative and RecoveryPark Farms utilize large vacant plots to produce and sell substantial produce amounts, providing employment and training opportunities, especially for disadvantaged residents.

Technology and Innovation:

- Hydroponics and Aquaponics: These technologies support year-round farming in controlled environments, crucial in Detroit's harsh winters, by using less water and bypassing soil dependency.
- Solar Energy: Some farms integrate solar panels, enhancing sustainability and reducing reliance on non-renewable energy.

Government and Community Support:

- Land Availability: Detroit facilitates urban farming through low-cost land leases and sales, spearheaded by local government initiatives.
- Educational and Financial Support: Non-profits and city programs offer essential resources and training for aspiring urban farmers.

Community Engagement and Impact:

- Skill Development and Employment: Urban farming initiatives provide essential job training and employment, focusing on hiring local residents facing employment challenges.
- Educational Programs: Collaborations with local schools deliver practical agricultural knowledge and promote healthy eating among students.

Challenges and Future Directions:

Despite its successes, urban farming in Detroit faces hurdles like needing more supportive city policies, scalability of farming projects, and consistent funding. Future plans aim to enhance technology use in farming, expand market reach for urban-produced goods, and improve local food system infrastructure.

12.6 Concluding Remark: Future Trends

As cities continue to innovate in this space, sharing best practices and technologies globally will be crucial for scaling up smart urban agriculture to meet growing food security challenges. As demonstrated by examples from around the world, there is no one-size-fits-all approach to smart urban agriculture. Cities must develop context-appropriate strategies based on their unique geographical, climatic, and

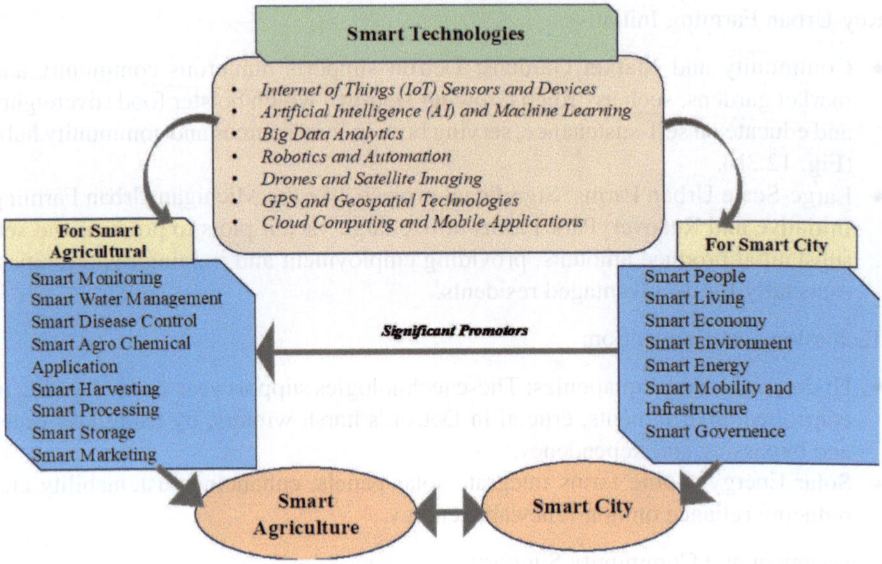

Fig. 12.4 Interlinkage between smart agriculture and smart cities

socio-economic conditions. This may involve a mix of high-tech indoor farming, rooftop gardens, community allotments, and peri-urban production. Looking ahead, several key areas will be crucial for advancing smart agriculture in smart cities:

- developing integrated data platforms to optimize urban food systems;
- scaling up circular economy approaches to close urban nutrient loops;
- leveraging AI and robotics to increase productivity of urban farms;
- creating supportive policy frameworks and incentives for urban agriculture;
- fostering innovation ecosystems and public–private partnerships;
- promoting education and workforce development for urban farming;
- ensuring equitable access to benefits of smart urban agriculture.

These key areas are interconnceted and there should have an integrated approaches of policy formulation, action planning and implementation of all these to maximize the benefits from smart agriculture in smart cities. However, to maximize the benefits it is necessary to ensure smart application of smart technologies both in agriculture and in city planning, development and management.

Figure 12.4 indicates different smart technologies that would serve smart cities and smart agriculture in diverse ways. Smart application and maintenance of all these smart technologies would contribute to each component of smart city and smart agriculture. A smart city is believed to be significant promotor for each component of smart agriculture and smart agriculture would enhance smart living, smart economy and smart environment in a smart city.

References

Adamson G (2023) Tech's impact on urban farming methods. West Tech Fest Blog. https://westtechfest.com/techs-impact-on-urban-farming-methods/2024

Aguilar L (2023) Detroit farms, gardens still fighting for right to flourish. The Detroit News. https://seas.umich.edu/sites/default/files/2023-08/Detroit%20farms%2C%20gardens%20still%20fighting%20for%20right%20to%20flourish.pdf. Accessed 25 July 2024

Cabinet Office of Japan (n.d.) Society 5.0, https://www8.cao.go.jp/cstp/english/society5_0/index.html. Accessed 10 Aug 2024

Dobele M, Zvirbule A (2020) The concept of urban agriculture–historical development and tendencies. Rural Sustain Res 43(338):20–26

De Zeeuw H, Van Veenhuizen R, Dubbeling M (2011) The role of urban agriculture in building resilient cities in developing countries. J Agric Sci 149(S1):153–163

Dos Santos MJPL (2016) Smart cities and urban areas—aquaponics as innovative urban agriculture. Urban for Urban Green 20:402–406

Engel E, Fiege K, Kühn A (2019) Farming in cities: potentials and challenges of urban agriculture in Maputo and Cape Town. Humboldt-Universität zu Berlin

EPA (n.d.) Climate change impacts on agriculture and food supply. https://www.epa.gov/climateimpacts/climate-change-impacts-agriculture-and-food-supply, Accessed 10 Aug 2024

FAO (2009) Global agriculture towards 2050. https://www.fao.org/fileadmin/templates/wsfs/docs/Issues_papers/HLEF2050_Global_Agriculture.pdf. Accessed 10 Aug 2024

Guttapulli, Ankitha (2024) https://www.archdaily.com/1012190/urban-agriculture-in-the-united-states-revitalizing-neighborhoods#:~:text=The%20UN%20Food%20and%20Agriculture,people%2C%20practicing%20urban%20agriculture%20worldwide. Accessed 25 June 2024

Halegoua G (2020) Smart cities. MIT press

Hall RE, Bowerman B, Braverman J, Taylor J, Todosow H, Von Wimmersperg U (2000) The vision of a smart city. https://www.osti.gov/servlets/purl/773961/

IBM (2023) What is smart farming?, https://www.maff.go.jp/e/policies/tech_res/smaagri/robot.html. Accessed 10 Aug 2024

Kalam A, Mustafa NMF (2024) Smart agriculture for smart cities. In: Smart cities: power electronics, renewable energy, and internet of things. CRC Press, pp 130–156

McDaniel A (2022) The intersection of smart cities and sustainability. How smart cities' sustainable infrastructure and design is pushing environmental and social progress forward, sustainly

Ministry of Land Infrastructure and Transport (2023) Smart City in Korea. Accessed from Seoul https://smartcity.go.kr/en/%EC%86%8C%EA%B0%9C/

Ministry of Agriculture, Forestry and Fisheries (MAFF) (2023) Smart agriculture. https://www.maff.go.jp/e/policies/tech_res/smaagri/robot.html. Accessed 10 Aug 2024

MIT D-Lab (n.d.) Internet of things: low cost sensors for agriculture. https://d-lab.mit.edu/research/mit-d-lab-cite/internet-things-low-cost-sensors-agriculture. Accessed 10 Aug 2024

Morchid A, El Alami R, Raezah AA, Sabbar Y (2023) Applications of internet of things (IoT) and sensors technology to increase food security and agricultural sustainability: benefits and challenges. Ain Shams Eng J:102509

OECD (2023) Environmental sustainability in agriculture 2023, https://openknowledge.fao.org/server/api/core/bitstreams/32da2942-3854-4736-af19-877b3ab22d35/content. Accessed 10 Aug 2024

Posivakova T, Svajlenka J, Hromada R, Korim P (2019) Ecological urban agriculture from the point of view basic elements of sustainability. IOP Conf Ser Mater Sci Eng 603(2):022022. IOP Publishing

Rajak P, Ganguly A, Adhikary S, Bhattacharya S (2023) Internet of Things and smart sensors in agriculture: scopes and challenges. J Agric Food Res 14:100776

Risalat Consultant (2023) Smart cities—the future of urban living experience here and now!. https://risalatconsultants.com/smart-cities/. Accessed 30 June 2024

Shea S (2020) Smart city. TecTarget. Sustainable development in Sub-Saharan Africa, paper no. 11 World Bank

Smit J, Ratta A, Nasr J (1996) Urban agriculture an opportunity for environmentally

Steel C (2013) Hungry city: how food shapes our lives. Vintage Books, UK, p 383

Union of Concerned Scientists (2019) Climate change and agriculture—a perfect storm in farm country. https://www.ucsusa.org/resources/climate-change-and-agriculture. Accessed 10 Aug 2024

United Nations (2021) UN/DESA policy brief #102: population, food security, nutrition and sustainable development. https://www.un.org/development/desa/dpad/publication/un-desa-policy-brief-102-population-food-security-nutrition-and-sustainable-development/. Accessed 10 Aug 2024

United Nations (2018) World urbanization prospects, 2018

Van Dijk M, Morley T, Rau ML, Saghai Y (2021) A meta-analysis of projected global food demand and population at risk of hunger for the period 2010–2050. Nat Food 2(7):494–501

Vishnivetskaya A, Alexandrova E (2019) "Smart city" concept: implementation practice. IOP Conf Ser Mater Sci Eng 497(1):012019. IOP Publishing

WWF (n.d.) Sustainable agriculture. https://www.worldwildlife.org/industries/sustainable-agriculture. Accessed 10 Aug 2024

Winkowska J, Szpilko D, Pejić S (2019) Smart city concept in the light of the literature review. Eng Manage Prod Serv 11(2):70–86

Chapter 13
Cost and Benefit of Green Infrastructures in Urban Heat Island Mitigation

Shazmin Shareena Ab. Azis and Shastitharran Baskaran

Abstract Anthropogenic heat from cooling systems fitted in buildings have increased surrounding temperature, which causes Urban Heat Island (UHI). UHI has increased energy consumption and greenhouse gas emissions. Hence, an immediate solution is required to reduce the impact of UHI. Green roofs and walls are proven to reduce UHI through evapotranspiration and heat insulators. According to past research, these components reduce 0.7–25% cooling energy demand and temperature by 0.2–20 °C. However, the monetary value of green infrastructure implementation remains undisclosed. Hence, this research is done to compare the economic worth of both green components using cost–benefit analysis. Green roof has been proven to perform better than green wall, as it is more capable in decreasing building energy consumption by 15% to 79% compared to green wall by 10% to 59%. Furthermore, the cost and benefit of a green roof is slightly better than a green wall whereby the ratio of minimum cost and benefits conveyed by a green roof in reducing cooling energy consumption is 1:1.19. Therefore, the construction of green infrastructure such as green roof is beneficial in terms of environment and economy for the property owners.

Keywords Cooling energy · Green roof · Green wall · Economic performance · Urban Heat Island (UHI)

13.1 Introduction

The occurrence of Urban Heat Island (UHI) is becoming more frequent in the urban city throughout the world. UHI is a situation where urban area has higher overall temperature compared to rural areas (Fernández et al. 2015; Fei and Zhao 2019). This phenomenon has caused various issues in terms of economy, environment, and

---

S. S. Ab. Azis (✉) · S. Baskaran
Faculty of Built Environment & Surveying, Universiti Teknologi Malaysia, Skudai, Johor, Malaysia
e-mail: shazmin@utm.my

M. G. Kibria et al. (eds.), *Urban Blue-Green Infrastructure Approach for Food Security and Climate Disaster Resilience*, Disaster Risk Reduction,
https://doi.org/10.1007/978-981-96-8746-6_13

social. Increasing migration of citizens to the urban area has increased the UHI impact toward the urban area where more buildings are constructed, which directly amplifies the impact of UHI. These also have increased the demand for cooling energy (Wang et al. 2018).

UHI causes cumulative issues where one activity that happens creates problems in different aspects. The high temperature of surrounding causes the building energy consumption to increase and concurrently increasing electricity cost for the property owners. A study showed that almost 10% of the energy usage around the world is used just to stabilize the high surrounding temperature caused by UHI impact (Akbari 2005). Other than that, previous studies also indicate that the demand for decreasing surrounding temperature increase by 30% with every 0.6 °C changes in the overall temperature (Akbari 2005). The roofs and walls of properties are the most dominant parts of the building that is exposed to the solar rays. These solar rays increase the overall temperature of the building during daytime which requires higher amount of air-conditioning to stabilize the temperature of the building which leads to higher building electricity usage. Therefore, there is a need to implement an appropriate solution to minimize UHI impact toward buildings such as introducing urban green infrastructure.

Green wall and green roof have high capability in decreasing heat projection toward the buildings through their natural life cycle process known as evapotranspiration (Olivieri et al.2013; Aflaki et al. 2017). These green infrastructures are capable of reducing the temperature of buildings at around 10% which leads to electricity savings up to 25% on average. Although this green infrastructure is prominent in reducing UHI impacts on buildings, the monetary value of its implementation is still undisclosed to date. Thus, this study is motivated to evaluate the cost and benefits of green roof and green wall implementation for the interest of property owners. The findings of this study act as a decision-making tool for property owners in considering the integration of green roof and green wall to their buildings for the purpose of reducing electricity demand together with electricity bills.

13.2 Literature Review

13.2.1 Urban Heat Island

Urban Heat Island (UHI) is instigated by heat produced in city settings that is trapped by metropolitan buildings and worsened by greenhouse gases and the absence of green spaces. It causes the urban area to be higher in temperature compared to countryside due to the high population and their activities (Takebayashi and Moriyama 2020; Oke 1982; Levermore et al. 2015). Moreover, temperature of urban area is warmer at night due to the sun's radiation being absorbed to the buildings (Levermore et al. 2015). Increasing effect of this phenomenon has caused various impact toward the humas in terms of environmental, economic, and social. Apart from that, these also have the tendency to disturb the natural ecosystem process causing harm to the biodiversity of mother nature in the long run.

The UHI phenomenon has been reported to increase its negative impact through the hot and sunny season where it increases the lack of comfortableness in citizens together with increasing the odds of getting disease that is related to increased heat projection from the surroundings (Desideri and Ȧsdrubali 2018). The intensity of UHI can be contributed by numerous confluences such as meteorological factors, the use of manmade building materials, and the lack of greenery which can be used as natural cooling impacts (Oke 1982; Mohajerani et al. 2017). Furthermore, in the context of real estate properties, the increasing heat will cause the electricity usage to increase which is needed to regulate the temperature of the buildings.

13.2.1.1 Urban Heat Island (UHI) Impact on Cooling Energy Consumption

UHI affects energy consumption of a building by increasing demand for cooling load by 19% (Kamboj and Ali 2020; Li et al. 2019). It was also found that this phenomenon increased electricity usage of buildings by 30% to 70% and 10% to 20%, respectively (Litardo et al. 2020). The usage of electricity in the urban areas has increased up to 4% for every one unit of increase in the surrounding temperature from the normal room temperature (Akbari 2005). The effect amplifies in the urban areas as there are many building structures together with decreasing green spaces that helps in climate regulation process. Table 13.1 shows the effects of UHI in various countries together with the statistics in changes in building electricity usage. From the studies, although the effect varies in a small scale to a large scale of electricity demand changes, the fact that the increasing temperature in the environment is indeed causing a rise in the electricity usage around the world.

According to Kumari et al. (2020), UHI temperature rates changing from 1.9 to 7.2 °C cause an increase of 11.4% in annual electricity consumption. Based on a projection by Yadav etl al. in 2017, a higher 3.0 °C caused by UHI would increase a 1856 GWh demand of electricity in Delhi. Research by Kumari et al. (2020) also shows an increase of 2600 GWh varying of an increase temperature reading by 1.9–7.2 °C. The study shows that 1 °C reduced UHI impact could lead to approximately USD 141 million savings in electricity costs in Delhi and with a 2 °C reduced UHI impact, the electricity savings can reach up to USD 200 million approximately between 2016 and 2017. Looking at the possible savings that could be generated, an appropriate action and solution need to be planned and executed to ensure the dominating effects of UHI can be reduced gradually. As the characteristics of properties is static and cannot be removed, solutions such as adapting features and asset that could be fitted onto the buildings which enhances the climate regulations need to be discovered, analyzed, and implemented. Increasing green area in the urban setting and buildings could be one of the ways to regulate and reduce the UHI affects.

Table 13.1 Urban Heat Island (UHI) impact on building electricity per 1 °C

Author	Research area	UHI impact		Urban Heat Island (UHI) impact on building electricity per 1 °C (%)
		Indoor air temperature (°C)	Cooling energy consumption	
Parkpoom and Harrison (2008)	Thailand	1	4.6%	4.6
Fung et al. (2006)	Hong Kong	1	2.4–9.2%	5.8
Franco and Stanstad (2008)	California	0.79–1.84	3.0%	3.0
Wangpattarapong et al. (2008)	Bangkok	1	7.49%	7.49
Amato et al. (2005)	Massachusetts	1	6.5%	6.5
EPA (2008)	United States	0.6	1.5–2.0%	2.9
Akbari (2005)	Los Angeles	1	2.0–4.0%	3.0
Kumari et al. (2020)	Delhi, India	1.9–7.2	11.4% (2600GWh)	2.5
Yadav et al. (2017)	Delhi, India	3.0	3.2% (1856GWh)	1.0
1 °C UHI =				**1.0–7.5%**

13.2.2 Green Vegetation Infrastructures

Building envelopes is also known as building components implemented or constructed as important component of buildings such as windows, walls, and roof, it features as the separator between the inner space and outer space of a building (Sadineni et al. 2011). In the context of Malaysia, the high amount of heat projection generated in the surroundings travels through the windows and building surfaces has increased the overall temperature of the building inner environment paired up with higher demand for cooling systems (Azis 2020). Green envelopes can reduce energy demands and consumption. Past researches have studied the types of green components that is available in the market which could be fitted on the surface of buildings to reduce electricity usage which includes green roof, green walls, and double skin façade (Azis 2020). Green components that are fitted into the buildings can act as a barrier and neutralizer of high heat projection from entering the buildings which can concurrently benefit in reducing the energy demand in buildings and enhancing sustainability and biodiversity.

13.2.2.1 Efficiency of Green Vegetation Infrastructure in Reducing Cooling Energy Consumption

A study by Azis (2020) regarding association between green infrastructure and building electricity consumption revealed that green infrastructures that are equipped with vegetations produced and delivered highest capability in promoting energy saving for the building. Residential properties that are fitted with green infrastructures have the capability to reduce annual energy consumption up to 25% on an annual period. These reduction in cooling energy consumption could yield up to approximately USD 170 savings annually. Apart from that, commercial properties that are equipped with green infrastructures could generate savings up to USD 37 annually with 4% reduction in electricity usage.

Green Roof

Green roof is a manmade roof structure which is constructed at roof surface of buildings. It consists of several layers such as water-proof platform, growing medium, and lastly the vegetation layer. Apart from that, it also has its accessory fittings such as with root barriers, drainage, and irrigation system. Past research has proven that green roof is impactive in reducing the use of air-conditioning in hot and humid situation which contributes to energy savings (Rumana and Hamdan 2009). Research conducted by Niachoue et al. (2001) found that the implementation of green roof can save energy by 37–48% when a 10 ACH of night ventilation is applied compared to a 4–7% in the basic scenario without night ventilation.

Table 13.2 shows the green roof efficiency in reducing cooling load. The green roof that is present in different countries has several layers present together with vegetations which will create a new ecosystem on the roof of buildings and the process of evapotranspiration will help in the climate regulation in the building surfaces. Furthermore, research has also proven that green roof integration provided 14.7% to 49% of annual cooling load reduction (Goerguelue and Ekren 2013; Tian et al. 2017). These show that green roof has the capability and ability to improve the surface and indoor temperature of the buildings making it an effective green infrastructure that can be implemented in the buildings as an initiative to combat UHI.

Green Wall

Green wall can be defined as the use of plants of vertical surface to reduce heat absorption on building structures which can cause a change in temperature. It evaporates water to the surface and caused a cooler temperature rather than exposed concrete surface (Jaafar et al. 2011). Table 13.3 shows the green wall efficiency in reducing cooling load. Green walls have the capability to reduce cooling load via decreasing surrounding temperature from 10 to 8% approximately which is almost half of the

Table 13.2 Green roof efficiency in reducing cooling load

Author	Research area	Percentage of efficiency (%)
Niachou et al. (2001)	Athens, Greece	37.0–48.0
Jiang and Tang (2017)	Shanghai, China	79.0
Tian et al. (2017)	Suzhou, China	27.7–35.8
Santamouris (2014)	Athens, Greece	20.0–40.0
Yang et al. (2018)	Singapore	31.0
Ebadati and Ehyaei (2020)	Tehran	16.3
Wong et al. (2003)	Singapore	15

Table 13.3 Green wall efficiency in reducing cooling load

Author	Research area	Percentage of efficiency (%)
Bakhtyari et al. (2024)	Iran	15
Wong et al. (2010)	Hong Kong, China	16
Coma et al. (2017)	Spain	58.9
Pan et al. (2018)	Hong Kong, China	2.0–42.0
Dahanayake and Chow (2017)	Hong Kong, China	13.0
Santamouris (2014)	Athens, Greece	10.0–40.0
Haggag et al. (2014)	Abu Dhabi	20

cooling load for the buildings. These show that green wall implementation onto the buildings can surpass the increasing temperature of UHI together with providing sustainability and aesthetics to the buildings. This green façade can maintain 5 °C lower temperature than an exposed wall which can improve energy performance of a building (Haggag et al. 2014). Sunakorn and Yimprayoon (2011). However, the efficiency of the green wall systems also depends on the type of plantations and vegetation that is present which ranges from 4 to 16 °C (Wong et al. 2010). Previous research has reported that green wall implementation could enhance the cooling of building temperature from 12 to 64% annually (Wong et al. 2010; Coma et al. 2017; Pan et al. 2018; Dahanayake and Chow 2017).

13.3 Methods

A total of three objectives were outlined to achieve the aim of this study. The three objectives of this study are to identify the cooling energy consumption impact from the Urban Heat Island (UHI) to analyze the efficiency of green wall and green roof in mitigating Urban Heat Island (UHI) and to evaluate the cost and benefit of green roof and green wall integration for property owner. The study adapted both quantitative and qualitative analyses to fulfill the objectives. Figure 13.1 shows the methodology process of this study. The first stage data was collected based on literature review and analyzed using systematic analysis. The second stage data was retrieved from literature review process using systematic analysis to finalize and tabulate the results.

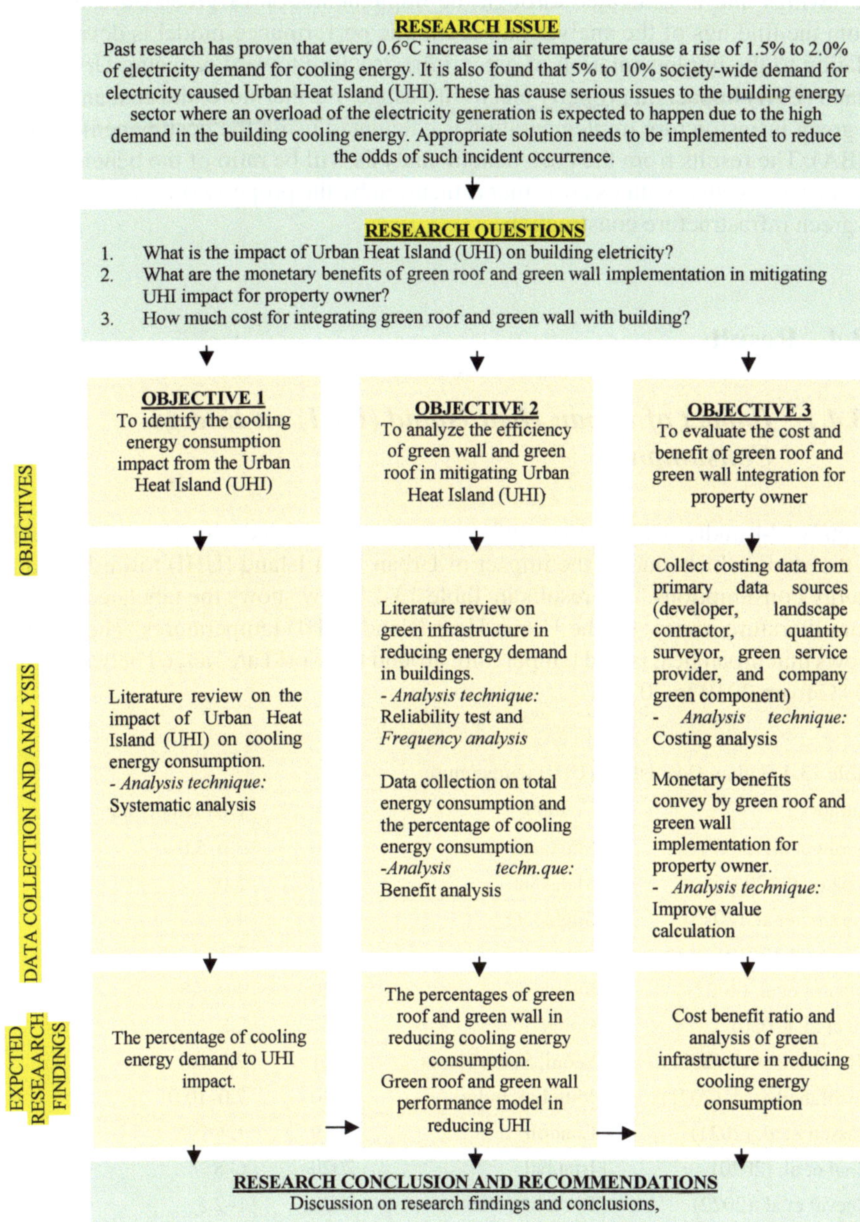

Fig. 13.1 Research methodology

Then, case study was conducted in an office building that is in Putrajaya. The data regarding the energy that is consumed by the office building were collected and the energy that is used particularly for the air-conditioning of the building is analyzed in the form of percentage. The data was used to carry out analysis that reveals the savings that is generated through the implementation of green infrastructures. From the findings of the analysis, an economic performance model is developed in relation to the implementation of green infrastructure toward reducing Urban Heat Island (UHI) impact on property electricity usage. Furthermore, the cost and benefit of green infrastructure implementation is calculated using the cost–benefit analysis (CBA). The results from the cost–benefit analysis will be ratio of the benefits that is associated together with its costs that is incurred by the property owners in the phase of green infrastructure construction.

13.4 Results

13.4.1 Impact of Urban Heat Island (UHI) on Energy Consumption

A total of 14 studies with various study areas were discovered which meets the criteria of the study which studies the impact of Urban Heat Island (UHI) toward building energy consumptions. The results in Table 13.4 below shows the tabulated findings from literature review on the Urban Heat Island (UHI) temperatures. The analysis shows that urban heat island temperature around the world are ranged between 1 and 10 °C from 2019 to 2021.

Table 13.4 Urban Heat Island (UHI) temperature

Authors	Country	Year	UHI temperature (°C)
Harun et al. (2020)	Malaysia	2020	3.0–5.0
Aziz et al. (2021)	Malaysia	2020	5.0
Borzino et al. (2020)	Singapore	2020	1.6–2.2
Aris et al. (2019)	Indonesia	2019	2.3
Saputra et al. (2022)	Indonesia	2020	4.9–7.8
Lu et al. (2020)	China	2020	2.8
Ngarambe et al. (2021)	Seoul, South Korea	2021	2.5–6.0
Shahfahad et al. (2021)	Mumbai, India	2020	7.0–10.0
Bassett et al. (2021)	London	2019	1.4–2.9
Unal et al. (2020)	Istanbul	2020	6–8
Cheval et al. (2022)	Romania, Europe	2020	1–2.5
Urban heat island temperature from 2019 to 2021			1–10

Collected data from thc above are calculated to derive the percentage of cooling energy consumption increment due to UHI temperature as shown in Eqs. 13.1 and 13.2 Hence, analysis from objective 1 shows that the energy needed to stabilize the temperature of building in the presence of heat island impact is between 1 and 7.5%.

$$\begin{aligned}&\text{Percentage of minimum cooling energy consumption increment due to minimum}\\&\text{UHI temperature} = \text{Percentage of minimum cooling energy demand increment}\\&\text{due to 1 Celsius temperature increment (\%)}\\&\times\ \text{Minimum UHI temperature in urban area (Celsius)}\\&= 1\% \times 1^{\circ}\text{C}\\&= 1\%\end{aligned} \tag{13.1}$$

$$\begin{aligned}&\text{Percentage of maximum cooling energy consumption increment due to maximum}\\&\text{UHI temperature} = \text{Percentage of maximum cooling energy demand increment}\\&\text{due to 1 Celsius temperature increment (\%)}\\&\times\ \text{Maximum UHI temperature in urban area (Celsius)}\\&= 7.5\% \times 10^{\circ}\text{C}\\&= 75\%\end{aligned} \tag{13.2}$$

13.4.2 Green Wall and Green Roof Performance Model in Reducing Urban Heat Island (UHI) Impact

A performance model is developed to evaluate the percentages of green roof and green wall efficiency in decreasing overall demand for cooling energy together with evaluating the savings that can be generated from the construction of green roof and green wall. The benefits in terms of monetary value that can be gained from the construction of green roof and green wall were identified using the Benefit Transfer Approach (BTA).

13.4.2.1 Green Vegetation Infrastructure Efficiency in Reducing Cooling Energy Consumption

Based on past research, green roof and green wall have the capability to efficiently reduce cooling energy consumption by 15–79% and 10–58.9% (Table 13.5).

Table 13.5 Green envelope component's efficiency in reducing cooling energy consumption

Green envelope components	Efficiency in reducing cooling energy consumption (%)	
	Min	Max
Green roof	15	79
Green wall	10	59

13.4.2.2 Cooling Energy Consumption for Residential Building in KL (RM)

Average percentage of cooling energy consumption for residential building in Kuala Lumpur was derived from the actual data collected from Tenaga Nasional Berhad (TNB) and the current average percentage of cooling energy consumption. According to Tenaga Nasional Berhad (TNB), total energy consumption for residential building in Kuala Lumpur is estimated around RM479889909380.03. Furthermore, average percentage of cooling energy consumption is 40% as shown in Table 13.6.

Hence, the average cooling energy consumption for residential building in Kuala Lumpur is RM342778506700.02 as shown in Eq. 13.3.

Cooling energy consumption in Kuala Lumpur

Table 13.6 Average percentage of cooling energy consumption

Citation	Area	Percentages (%)
Kubota et al. (2011)	Malaysia	29
Umbark et al. (2020)	Libya	21
Al-mudhafar et al. (2021)	Iraq	50
International Energy Agency (2018)	United Kingdom	20
Seo et al. (2013)	Queensland, Australia	73
Kim and Suh (2021)	South Korea	20
Ürge-Vorsatz et al. (2015)	Europe	25
Li et al. (2021)	US	35
Yun and Steemers (2011)	US	47
DoE (2011)	US	37
Jelle et al. (2012)	New York	60
Hu et al. (2019)	China	46
Santamouris (2016)	Europe	61
Hassan et al. (2014)	Malaysia	50
Sukarno et al. (2015)	Indonesia	30
Abdollah et al. (2021)	Egypt	50
Average percentages of cooling energy consumption		40

Table 13.7 Average cooling energy consumption with UHI impact in Kuala Lumpur

UHI impact on cooling energy consumption	Percentages	Average cooling energy consumption with UHI impact
Min UHI impact	0.1	RM34277850670.00
Max UHI impact	0.75	RM257083880025.02

$$= \text{Average percentages of cooling energy consumption (\%)}$$
$$\times \text{ Total energy consumption of residential building in Kuala Lumpur (RM)}$$
$$= 40\% \times \text{RM}479{,}889{,}909{,}380.03$$
$$= \text{RM}342{,}778{,}506{,}700.02 \quad (13.3)$$

13.4.2.3 Average Cooling Energy Consumption with UHI Impact in Kuala Lumpur

Cooling energy consumption with Urban Heat Island (UHI) impact in Kuala Lumpur was calculated based on the current minimum and maximum UHI impact on cooling energy consumption (1–75%) and the cooling energy consumption for residential building in Kuala Lumpur (RM342778506700.02). Table 13.7 shows the tabulated results of the calculation.

13.4.2.4 Cooling Energy Consumption Reduction in Reducing UHI Impact

Cooling energy consumption reduction of green vegetation infrastructures in reducing UHI impact is as tabulated in Table 13.8. Green roof can reduce cooling energy consumption due to UHI impact by RM5141677600.50 to RM203096265219.76 depending on the minimum and maximum impacts. Besides, green wall was also proven to reduce cooling energy consumption by RM3427785067.00 to RM151679489214.76.

Dewan Bandaraya Kuala Lumpur annual report 2021 stated that total residential area in Kuala Lumpur is 45,217,425 sq. m with 69% occupancy rate. Hence, this data was used to estimate the overall cooling energy consumption reduction due to green envelope component implementation (RM/sq. m). Table 13.9 shows the results of cooling energy consumption reduction in reducing UHI impact (RM/sqm).

Based on the table above, green roof can save cost by RM113.71 per sq.m due to its 15% efficiency in combating a minimum UHI impact of 1%. However, a higher saving cost (RM598.87 per sq.m) can be derived if the green roof provides maximum efficiency of 79%. These results can also be seen on the maximum UHI impact where low green roof efficiency is able to save RM852.83 per sq.m compared to RM4491.55 per sq.m saving for higher green roof efficiency. Green wall can save

Table 13.8 Cooling energy consumption reduction in reducing UHI impact (RM)

Green vegetation infrastructure	Efficiency (%)	Cooling energy consumption reduction in reducing UHI impact (RM)	
		Min	Max
		Cost saving (RM)	Cost saving (RM)
Green roof			
Min	0.15	RM5141677600.50	RM38562582003.75
Max	0.79	RM27079502029.30	RM203096265219.76
Green wall			
Min	0.1	RM3427785067.00	RM25708388002.50
Max	0.59	RM20223931895.30	RM151679489214.76

Table 13.9 Cooling energy consumption reduction in reducing UHI impact (RM/sqm)

Green vegetation infrastructure	Efficiency (%)	Cooling energy consumption reduction in reducing UHI impact (RM/sqm)	
		Minimum	Maximum
		Cost saving (RM)	Cost saving (RM)
Green roof			
Minimum	15.0	RM113.71	RM852.83
Maximum	79.00	RM598.87	RM4491.55
Green wall			
Minimum	1.00	RM75.81	RM568.55
Maximum	58.90	RM447.26	RM3354.45

cost by RM75.81 per sq.m to RM568.55 per sq.m for minimum green wall efficiency in reducing cooling energy consumption while RM447.26 per sq.m to RM3354.45 per sq.m can be saved for maximum green wall efficiency.

The result shows that in square meter measurement, green roof yielded slightly higher cost saving than green wall. This is due to the efficiency in reducing cooling energy consumption of green wall which is lower than green roof. Henceforth, this model has proved that implementation of green envelope components is impactive in reducing cooling energy consumption due to Urban Heat Island (UHI) impact. Therefore, green roof and green wall must be implemented and utilized to its best condition to achieve highest efficiency and thus highest cost saving.

13.4.3 *Cost Benefit of Green Roof and Green Wall Implementation for Property Owner*

The third objective is to evaluate the performance of green roof and wall in reducing UHI impact through cost–benefit analysis. To evaluate the performance of green roof and wall in reducing UHI impact through cost–benefit analysis, cost–benefit analysis approach is adopted. Cost–benefit approach is widely used to understand the investment worth of a decision that is being made (Koopmans and Mouter 2020). In the context of this study, the worth of implementing green roof and green wall is analyzed for the benefits in reducing UHI effects. The benefit represents by monetary saving gain by property owner due to application of green roof and wall (as derived in objective 2). Meanwhile cost is represented by the costing data that is retrieved from green roof and green wall contractors to integrate green roof and wall with building. Both the costs and benefits of green roof and green wall are identified and analyzed to identify whether, in overall, the implementation would give a positive or negative outcome.

13.4.3.1 Cost of Green Vegetation Infrastructure (RM)

The costs of the green vegetation infrastructure are collected in the form of quotation by the contractors. Cost of green roof implementation was collected from three companies which are Sika Kimia Sdn. Bhd., Aathaworld Sdn. Bhd., and Belalang Inovasi. According to the contractors, the cost for the implementation of green roof is influenced by several attributes such as type of moisture barrier, green roof system, plant material, irrigation system, and professional installation. The total cost of a green roof construction is estimated by RM3,770 per sq.m to RM5,060 per sq.m. The cost details for each attribute are as shown in Table 13.10.

As for the green wall implementation, the cost data was collected from Vertical Green Sdn. Bhd., Belalang Inovasi Sdn. Bhd., and Green Planet Midori Sdn. Bhd. Besides, the implementation of green wall also includes similar attributes. The total

Table 13.10 Green roof implementation cost

Green roof attributes	Total cost per sqft. (RM)		
	Sika Kimia Sdn. Bhd	Aathaworld Sdn. Bhd	Belalang Inovasi
Moisture barrier	8	15	10
Green roof system	257	250	180
Plant material	93	80	70
Irrigation system	48	50	40
Installation	84	70	50
Total costs (RM/sqft)	**470**	**465**	**350**
Total costs (RM/sqm)	**5,060**	**5,000**	**3,770**

Table 13.11 Green wall implementation cost

Green wall attributes	Total cost per sqft. (RM)		
	Vertical green Sdn. Bhd	Belalang Inovasi	Green Planet Midori
Moisture barrier	20	20	10
Green wall system	300	250	347
Plant material	100	90	125
Irrigation system	60	50	64
Installation	80	70	114
Total costs (RM/sqft)	**560**	**480**	**660**
Total costs (RM/sqm)	**6,030**	**5,170**	**7,100**

cost of green wall implementation is estimated by RM5,170 per sq.m to RM7,100 per sq.m. From the average costs of green roof and green wall, green wall is more costly than green roof as green wall is implemented on various types of wall surfaces such as angled and curved surfaces which require special design of panels and type of plants according to the wall's location in the building. On the other hand, green roof is placed on a flat or slanted roof which does not require custom panels. Apart from that, green wall tends to contain a more expensive and aesthetic looking plants and vegetation compared to green roof which makes it more costly as well (Table 13.11).

13.4.3.2 Cost–Benefit Relatio of Green Vegetation Infrastructure

The green roof and green wall costings and monetary benefits in terms of energy savings are used as an input to in the cost–benefit analysis. This analysis is conducted to assess the worth of implementing both types of green envelope components in a ratio form. The cost–benefit analysis ratio is calculated using formula as follows:

$$\text{Cos-benefit ratio of Green Vegetation Infrastructure} = \frac{\text{Green Vegetation Infrastructure benefit (RM)}}{\text{Green Vegetation Infrastructure cost (RM)}} \quad (13.4)$$

Table 13.12 shows the cost–benefit analysis that is done for the green roof and green wall infrastructure's minimum and maximum cost–benefit ratio for green roof is between 1: 0.02 and 1: 1.19. This ratio indicates that the benefit conveyed by green roof is around 1.19 times higher than minimum cost that is associated with integrating green roof with building. However, the results also show that the benefits conveyed by green roof are lower than the implementation cost by 0.02 to 0.89 when it was compared to the maximum cost of green roof implementation. This is because, green types and size of the area of implementation influence the cost of green roof which involves higher maintenance costs for bigger green roof as well.

This study proves that the implementation of green roof to reduce Urban Heat Island (UHI) impact is only worthy with minimum implementation cost. Although

Table 13.12 Cost–benefit analysis for green vegetation infrastructure

Green vegetation infrastructure	Benefits (RM)	Cost (RM)		Cost–benefit analysis	
		Min	Max	Min	Max
Green roof					
Minimum	113.71	3770	5060	0.03	0.02
Maximum	4,491.55			1.19	0.89
Green wall					
Minimum	75.81	5170	7100	0.01	0.01
Maximum	3,354.45			0.65	0.47

the benefits gained are less when compared to the highest cost of implementation on green roof, it is a good investment as the green roof that is implemented has the ability to provide benefits not only in terms of UHI reduction but also in terms of increasing roof lifespan, mitigating stormwater, and providing sound insulations. Green roof implementation will be effective and beneficial if proper planning and budgeting is made in terms of the cost of implementation and ensuring all attributes is taken into consideration.

As for the cost benefits of green wall, it was found that the implementation of green wall in reducing Urban heat Island (UHI) impact is not worthy due to low efficiency of green wall in reducing cooling energy consumption and a high implementation cost. The minimum and maximum cost–benefit ratio for green wall is between 1: 0.01 and 1: 0.65. These results proved that the even minimum cost of green wall implementation is higher the benefits convey. This shows that green wall implementation leans its benefits more toward the aesthetics side of the building compared to green roof which provides aesthetics and benefit in terms of environment.

From the cost–benefit analysis, green roof shows better efficiency in terms of costs and benefits compared to green wall. Green roof is a better green infrastructure that can be implemented for the mitigation of UHI in urban areas. However, green wall also showed positive results in energy savings in the building but due to its high cost it becomes less effective than green roof generally. Therefore, it is recommended that the both green vegetation infrastructures must be installed according to the suitability to climate so the building can achieve highest and best use to increase the efficiency. Besides, it is also suggested that the installation cost of green roof and green wall must be controlled to secure the worth of the implementation.

13.5 Conclusion

In conclusion, the green component such as green roof and green wall integration to the residential and commercial buildings in the urban city is a vital solution that need to be focused to reduce the impact of Urban Heat Island (UHI).These components have great capabilities in providing benefits in the environmental and economic sectors. Green roof has been proven to perform better than green wall as it is more efficient in reducing cooling energy consumption by 15% to 79% compared to green wall by 10% to 59%. Besides, this study also discovered that implementation of green roof is a better solution in terms of its cost and benefits compared to green wall where the cost–benefit ratio is 1:1.19.The property owners could benefit from the ability of green roof as a tool that can reduce cooling energy consumptions of building which is directly related to reduced electricity bill costs. Local government, residential and commercial property owners should consider fitting this green infrastructure to the buildings to reduce energy consumption together with enhancing sustainability in urban areas. Future research can focus on identifying the impacts of these green infrastructure toward other economic and environmental issue that is developing due to urbanization and also development.

References

Abdollah MAF, Scoccia R, Filippini G, Motta M (2021) Cooling Energy use reduction in residential buildings in Egypt accounting for global warming effects. Climate 9(3):45

Aflaki A, Mirnezhad M, Ghaffarianhoseini A, Ghaffarianhoseini A, Omrany H, Wang ZH, Akbari H (2017) Urban heat island mitigation strategies: a state-of-the-art review on Kuala Lumpur, Singapore and Hong Kong. Cities 62:131–145

Akbari H (2005) Energy saving potentials and air quality benefits of urban heat island mitigation. http://www.osti.gov/bridge/purl.cover.jsp;jses-sionid=1BC14739097AA40B8236D530 7BC8FC0C?purl=/860475-UlHWIq/

Al-mudhafar AH, Hamzah MT, Tarish AL (2021) Potential of integrating PCMs in residential building envelope to reduce cooling energy consumption. Case Stud Therm Eng 27:101360

Amato AD, Ruth M, Kirshen P, Horwitz J (2005) Regional energy demand responses to climate change: methodology and application to the commonwealth of Massachusetts. Clim Change 71(1):175–201

Aris A, Syaf H, Yusuf DN (2019) Analysis of urban heat island intensity using multi temporal landsat data; case study of Kendari City, Indonesia. IOP Conf Ser Earth Environ Sci 389(1):012002. IOP Publishing

Azis SSA (2020) Improving present-day energy savings among green building sector in Malaysia using benefit transfer approach: cooling and lighting loads. Renew Sustain Energy Rev 137:110570

Aziz NAA, Fuazi MF, Azmi A, Zakaria MZ, Patah NA, Hashim SAM (2021) Study On urban heat island around Kuala Lumpur during Malaysia's movement control order using multi-temporal remote sensing data. In: IEEE workshop on geoscience and remote sensing.

Bakhtyari V, Fattahi K, Movahed K, Franz A (2024) Investigating the effect of living walls on cooling energy consumption in various urban microclimates, building heights, and greenery coverage areas. Sustainability 16(2):920. https://doi.org/10.3390/su16020920

Bassett R, Janes-Bassett V, Phillipson J, Young PJ, Blair GS (2021) Climate driven trends in London's urban heat island intensity reconstructed over 70 years using a generalized additive model. Urban Clim 40:100990

Borzino N, Chng S, Mughal MO, Schubert R (2020) Willingness to pay for urban heat island mitigation: a case study of Singapore. Climate 8(7):82

Cheval S, Dumitrescu A, Irașoc A, Paraschiv MG, Perry M, Ghent D (2022) MODIS-based climatology of the surface urban heat island at country scale (Romania). Urban Clim 41:101056

Coma J, Pérez G, de Gracia A, Burés S, Urrestarazu M, Cabeza LF (2017) Vertical greenery systems for energy savings in buildings: a comparative study between green walls and green facades. Build Environ 111:228–237

Dahanayake KKC, Chow CL (2017) Studying the potential of energy saving through vertical greenery systems: using EnergyPlus simulation program. Energy Build 138:47–59

Department of Energy (2011) Increasing efficiency of building systems and technologies (Chapter 5). In: Quadrennial technology review: an assessment of energy technologies and research opportunities. Retrieved from U.S. Department of Energy website

Desideri U, Asdrubali F (eds) (2018) Handbook of energy efficiency in buildings: a life cycle approach. Butterworth-Heinemann

Ebadati M, Ehyaei MA (2020) Reduction of energy consumption in residential buildings with green roofs in three different climates of Iran. Adv Build Energy Res 14(1):66–93

EPA A (2008) Reducing urban heat islands: compendium of strategies. Heat Island Reduct Act:1–23

Fei W, Zhao S (2019) Urban land expansion in China's six megacities from 1978 to 2015. Sci Total Environ 664:60–71

Fernández FJ, Alvarez-Vázquez LJ, García-Chan N, Martínez A, Vázquez-Méndez ME (2015) Optimal location of green zones in metropolitan areas to control the Urban heat Island. J Comput Appl Math 289:412–425

Franco G, Sanstad AH (2008) Climate change and electricity demand in California. Clim Change 87(1):139–151

Fung WY, Lam KS, Hung WT, Pang SW, Lee YL (2006) Impact of urban temperature on energy consumption of Hong Kong. Energy 31(14):2623–2637

Goerguelue S, Ekren N (2013) Energy saving in lighting system with fuzzy logic controller which uses light-pipe and dimmable ballast. Energy Build 61:172–176

Haggag M, Hassan A, Elmasry S (2014) Experimental study on reduced heat gain through green facades in a high heat load climate. Energy Build 82:668–674

Harun Z, Reda E, Abdulrazzaq A, Abbas AA, Yusup Y, Zaki SA (2020) Urban heat island in the modern tropical Kuala Lumpur: comparative weight of the different parameters. Alex Eng J 59(6):4475–4489

Hassan JS, Zin RM, Abd Majid MZ, Balubaid S, Hainin MR (2014) Building energy consumption in Malaysia: an overview. Jurnal Teknologi (Sci & Eng) 70(7)

Hu Y, Hou M, Jia G, Zhao C, Zhen X, Xu Y (2019) Comparison of surface and canopy urban heat islands within megacities of eastern China. ISPRS J Photogramm Remote Sens 156:160–168

IEA (2018) Global Energy Review. OECD/IEA, Paris

Jaafar B, Said I, Rasidi MH (2011) Evaluating the impact of vertical greenery system on cooling impact on high rise buildings and surroundings: a review. RUAS (Rev Urban Archit Stud) 9(2):1–9

Jelle BP, Hynd A, Gustavsen A, Arasteh D, Goudey H, Hart R (2012) Fenestration of today and tomorrow: a state-of-the-art review and future research opportunities. Sol Energy Mater Sol Cells 96:1–28

Jiang L, Tang M (2017) Thermal analysis of extensive green roofs combined with night ventilation for space cooling. Energy Build 156:238–249

Kamboj S, Ali S (2020) Urban sprawl of Kota city: a case study of urban heat island linked with electric consumption. Mater Today Proc

Koopmans C, Mouter N (2020) Cost-benefit analysis. In: Advances in transport policy and planning, vol 6. Academic Press, pp 1–42

Kubota T, Jeong S, Toe DHC, Ossen DR (2011) Energy consumption and air-conditioning usage in residential buildings of Malaysia. 国際協力研究誌, 17(3):61–69

Kumari P, Kapur S, Garg V, Kumar K (2020) Impact of surface temperature on energy consumption in a calibrated building: a case study of Delhi. Climate 8(6):71

Levermore GJ, Parkinson JB, Laycock PJ, Lindley S (2015) The urban heat island in Manchester 1996–2011. Build Serv Eng Res Technol 36(3):343–356

Li X, Yao R (2021) Modelling heating and cooling energy demand for building stock using a hybrid approach. Energy Build 235:110740

Li X, Zhou Y, Yu S, Jia G, Li H, Li W (2019) Urban heat island impacts on building energy consumption: a review of approaches and findings. Energy 174:407–419

Litardo J, Palme M, Borbor-Cordova M, Caiza R, Macías J, Hidalgo-Leon R, Soriano G (2020) Urban heat island intensity and buildings' energy needs in Duran, Ecuador: simulation studies and proposal of mitigation strategies. Sustain Cities Soc 62:102387

Lu L, Weng Q, Xiao D, Guo H, Li Q, Hui W (2020) Spatiotemporal variation of surface urban heat islands in relation to land cover composition and configuration: a multi-scale case study of Xi'an, China. Remote Sens 12(17):2713

Mohajerani A, Bakaric J, Jeffrey-Bailey T (2017) The urban heat island impact, its causes, and mitigation, with reference to the thermal properties of asphalt concrete. J Environ Manage 197:522–538

Ngarambe J, Oh JW, Su MA, Santamouris M, Yun GY (2021) Influences of wind speed, sky conditions, land use and land cover characteristics on the magnitude of the urban heat island in Seoul: an exploratory analysis. Sustain Cities Soc 71:102953

Niachou A, Papakonstantinou K, Santamouris M, Tsangrassoulis A, Mihalakakou G (2001) Analysis of the green roof thermal properties and investigation of its energy performance. Energy Build 33(7):719–729

Oke TR (1982) The energetic basis of the urban heat island. Q J R Meteorol Soc 108(455):1–24

Olivieri F, Di Perna C, D'Orazio M, Olivieri L, Neila J (2013) Experimental measurements and numerical model for the summer performance assessment of extensive green roofs in a Mediterranean coastal climate. Energy Build 63:1–14

Pan L, Wei S, Chu LM (2018) Orientation impact on thermal and energy performance of vertical greenery systems. Energy Build 175:102–112

Parkpoom S, Harrison GP (2008) Analyzing the impact of climate change on future electricity demand in Thailand. IEEE Trans Power Syst 23(3):1441–1448

Rumana R, Hamdan MA (2009) The passive cooling impact of green roof in high-rise residential building in Malaysia. Sustain Archit Urban Dev 1:271–282

Sadineni SB, Madala S, Boehm RF (2011) Passive building energy savings: a review of building envelope components. Renew Sustain Energy Rev 15(8):3617–3631

Santamouris M (2014) Cooling the cities–a review of reflective and green roof mitigation technologies to fight heat island and improve comfort in urban environments. Sol Energy 103:682–703

Santamouris M (2016) Cooling the buildings–past, present and future. Energy Build 128:617–638

Saputra A, Ibrahim MH, Shofirun S, Saifuddin A, Furoida K (2022) Assessing urban heat island in Jakarta, Indonesia during the pandemic of Covid-19. IOP Conf Ser Earth Environ Sci 986(1):012069. IOP Publishing

Seo S, Wang CH, Grozev G (2013) Cooling energy consumption and reduction effect for residential buildings in South East Queensland, Australia. Build Environ 59:408–416

Shahfahad, Rihan M, Naikoo MW, Ali MA, Usmani TM, Rahman A (2021) Urban heat island dynamics in response to land-use/land-cover change in the coastal city of Mumbai. J Indian Soc Remote Sens 49(9):2227–2247

Sukarno I, Matsumoto H, Susanti L, Kimura R (2015) Urban energy consumption in a city of Indonesia: general overview. Int J Energy Econ Policy 5(1):360–373

Sunakorn P, Yimprayoon C (2011) Thermal performance of biofacade with natural ventilation in the tropical climate. Procedia Eng 21:34–41

Takebayashi H, Moriyama M (2020) Background and purpose. In: Adaptation measures for urban heat islands. Academic Press, pp 1–8

Tian Z, Lei Y, Gu X (2017) Building energy impacts of simple green roofs in the hot summer and cold winter climate zone: Suzhou as a study case. Procedia Eng 205:2918–2924

Umbark MA, Alghoul SK, Dekam EI (2020) Energy consumption in residential buildings: comparison between three different building styles. Sustain Dev Res 2(1):p1–p1

Ünal YS, Sonuç CY, İncecik S, Topcu HS, Diren-Üstün DH, Temizöz HP (2020) Investigating urban heat island intensity in Istanbul. Theoret Appl Climatol 139(1):175–190
Ürge-Vorsatz D, Cabeza LF, Serrano S, Barreneche C, Petrichenko K (2015) Heating and cooling energy trends and drivers in buildings. Renew Sustain Energy Rev 41:85–98
Wangpattarapong K, Maneewan S, Ketjoy N, Rakwichian W (2008) The impacts of climatic and economic factors on residential electricity consumption of Bangkok metropolis. Energy Build 40(8):1419–1425
Wang Y, Li Y, Sabatino SD, Martilli A, Chan PW (2018) Impacts of anthropogenic heat due to air-conditioning systems on an extreme high temperature event in Hong Kong. Environ Res Lett 13(3):034015
Wong NH, Tan AYK, Chen Y, Sekar K, Tan PY, Chan D, Wong NC et al (2010) Thermal evaluation of vertical greenery systems for building walls. Build Environ 45(3):663–672
Wong NH, Chen Y, Ong CL, Sia A (2003) Investigation of thermal benefits of rooftop garden in the tropical environment. Build Environ 38(2):261–270
Yadav N, Sharma C, Peshin SK, Masiwal R (2017) Study of intra-city urban heat island intensity and its influence on atmospheric chemistry and energy consumption in Delhi. Sustain Cities Soc 32:202–211
Yang J, Pyrgou A, Chong A, Santamouris M, Kolokotsa D, Lee SE (2018) Green and cool roofs' urban heat island mitigation potential in tropical climate. Sol Energy 173:597–609

Chapter 14
Financing Urban BGI—Approaches and Challenges

Shubham Sharma, Shashi Kant Sharma, and Subir Sen

Abstract Traditionally, infrastructure development often comes at the expense of the environment, and such directly or indirectly contribute to climate change. However, modern engineering is prioritizing sustainable infrastructure that protects and preserves environmental resources. This shift has led to the rise in green and blue infrastructure (BGI), particularly relevant for urban areas, which contribute more than half of the world's GDP and responsible for 70 percent of global greenhouse gas emissions and 60 percent of resource use. As urbanization continues to grow, with 68 percent of the world's population projected to live in urban areas by 2050, the importance of BGI has increased. An allied question is regarding the funding requirements and the source of such funding? BGI focuses on climate change, health, and urban resilience, aligning with Sustainable Development Goals (SDGs) on land (SDG 15), water (SDGs 6 and 14), and climate change (SDG 13). Therefore, the scope for finance is wide. In this chapter, we discuss these issues and with reference to a few noteworthy global BGI initiatives including Singapore's "Active, Beautiful, Clean Waters" (ABC) Programme, the U.S. "Grey to Green Initiative," and China's "Sponge City Programme." In India, the Smart Cities Mission and AMRUT reflect efforts toward sustainable urban infrastructure. However, financial constraints, including limited funding, high costs of land acquisition, and the complexities associated with economic valuation of non-market benefits, make it difficult to secure sustained financial support for BGI projects.

Keywords Blue and green infrastructure · Sustainable urban development · Resilience · Adaptation financing

---

S. Sharma · S. K. Sharma · S. Sen (✉)
Department of Humanities and Social Sciences, Indian Institute of Technology Roorkee, Roorkee, Uttarakhand, India
e-mail: subir.sen@hs.iitr.ac.in; subir.econ@gmail.com

M. G. Kibria et al. (eds.), *Urban Blue-Green Infrastructure Approach for Food Security and Climate Disaster Resilience*, Disaster Risk Reduction,
https://doi.org/10.1007/978-981-96-8746-6_14

14.1 Introduction

Urban areas globally are experiencing increased environmental, social, and economic challenges, particularly with the expansion of the population density of the cities. This rapidly increasing rate of urbanization has led to major changes in the natural landscapes with frequent rise of impervious surfaces, declining green cover, and deterioration of natural water systems (Udas-Mankikar and Driver 2021; Plieninger et al. 2022). These changes have aggravated various urban challenges, including but not limited to air and water pollution, urban heat island effects, urban flooding, and loss of biodiversity from deteriorating flora and fauna. To address these challenges, a series of policies and strategies, aligning with the sustainable development, have been advocated, in order to enhance the liveability and resilience of the urban regions. One such strategy is the integration of the **Urban Blue-Green Infrastructure (BGI)**.

Urban Blue-Green Infrastructure is a system of natural and semi-natural spaces that exists in and around urban centers or cities. It constitutes the elements of both "blue" and "green components." The **blue** element represents rivers, lakes, ponds, wetlands, and stormwater management systems, whereas the parks, gardens, street trees, green roofs, green walls, and urban forests constitute the **green** element (Well and Ludwig 2020). The BGI represents a system in which both the blue and green elements coexist to create a multi-purpose infrastructure that serves various functions ranging from managing the environment to providing social and economic benefits (Kaur and Gupta 2022). The objective of the BGI approaches is to build multifunctional landscapes that means bioswales, rain gardens, green roofs, and wetlands, all acting in unison to enhance ecosystem services and reduce the impacts of urban heat islands, stormwater runoff, flooding, and improving air quality, among others (Well and Ludwig 2020). The integrated system helps in mitigating the urban challenges and improves the urban biodiversity. The process acknowledges the role of interconnectedness of blue and green infrastructure in providing climate-resilient urban infrastructures and in mitigating the urban stress from climate change and environmental degradation.

Box 14.1 Successful Urban-BGI projects worldwide

- **Singapore**: Active, Beautiful, Clean Waters Program
- **New York City**: Green Infrastructure Initiatives
- **Melbourne**: Water Smart City Model
- **Copenhagen**: Cloudburst Management Plan
- **Toronto**: Green Roof Subsidy Program
- **Canada**: Rain City Strategy
- **China**: Sponge City Programme.

The Urban BGI is preceded by traditional urban infrastructure solutions, such as **Green** and **Gray Infrastructure**. Green Infrastructure program comprises of a set of water management practices, such as *green roofs, rain gardens, permeable pavements*, and *wetlands*, to manage urban environment and enhance the sustainability by improving the air and water quality, and stormwater management (Abbott et al. 2013). It was further integrated with the Gray Infrastructure initiatives to enhance the efficiency of stormwater management and wastewater treatment to deal with the likelihood of urban flooding and water pollution. It included the construction of structures, such as *water treatment plants, wastewater treatment plants*, and *dams and reservoirs* (Xu et al. 2019). Both the initiatives were combined together along with the component of Blue Infrastructure in the BGI with the aim to protect multiple ecosystem services by making it more sustainable and economically efficient.

The Urban BGI makes a holistic attempt to mitigate urban flooding, improve the water quality, ensure advanced stormwater management, enhance urban biodiversity, and reduce urban heat island effects (Lawson et al. 2014; Xu et al. 2019). Although there are several benefits of the Urban BGI, yet its implementation is challenging. For example, policy and institutional barriers (conflicting priorities among competing governance institutions) (Deely et al. 2020), technical barriers (lack of adequate technical expertise) (Keeley et al. 2013; Johns, 2019), low community engagement and social acceptance (due to lack of awareness about the benefits of the BGI) (Keeley et al. 2013; Furlong et al. 2018), and possibly the most important of all, the financial constraints (high costs in the planning, design and implementation) (Dupras et al. 2015; Di Marino et al. 2019).

Despite the role of BGI in mitigating urban challenges, such as environmental degradation, increasing population density, and climate change, the implementation of the Urban BGI projects is plagued with severe financial bottlenecks. These include a diversified range of funding avenues needed to finance the cost of the initial capital infrastructure, and operation and maintenance at later stages, thereby ensuring their successful implementation in the urban landscape. The funding mechanisms may include public–private partnerships (PPPs), Grants from the government (national or sub-national), and environmental impact bonds, among other green infrastructure financing tools (Merk et al. 2012). Cities and urban centers can also enhance their funding portfolios with private investments through new financing models sharing both risks and rewards associated with BGI projects. It demonstrating both, possible cost savings and environmental benefits such mechanism could yield. This illustrates the mixed sources of financial support for Urban BGI projects. The role of the community-led initiatives cannot be isolated from the financing equation. In many cases, the local stakeholders play an important role in promoting the BGI projects and their involvement helps in enhancing the perceived value and acceptance of investments in BGI (Deely et al. 2020). Community-led initiatives and funding programs create ownership and accountability—which are the necessary pre-requisites needed for long-term sustainability of BGI solutions.

Given the increasingly stressful environmental challenges that cities are increasingly facing, financing Urban Blue-Green Infrastructure is much more than an engineering and planning pursuit. It is strategic resource management. Effective financing strategies are essential in facilitating the uptake of sustainable urban practices to enhance resilience and quality of life in an urban setting. Therefore, comprehensive financial planning and investment are more urgent than ever.

The remainder of the chapter is organized as follows: Sect. 14.2 presents the overall understanding of the BGI. Section 14.3 discusses the various challenges and barrier in the implementation of BGI. Section 14.4 illustrates the different financing mechanisms available to fund the BGI. The final section concludes the chapter.

14.2 Understanding the Blue-Green Infrastructure (BGI)

14.2.1 The Need for Urban Blue-Green Infrastructure

The world has observed a rapid increase in the rate of urbanization in the past few decades. The current demographic landscape reveals that 55% of the global population resides in urban areas, and projections by the United Nations indicate a substantial increase in this proportion, anticipated to reach 68% by the year 2050 (UN DESA [1] 2015). This rise in the growth of cities and urbanization often results from the loss of natural environments (Padigala 2012). The rural–urban migration further increases the burden on the urban resources and contributes to environmental damage (Kundu 2011). Urbanization frequently results in the physical expansion of cities, which can encroach upon and diminish natural landscapes (Kaur and Gupta 2022). This phenomenon, where urban development supersedes the preservation of natural environments, poses a significant challenge to maintaining ecological balance and biodiversity. The rapid expansion of the concrete structures impacts the permeability of the land which in turn impacts ground water recharge and water retention capacity, causing surface water runoff and flooding, respectively. These disrupt the hydrological cycle of the urban areas (Du et al. 2015). In addition, the scarcity and poor quality of water, and overflowing sewers due to lack of adequate drainage infrastructure, are common concern of the urban areas (Li et al. 2017). Hence, the shift towards urbanization underscores the need for comprehensive planning and sustainable development strategies to accommodate the growing urban population while at the same time, mitigating potential challenges associated with rapid urbanization.

Climate change is the by-product of the interactions between humans and environment where humans have excessively drawn resources from the environment to meet their wants (IPCC 2014). However, we may not be prepared for the adversaries following the change in the rate of the climate change (Plieninger et al. 2022). The increased occurrences of droughts, floods, wildfires, and heat waves among others in the recent decades reveal the vulnerabilities that the ecosystem and the humans may be exposed to as result of the climate change (Ghofrani et al. 2017). There may be a significant loss to biodiversity following such climate extremes. The impact of such variability in the climate is further aggravated by the lack of awareness and

[1] United Nations. Department of Economic and Social Affairs.

preparedness of the communities and the concerned agencies responsible for urban planning (Faggian and Sposito 2009). This has caused climate change to be a subject of debate among the policymakers and environmental agencies at the global level (Göpfert et al. 2019). There are two primary responses in the form of adaptation and mitigation. Steps can be taken to adapt to the changing climatic conditions and minimize it as much possible or do not take necessary steps at the earliest that may subject us to more vulnerabilities and dreadful consequences (Leary et al. 2012).

The first step toward the adaptation is the mitigation of vulnerabilities and minimization of exposure to current climatic fluctuations (Ghofrani et al. 2017). With increasing awareness about climate change and the associated vulnerabilities, there has been growing support for nature-based solutions to address the climatic risks (Foster et al. 2011). This requires promotion of sustainable engineering solutions such as Blue-Green Infrastructure (BGI) (Sánchez and Govinadarajulu 2023). Blue-green infrastructure (BGI) can be referred to as an inter-connected system consisting of natural and engineered landscape elements, encompassing water bodies such as rivers, streams, lakes, and green spaces including parks and marshlands (Abbott et al. 2013; Xu et al. 2019). This integrated network facilitates harmony in the joint existence of humans and environment without significant loss of biodiversity within the urban setup. The adaptation strategies though BGI provide resilience against future extreme climatic events and assure mental, social, and economic well-being without compromising the environmental quality (Gentle et al. 2001; Foster et al. 2011; IPCC 2014; Connop et al. 2016). Given the significant benefits received from the BGI in the restoration of biodiversity and its growing awareness, a comprehensive review of the approaches and challenges involving BGI is extremely relevant.

14.2.2 Types of Urban Blue-Green Infrastructure

There have been serious discussions considering the factors to classify the typology of the BGI (Koc et al. 2017). The factors include considerations such as whether to it making technical, man-made intervention, or natural and ecologically driven (Coutts 2016; Connop et al. 2016), whether it should address local challenges or should cover wider landscapes (Hansen et al. 2019). Addressing such considerations require a deeper understanding of the local and global socio-economic and environmental, and availability of the necessary infrastructures to support the BGI intervention (Mell and Scott 2023).

The type of Urban BGI can be categorized as micro, meso, and macro depending on the scale of the intervention worldwide (Sehrawat and Shekhar 2023). The micro-scale solutions are the ones that are adopted by the individual households or applied along the streets. They are useful in addressing the urban heat island effect and graywater management. Examples of the micro-scale solutions include rainwater harvesting, rain gardens and bioswales, green roofs, and vertical gardens. The meso-scale solutions pertain to more extensive and focused strategies if implementing BGI, operating between micro-scale and macro-scale systems. They are mostly implemented within the geographical boundary of the city, in the nearby localities, towns, or specific areas. Examples of the meso-scale solutions include, urban wetlands and ponds, green corridors, permeable pavements, and green streets. Macro-scale solution pertains to the wider landscapes and citywide or beyond city strategies to implement the BGI. They largely cater to the regional or zonal environmental concerns. Examples of macro-scale solutions are sponge cities, flood plain restoration and river reclamation projects, and integrated urban water management plans (Table 14.1).

Table 14.1 Scale-wise typology of BGI implementation strategies

Scale	Solutions	Benefits
Micro	Rainwater harvesting	Provides access to freshwater for non-potable purposes
	Rain garden	Enhances stormwater management and adds aesthetic value
	Bioswales	Helps in effective stormwater management and minimizes soil erosion and flooding
	Green roofs	Improves air quality and reduces heat island effect
	Vertical gardens	Helps in mitigating island heat effect
Meso	Urban wetlands and ponds	Helps in flood control and water purification
	Urban parks	Enhances water retention capacity controlling urban flooding, enhances vegetation and green cover of the locality
	Street trees	Habitat creation, flood mitigation, and aesthetic greening
	Green corridors	Restoring natural habitats
	Permeable pavements and green streets	Curb surface water runoff and enhance stormwater infiltration
Macro	Sponge cities	Helps in natural water retention and the risk of flooding
	Flood plain restoration and river reclamation	Helps in water storage, reduced urban flooding, and enhance ecological value of water bodies
	Integrated urban water management	Improve water footprints of the city/region by wastewater treatment, water reuse, and stormwater management
	River corridors	Helps in water storage addressing water scarcity, helps in drought control by providing access to water in water scant areas, adds aesthetic value to the water bodies
	Formal green belts	Reduces urban heat, flooding, aesthetic greening, restores natural habitat

Source Authors' own elaboration

14.3 Challenges and Barriers to Urban BGI Implementation

14.3.1 Technical and Topographical Challenges

A major hindrance to the effective implementation of the BGI is related to the uncertainties around the design, planning, site of the project, construction, and maintenance of the project (Govindarajulu 2014; Sehrawat and Shekhar 2023). It is primarily related with the concerned departments or ministries that are required to undertake the project. The managers handling the operations of the BGI project often report limited capability in the designs and lack of technical expertise and experience in handling such projects as a cause of concern (NRC 2009; Sun et al. 2020). There is also lack of sufficient data about the performance of the existing BGI projects as its benefits are intangible and are visible mostly in the long run (Mguni et al. 2016; Kaur and Gupta 2022). Such limitations tend to raise questions about the effectiveness of the project.

The biophysical or topographical challenges mainly include the terrain of the project site, the soil characteristics, the physical characteristics of the area, and the climatic zone of the site, among others (Pataki et al. 2011; Liao et al. 2017). Reviewing all the challenges discussed, one of the most prominent ones is the lack of physical space which is a major concern in the cities as BGI projects need greater land cover as compared to its traditional alternatives, such as gray or green infrastructures (Furlong et al. 2018; Sehrawat and Shekhar 2023). Technologies should be designed considering the local conditions and addressing the challenges of the specific localities. This may help in successful implementation, operation, and maintenance of the project (Li et al. 2017). The uncertainties around the design of the BGI may negatively impact the induction of the project. An uncertainty about the design is regarding the health benefits from the improved ecosystem services that is promoted but the demonstration of the same is not clearly visible (Deely et al. 2020). In addition, there is no clarity about the specific health benefits from the project. In the absence of adequate knowledge about the design, the project may cause harm than good, for instance, some plant species may cause negative impact on health (Cosgrove and Loucks 2015; O'Donnell et al. 2017). The designing includes the engineering of the project and the selection of appropriate technologies and methods tailored to suit the physical conditions of the site, among others.

A major design-related challenge is the construction of the infrastructure, particularly incorporating the already existing alternative technologies to the relatively new BGI (Evans et al. 2017). However, it is often argued that integrating the present infrastructures including the green and gray infrastructures can be more cost-saving than considering the BGI in isolation (Liu et al. 2016; Fryd et al. 2013).

14.3.2 Regulatory Challenges

The presence of an appropriate regulatory mechanism is an important aspect in the effective implementation of the urban BGI (Ahmed and de Oliveira 2017; Mansur et al. 2022). Gersonius et al. (2016) state that clarity in the legal arrangements is a pre-requisite for a sustained and long-term commitment in the new responsibilities. A relatively poor amalgamation of the rules and regulations into the new responsibilities often results in the failure to utilize the full potential of an intervention (Kremer et al. 2016; Dhakal and Chevalier 2016). An example from China is the consideration of the stormwater infrastructure system to only consider the graywater management by the national regulators and ignore other forms of wastewater treatments (Li et al. 2017).

Another major concern is the coordination among various regulatory authorities and ministries to extend hands of cooperation and remain inter-connected through a single channel (Borgström 2019; Udas-Mankikar and Driver 2021). An example of the same is the "Santiago Verde" or "Green Santiago" plan that was proposed by Santiago during 2006–2011 to create an inter-connected network of the BGI at the regional level. However, the program could not be finalized and remained only at the conceptualization stage due to their failure to create a joint network of multiple jurisdictions and sectors (Vásquez et al. 2016). However, China too introduced such an initiative named, "Sponge City" program, following the severe flooding in 2012. This program was jointly supported by its Ministry of Water, Resources, the Ministry of Housing and Rural–Urban Development, and the Ministry of Finance (Li et al. 2017).

Another major hindrance in the effective implementation of the urban BGI is the discrepancies in the property rights (Baptiste et al. 2015). For instance, in the US, the plantation of trees do not have to be dealt with any property right regulations, however, the protection of the existing tress from being cut-down is a challenge. Similar rules are practiced in Australia regarding afforestation and deforestation (Furlong et al. 2018).

In order to bring the BGI to the mainstream, change in the paradigm of Governance is a key, say, from centralized to the decentralized governance. Dhakal and Chevalier (2016) proposed the development of a new unit, "local hydrological district" in the US that will be responsible in the identification, management, and operation of the green infrastructures in the wastewater management. The regulatory authorities at the regional level may be present only to ensure vigilance of the local hydrological districts. However, it was observed that there was little interest on part of the agencies to decentralize governance at the district level owing to lack of technical support and financial constraints. This highlights that although some ideas may be conceptualized from the theoretical perspective, however, their transition to reality may be plagued by technical, institutional, and financial constraints.

14.3.3 Socio-Cultural Challenges

The promotion of social equity is a major component of every initiative, especially the ones inducted by the Government. However, ensuring equitable distribution of resources is a major concern (Barnhill and Smardon 2012; Williams et al. 2019). A common phenomenon in the cities is that the wealthier or high-income group citizens have greater access to green vegetation and green space as compared to the low-income group or marginalized citizens (Hoang and Fenner 2016). Access to such green space is associated with positive impacts on the mental, physical, and social well-being of the citizens (Chiesura 2004; Depietri and McPhearson, 2017). This disparity in the access to the ecosystem services acts as an environmental injustice to the marginalized (Smiley et al. 2016). Arshad and Routray (2018) state that marginalized community has lack of access to resources and in all possibilities, this also means lack of access to the BGI.

An important factor to be considered in the intervention programs such as BGI is the presence of basic infrastructures needed to support further development projects (O'Donnell et al. 2017). Such infrastructures are generally located in the richer neighborhoods and hence they are the early beneficiaries of BGI projects as compared to their poorer counterparts (Parr et al. 2016). The implementation of such projects may increase the standard of living and the aesthetics of nearby areas, and hence raise the property prices that may further make it unaffordable for the poor who may be then forced to migrate to the outskirts of the city owing to the "rent-gap" problem (Smith 1984; Gulsrud et al. 2018). Hence, attempts should be made to eliminate disparity in the provisioning of such interventions.

The perspectives of the various stakeholder's group should also be considered while considering the BGI projects. The stakeholder's group primarily involves the manager's group and the resident's group (Deely et al. 2020). The manager's group is often plagued by the "technocratic path dependence" barrier (Matthews et al. 2015; Wihlborg et al. 2019). To put, in other words, it simply implies that managers are hesitant to adopt new strategies or technologies to deal with an issue as they are more comfortable and used to the previously used technologies or methods (David 2001). Hence, there may be reluctance to adopt or invest in a relatively new BGI regime. Similarly, the resident's group may show unwillingness to adopt to new BGI projects and alter their existing practices (Johns 2019). In order to ensure an intervention to be successful, active involvement and adaptation of the citizens is an important aspect (Derkzen et al. 2017). This remains true for the success of the BGI projects as well. If people are not convinced that BGI projects may be vital in dealing with the associated concerns, successful implementation of the same cannot be ensured.

A clear understanding and knowledge of the functionality and the benefits of the BGI is rudimentary for its effective implementation (Schiappacasse and Müller 2015; Evans et al. 2017). Equally important is the awareness about its alternative technologies, for instance, gray infrastructures and its impact on the biodiversity. The lack of clarity among the stakeholders may be the reason for limited or sometimes no awareness of its benefits and the ecosystem services that it ensures and restores

(Deely et al. 2020). Lack of information about the proposed BGI project may be a reason for its low acceptance as a potential solution to address the environmental concerns (Sussams et al. 2015). However, another factor could be lack of culture in the society of placing high values to the ecosystem services provided by the project.

In some other cases, few communities may be willing to support the BGI projects, however, under representation of their community may lead to under-valuation and delay in the BGI process (Hoyle et al. 2017; Finewood et al. 2019). Hence, a major cause of inequitable development of BGI projects and their subsequent benefits could be the lack the equitable representation of the community. Hence, joint participation of the community and the organizations is also key to successful implementation of the BGI projects (Sharma and Tomar 2010; Chu et al. 2016). Avalon Green Alley Network project in Los Angeles (LA) is an example of such joint cooperation between the community and the regulatory authority. The area composed six valleys that experienced severe flooding every year. However, the project was jointly inducted and adapted by the communities and the regulatory agencies in LA that not only reduced the flooding but also improved the living conditions of the neighborhood.

The ownership of the BGI projects, public or private, is also an important subject of discussion. Since there are positive externalities from the BGI projects to the other residents as well who have not paid for the project, private entrepreneurs may have little interest in the BGI projects (Di Marino et al. 2019). Private entities may only be interested in such projects if they can be made not excludable and non-rival in consumption. However, in such an instance, the benefits of such ecosystem services may not be able to meet its socially desirable objectives.

14.3.4 Institutional Challenges

Institutional barriers arising due to poor governance may be cited as one of the prominent barriers in relation to the BGI (Kabisch et al. 2017). It may cause lack of effective policies concerning BGI process and lack of strong leadership that may further impede the pace of induction of the BGI projects and their operation and maintenance (Borelli et al. 2018; Romero, 2016). Joint operation of the concerned departments and agencies are essential to ensure effective implementation of the BGI (Borgström 2019; Udas-Mankikar and Driver 2021). Currently, this may be lacking as the environmental agencies managing various ecosystem services function separately as different departments under different ministries. Lack of communication among such different managerial groups may limit the benefits that may arise from the BGI projects (Thorne et al. 2015). Another cause of concern in this regard is the long-term commitment from the managers (as the projects may take a long time), especially if they are unsure about its completion (Bissonnette et al. 2018; Stefanakis et al. 2021). Political will and consciousness is also equally important (Thorne et al. 2015). A deterrence in that may hinder the motivation to modify the legislation in accordance to the changing needs. Poor leadership may give rise to lack of clarity about the roles and responsibilities of the respective stakeholders (Romero 2016). This may further cause delay in the induction of the BGI projects and operation and maintenance of the existing ones.

Long-term vision and monitoring is vital in the successful implementation of any project (Dhakal and Chevalier 2016). BGI projects may require a long term to be completed and hence require continued investment and monitoring during the period (Bissonnette et al. 2018; Stefanakis et al. 2021). Neglecting this may impact the sustainability of the project. Lack of visionary approach may hinder the progress of the project. An example of the visionary approach could be the inclusion of the additional ecosystem services to the already existing BGI projects according to the changing requirements (Sehrawat and Shekhar, 2023). Focus should not be limited to any specific characteristic, rather, BGI projects that are designed to cater to multiple benefits require that all of its functionalities are monitored with equal focus. Failure to do so may prevent realization of the full potential of such interventions.

In addition to legislation, defining clear policy is an important component for the overall development of the new BGI projects and effective implantation of the existing ones (Deely et al. 2020). Currently, there is lack of prominent policies that have been defined bearing in mind the environmental factors or climate concerns, at the local or regional level (Furlong et al. 2018). Lack of such policies undervalue the need of BGI projects and also undermine their efficiency. However, some efforts have been made to incorporate climate science and environmental concerns in policymaking (Pickett et al. 2016). An example of the same is from the European Union (EU), where the EU is promoting the implementation of green infrastructures (GI) as an adaptation strategy in dealing with the climate change.

A major barrier of the institutions in the implementation of the BGI is the limited availability of resources and competing uses of the available ones. This limitation is similar to other infrastructural investment projects in the low- and middle-income countries, where a thorough cost–benefit analysis is conducted and projects are granted approval only if the benefits arising from the same outweigh the costs incurred (Hamel and Tan 2021). Since the benefits arising from the BGI projects are mostly intangible and are visible only in the long run, seeking investments for the same may be a difficult task (Zuniga-Teran et al. 2019). This is also because there is lack of mechanism to effectively calculate the benefits of the non-market environmental goods. Failure to evaluate the non-market ecosystem benefits from the BGI projects often leads to long delays in their implementation (Keeley et al. 2013).

14.3.5 Financial Challenges

The financial assistance is a major component of any infrastructure project. Hence, continued financial support is also an inherent component of the BGI without which its implementation cannot be successful (Hoyle et al. 2017; Di Marino et al. 2019). The financial aspect of the BGI impacts the associated stakeholders, namely, the residents, resource-owners, and managers differently (Deely et al. 2020). The residents are the ones who pay for the ecosystem services ensured from the project. The resource-owners are the ones who receive or are remunerated for the provisioning of the ecosystem services from the BGI intervention. The mangers may be the governing or regulating agencies that act as intermediaries between the beneficiaries and the providers of the ecosystem services.

In addition to the regulating the project, a major objective of the managers would be maximizing the social welfare arising from the BGI intervention. The funding of the BGI may not be limited to the private entrepreneurs, rather Government may also initiate steps toward the BGI interventions. This is particularly common in low-income and developing economies where private entrepreneurs may show limited interest in the BGI projects where there is lack of clarity about the returns from such investments (Steadman et al. 2023). However, a common concern as with most of the public investment is the limited funding with the Government and the competing used of the available funds (Toxopeus and Polzin 2021). This leads to discussions and debates about the selection of the project and eventually the finalized project is the one that ensures maximum social output. It has also been stated that BGI being a relatively new concept requires additional training and recruitment of experts that further enhance the financial burden of the project (Johns 2019; Meran et al. 2021) as compared to the gray infrastructures. There are notable differences in the gray infrastructures and the BGI as the latter requires more land, technical expertise, and funds to ensure resource-owners to participate in "payment for ecosystem services" (PES) schemes (Deely et al. 2020).

The cost–benefit analysis of the project is a major component that plays a vital role in the decision-making of the BGI project (Levy et al. 2014; Hamel and Tan 2021). The benefits associated to the BGI include improved aesthetic value, stormwater management, availability and improved quality of drinking water, protection from flood and drought, and improved vegetation and green cover, among others (Roberts et al. 2012; Hamel and Tan 2021). The economic valuation of these benefits is difficult and may require complex econometric analysis (Connop et al. 2016; Hamin et al. 2018; Finewood et al. 2019). Revenue can be generated from the BGI only if people can be made aware and be educated about the multiple benefits of the BGI (Sehrawat and Shekhar 2023).

There are various costs associated to the BGI, one of the most prominent is the cost of procurement of land (Pontee et al. 2016). There are two major categories of costs involved in the acquisition of land, the opportunity cost, and the transaction cost (Deely et al. 2020). The opportunity cost may be regarded as the loss of the alternative opportunities of using a good or service. In terms of land acquisition, it may be said that the landowner may not be able to use the land for further crop production and hence may be giving up on his forthcoming revenues from the same (Ryan et al. 2016). Moreover, he/she may assume that the land prices would rise in the future and he/she may gain if the land is held back. These assumptions make the acquisition of land, a difficult process. In such cases, they must be compensated with at least that the next best alternative may offer. In some other circumstances, the landowners may not be willing to sell the land at any rate owing to the emotional and cultural connect with the land (McGurk et al. 2019).

The other type of costs may be referred to as transaction costs. It may be regarded as the amount that is necessary to ensure that the buyers and sellers may enter into an agreement (Deely et al. 2020). This amount may be different for developed and developing economies, based on certain factors. The benefits accrued from the ecosystem services and the costs incurred in monitoring and maintenance of the BGI may be regarded as important factors determining the transaction costs for developed economies (Wunder 2005). For developing economies, the transaction costs may further consider genuineness of the land ownership, certainty regarding remunerations and cultural factors (Schomers and Matzdorf 2013).

In addition to the economic valuation of the ecosystem services provided by the BGI, an important concern is determining the link between the beneficiaries and the service providers. This is important to generate revenues from the beneficiaries that are mostly in the form of tax credits or some form of cess, user charges, or taxation (Mell 2018). However, as there are positive externalities such of the BGI that are spread across people, it is difficult to identify the beneficiaries and tax them accordingly (Toxopeus and Polzin 2021). Hence, there may be a free-rider problem[2] with respect to the provisioning of the BGI (Obeng et al. 2018). This prevents the provisioning of the services at the socially desirable level as the provider may be incurring losses.

From the purview of the beneficiaries or the citizens, they are sometimes required to pay to restore the services that were not harmed by them (Ferraro and Pattanayak 2006; O'Brien et al. 2019). For instance, the waterbodies in a city may be polluted from the by-products or waste discharged from the nearby factories but the entire city may be taxed to restore the quality of such loss of biodiversity. These steps discourage public support and participation, and also reduce their willingness to pay for genuine BGI projects as well (Toxopeus and Polzin 2021). Hence, surveys should be conducted so that the projects tax the targeted participants who are the real beneficiaries of the services (Engel et al. 2008).

14.4 Financing BGI

Blue-Green Infrastructure (BGI) is gaining recognition worldwide as a sustainable approach to urban development, encompassing a range of natural and engineered systems designed to manage water, enhance biodiversity, and promote ecological resilience (Sanchez and Govindarajulu 2023; Udas-Mankikar and Driver 2021). As cities grapple with the challenges of rapid urbanization and climate change, financing BGI projects has emerged as a critical issue on the global agenda (Dhanya et al. 2022). From innovative financing mechanisms to public–private partnerships, various strategies are being employed to fund BGI initiatives and create more resilient and

[2] The problem of "free-riders" refers to a situation of a market failure where an individual or group of individuals benefit from the goods or services but are unwilling to pay for it (Musgrave and Musgrave 1973).

livable urban environments (Chakraborty 2023). In the context of India, the notion of blue-green infrastructure is relatively nascent and is gradually being defined through its incorporation into various policies at the central, state, and sub-regional levels (Sengupta et al. 2022). For example, the Delhi Development Authority,[3] tasked with formulating a blue-green masterplan for the city, delineates blue and green infrastructure distinctly within the realm of urban planning. Blue infrastructure encompasses water bodies such as rivers, canals, ponds, wetlands, floodplains, and water treatment facilities, while green infrastructure comprises elements like trees, lawns, hedgerows, parks, fields, and forests.

As policymakers define and refine the concept of blue-green infrastructure within the Indian context, there is a growing need to align financing mechanisms with the objectives of promoting sustainable urban development (Mell 2018; Udas-Mankikar and Driver 2021). Public funding, private investments, and innovative financing models can play pivotal roles in mobilizing resources for blue-green infrastructure projects, ensuring their successful implementation and long-term viability. These components of financing are explained in the next sub-sections.

14.4.1 Public Funding

Public funding serves as a cornerstone for financing BGI projects globally, with governments at all levels committing resources to promote environmental sustainability and resilience in urban areas (Peters and Landström 2022). Initiatives such as the European Union's Horizon 2020 program and the United States' Clean Water State Revolving Fund demonstrate the importance of public investment in supporting BGI initiatives (Wilbers et al. 2022). Through budget allocations, grants, and subsidies, governments aim to incentivize the adoption of sustainable practices and technologies, driving the implementation of BGI projects across diverse urban landscapes.

14.4.1.1 Smart City Missions (SCMs) and AMRUT: Indian Context

Government-led initiatives play a crucial role in financing BGI projects, with budgetary allocations supporting a wide range of initiatives aimed at enhancing urban resilience. Examples from cities like Copenhagen, where investments in green infrastructure have transformed the urban landscape, highlight the positive impact of government funding on BGI implementation. However, challenges such as bureaucratic delays and administrative hurdles can hinder the efficient use of public funds, underscoring the need for streamlined processes and effective governance mechanisms.

---

[3] Landscape and Environment Planning Department. More information can be accessed at https://dda.gov.in/roles.

One of the primary sources of financing for blue-green infrastructure (BGI) projects in India is through government budget allocations, particularly through initiatives like the Atal Mission for Rejuvenation and Urban Transformation (AMRUT)[4] and the Smart Cities Mission (SCM).[5] These missions, spearheaded by the central government, allocate significant funds toward urban development projects, including those aimed at enhancing environmental sustainability through the implementation of BGI. AMRUT, launched in 2015, focuses on improving basic urban infrastructure in cities across India, with an emphasis on water supply, sewerage, and urban green spaces (Smith and Pathak 2018). Under this mission, substantial budgetary allocations are earmarked for the development of green infrastructure projects such as parks, gardens, and tree plantation drives. Similarly, the SCMs, initiated in 2015, allocate funds to selected cities to undertake transformative projects aimed at improving urban livability and sustainability (Shruti et al. 2020; Singh and Mishra 2023). BGI projects, including those related to water management, green spaces, and sustainable transportation, receive funding under this mission (Randhawa and Kumar 2017).

14.4.1.2 Grants and Subsidies

Grants and subsidies play a vital role in financing BGI projects globally, providing financial incentives for the adoption of sustainable practices and technologies (Myong Koh 2018). In countries like Australia, government-funded programs support initiatives such as green roof installations and rainwater harvesting systems. While grants and subsidies can accelerate the uptake of BGI measures, challenges related to fund disbursement and allocation inefficiencies may arise, necessitating improved coordination and monitoring mechanisms to ensure effective implementation. In addition to budget allocations, grants and subsidies provided by the government play a crucial role in financing BGI projects in India. Various central and state government agencies offer financial assistance in the form of grants and subsidies to promote the implementation of sustainable urban infrastructure projects (Malhotra et al. 2022). These grants often target specific components of BGI, such as rainwater harvesting systems, green roof installations, and wastewater treatment plants. Additionally, subsidies may be provided to incentivize private sector participation in BGI initiatives, thereby lowering the financial burden on project implementers.

---

[4] More information on AMRUT is available at https://mohua.gov.in/upload/uploadfiles/files/AMRUT-Operational-Guidelines.pdf.

[5] More information on SCM can be accessed at https://smartcities.gov.in/sites/default/files/2023-09/SCM_UN_Report%20.pdf.

14.4.2 *Private Sector Engagements*

Private sector involvement in financing BGI projects is essential for unlocking additional resources and driving innovation in urban sustainability (Trencher et al. 2013). Venture capital investments, green bonds, and alternative financing models offer promising avenues for mobilizing private funds toward BGI initiatives. Examples from cities like Singapore, where public–private partnerships have successfully implemented BGI projects, demonstrate the potential of private sector engagement in advancing urban resilience (Shamsuzzoha et al. 2021). However, challenges such as market volatility and investment risks can deter private investors, highlighting the importance of regulatory certainty and risk mitigation strategies.

14.4.2.1 Green Bonds and Venture Capital

Green bonds[6] have emerged as a popular financing mechanism for BGI projects, attracting investors seeking to support environmentally sustainable initiatives. Green infrastructure is primarily viewed as the responsibility of the local government as it is a public good (Toxopeus and Polzin 2021). However, the local governments find it extremely challenging to fund such costs/investments using local taxation due to the low willingness of the people to pay. In contrast, countries like Sweden have successfully used green bonds to fund projects such as urban green spaces and stormwater management systems (Maltais and Nykvist 2020). While green bonds provide opportunities for diversifying investment portfolios, challenges such as market liquidity and bond pricing can affect their appeal to investors, highlighting the need for standardization and transparency in bond issuance. In the UK, local planning authorities are considering public-based crowdfunding to finance their green investment projects. Notably, the Swindon Borough Council has become the first to secure funding for solar park development through crowdfunding, with other councils contemplating similar approaches (Davis and Cartwright 2019). Alternatives to green bonds include co-financing investment projects among interested stakeholders or partners (O'Donnell et al. 2017; Tubridy 2021). These mechanisms, which encompass green bonds and other debt-financing sources, can also address shortfalls in investment typically covered by traditional funding sources like taxes and fees (Cousins and Hill 2021). Besides taxes and fees, local authorities can finance their infrastructure through asset sales, park trusts, and other means. Community-led funding initiatives include guerrilla gardening and community asset transfers, while private financing sources consist of corporate sponsorships and private management of public utilities (Mell 2018).

In addition to this, private sector investments, particularly through venture capital funding, play a crucial role in accelerating BGI projects worldwide. From green infrastructure startups to sustainable development ventures, venture capital firms

[6] Green bonds are fixed income securities used for raising capital through environmental-friendly projects (Sartzetakis2021).

are increasingly recognizing the potential for high returns in the BGI sector (Lin 2022). Countries like China are making use of these instruments in developing business cases for urban sustainability (Cheng et al. 2019). However, challenges such as regulatory barriers and market uncertainties may impede investment flows, emphasizing the need for supportive policy frameworks and risk mitigation mechanisms (Toxopeus and Polzin 2021). Private sector investment, including venture capital funding, represents another significant source of financing for BGI projects in India. With the increasing recognition of the economic and environmental benefits associated with BGI, private investors are increasingly showing interest in funding projects in this domain. Venture capital firms specializing in environmental sustainability and infrastructure development may provide funding for startups and enterprises involved in the design, implementation, and management of BGI projects.

14.4.2.2 Alternative Finance

Alternative financing models, including Corporate Social Responsibility (CSR) initiatives and Payment for Ecosystem Services (PES) schemes, offer innovative pathways for funding BGI projects globally. Examples from cities like Kolkata, where CSR contributions have funded large-scale green infrastructure projects, demonstrate the potential of these models in promoting collaboration between public and private stakeholders. However, regulatory compliance and resource allocation challenges may pose obstacles to the successful implementation of alternative financing models, underscoring the importance of stakeholder engagement and capacity-building efforts.

14.4.3 Mixed Approach: PPP Scheme

Public–private partnerships (PPPs) present a collaborative approach toward financing and implementing BGI projects in India. Delhi's collaboration with private developers to create green spaces and water bodies within urban development's showcases the potential for PPPs to enhance urban sustainability. Similarly, Singapore's successful implementation of BGI projects through partnerships between government agencies and private stakeholders underscores the importance of collaborative efforts in realizing the vision of sustainable urban development. In addition, cities like Copenhagen and Rotterdam have leveraged PPPs to develop innovative stormwater management systems that integrate green spaces into urban landscapes, mitigating flood risks while enhancing public amenities (Wang and Ma 2021). Furthermore, in countries like the United States and Australia, PPPs have played a crucial role in rejuvenating degraded urban environments by transforming abandoned industrial sites into vibrant green corridors and recreational areas, revitalizing communities while promoting environmental sustainability. These examples underscore the versatility and scalability of PPPs in catalyzing transformative change in urban landscapes, demonstrating their potential to create resilient, liveable cities through collaborative efforts between public authorities, private developers, and local communities (Table 14.2).

Table 14.2 Pros and cons of blue-green infrastructure finance

Pros	Cons
Environmental benefits: Enhances biodiversity, reduces urban heat, improves air and water quality, and mitigates the impact of extreme weather events	*Substantial investment*: BGI projects require significant upfront investments, which can be a barrier for governments, particularly for developing economies
Resilience: Increases urban resilience to climate change and extreme weather events, helping cities adapt to future environmental challenges	*Complex measurement for return on investment*: The financial benefits, such as cost savings from flood prevention, are often indirect and harder to measure, making it challenging to attract traditional investors
Social benefits: Provides recreational spaces, enhances mental well-being, improves public health, and can increase property values	*Long payback periods*: Returns on BGI investments may be realized over a long period, making it less appealing to investors looking for quicker returns
Cost-effective: Can reduce infrastructure repair costs, lower flood damage expenses, and save on water treatment costs over time	*Maintenance requirements*: BGI often requires ongoing maintenance to remain effective, which can be a significant operational cost over time
Employment opportunities: Creates employment opportunities in different areas such as landscaping, construction, and maintenance	*Risk of gentrification*: Increased property values in areas with BGI can lead to higher rents and living costs, potentially displacing low-income residents
Funding availability Growing support from green bonds, climate finance funds, and sustainable investment vehicles increases the potential for BGI financing	*Land use constraints*: Implementing BGI can require large spaces, and finding available urban land can be challenging. Additionally, it may conflict with existing land uses
Multifunctionality: Combines infrastructure with ecological functions, supporting water management, recreation, and carbon sequestration in a single investment	*Regulatory constraints*: BGI projects can face bureaucratic challenges, as they may require changes in zoning, environmental approvals, and interagency coordination
Community engagement: Involving communities in BGI projects can lead to better social outcomes and community ownership of green spaces	*Financial risk and investor hesitancy*: BGI is still relatively new, and investors may perceive it as risky or uncertain compared to conventional infrastructure investments

Source Author's own elaboration

14.5 Conclusion

This chapter illustrates the growing significance of BGI in the urban landscape, highlighting various successful BGI initiatives across the world, such as Singapore's Active, Beautiful, Clean Waters Programme, the Grey to Green Initiative in the United States, Canada's Rain City Strategy, and China's Sponge City Programme. These examples demonstrate the potential of BGI to enhance urban resilience and address climate challenges through innovative nature-based solutions. The accelerated pace of urbanization is having significant ecological impacts, including loss of natural habitats, increased surface water runoff, reduced groundwater recharge, and degradation of overall water quality. These challenges underscore the need for sustainable urban development strategies that integrate natural and engineered solutions to enhance urban resilience. BGI, as an inter-connected system of water bodies and green spaces, offers a comprehensive approach to mitigate the environmental impacts of urbanization and climate change. By incorporating natural elements such as rivers, lakes, parks, and engineered components like green roofs, permeable pavements, and bioswales, BGI promotes sustainable urban environments. These infrastructures not only enhance the aesthetic value of urban areas but also provide multiple benefits such as improved air quality, reduced urban heat island effects, stormwater management, and flood control. Moreover, BGI supports biodiversity conservation, water conservation, and recreational opportunities, contributing to the overall well-being of urban communities. Given the urgent need for adaptive responses to climate change and urbanization, promoting BGI is crucial for creating resilient urban landscapes. The adoption of BGI not only mitigates current vulnerabilities but also prepares urban areas for future climatic extremes, fostering sustainable, liveable, and ecologically balanced cities. This illustrates the importance of incorporating BGI into urban planning and policymaking to ensure sustainable development and ecological harmony in the rapidly urbanizing world.

The implementation of urban BGI faces numerous challenges across various domains, including technical, regulatory, socio-cultural, institutional, and financial aspects. Technical and topographical challenges primarily revolve around uncertainties in design, planning, construction, and maintenance, coupled with a lack of technical expertise and adequate data on existing BGI projects. These issues are compounded by site-specific biophysical constraints such as terrain, soil characteristics, and climate, as well as the complexities of integrating BGI with existing infrastructure. Regulatory challenges include the absence of clear legal frameworks, poor coordination among multiple regulatory bodies, and inconsistencies in property rights, all of which hinder the effective implementation and integration of BGI. Socio-cultural challenges, such as unequal access to green spaces and resistance from stakeholders due to technocratic path dependence or lack of awareness, further impede the adoption of BGI. The success of these projects often relies on effective stakeholder engagement and the promotion of social equity that remains a significant hurdle in urban settings. Institutional and financial barriers also play a crucial role in limiting the potential of BGI projects. Institutional challenges are primarily driven by poor governance, lack of inter-departmental cooperation, insufficient long-term commitment, and inadequate policy frameworks that fail to prioritize environmental and climate concerns. These issues often lead to delays in project initiation, operation, and maintenance. Financial constraints, including limited funding, high costs of land acquisition, and the complexities associated with economic valuation of non-market benefits, make it difficult to secure sustained financial support for BGI projects. Moreover, lack of a clear mechanism to link beneficiaries with service providers complicates revenue generation, further exacerbating financial challenges. Overcoming these barriers requires a concerted effort to enhance technical capabilities, establish robust regulatory and policy frameworks, promote social inclusivity, and secure long-term financial investment to ensure the sustainability and effectiveness of urban BGI initiatives.

Financing BGI is pivotal to promoting sustainable urban development in the face of rapid urbanization and climate change. This requires a diversified approach that leverages public funding, private sector investments, and alternative financing models. Public funding remains foundational, with government initiatives such as the SCM and AMRUT playing crucial roles in supporting BGI projects in India. These initiatives have demonstrated the potential to transform urban landscapes by allocating substantial budgetary resources and offering grants and subsidies to promote sustainable practices. However, the effectiveness of public funding is often challenged by bureaucratic delays and administrative inefficiencies, necessitating streamlined processes and enhanced governance frameworks to ensure the optimal use of resources. Moving forward, aligning financing mechanisms with strategic urban development goals and fostering a collaborative environment among stakeholders will be essential to overcoming the existing challenges and unlocking the full potential of BGI as a tool for sustainable urban transformation.

References

Abbott J, Davies P, Simkins P, Morgan C, Levin D, Robinson P (2013) Creating water sensitive places–scoping the potential for water sensitive urban design in the UK. CIRIA, London

Ahmed A, Puppim de Oliveira JA (2017) Integration of biodiversity in urban planning instruments in developing countries: the case of Kumasi metropolitan assembly, Ghana. J Environ Plann Manage 60:1741–1764

Arshad HSH, Routray JK (2018) From socioeconomic disparity to environmental injustice: the relationship between housing unit density and community green space in a medium city in Pakistan. Local Environ 23(5):536–548

Barnhill K, Smardon R (2012) Gaining ground: green infrastructure attitudes and perceptions from stakeholders in Syracuse, New York. Environ Pract

Baptiste AK, Foley C, Smardon R (2015) Understanding urban neighborhood differences in willingness to implement green infrastructure measures: a case study of Syracuse, NY. Landsc Urban Plan 136:1–12

Borelli S, Conigliaro M, Quaglia S, Salbitano F (2018) Urban and Peri-urban agroforestry as multifunctional land use. Agrofor Anecdotal Mod Sci:705–724

Borgström S (2019) Balancing diversity and connectivity in multi-level governance settings for urban transformative capacity. Ambio 48:463–477

Bissonnette JF, Dupras J, Messier C, Lechowicz M, Dagenais D, Paquette A, Jaeger JAG, Gonzalez A (2018) Moving forward in implementing green infrastructures: stakeholder perceptions of opportunities and obstacles in a major North American metropolitan area. Cities

Chakraborty D (2023) Challenges and opportunities of green infrastructure in urban India. In: Fifth world congress on disaster management, vol V. Routledge, pp 3–9

Cheng C, Hua Y, Tan D (2019) Spatial dynamics and determinants of sustainable finance: evidence from venture capital investment in China. J Clean Prod 232:1148–1157

Chiesura A (2004) The role of urban parks for the sustainable city. Landsc Urban Plan 68(1):129–138

Chu E, Anguelovski I, Carmin JA (2016) Inclusive approaches to urban climate adaptation planning and implementation in the global South. Clim Policy 16(3):372–392

Connop S, Vandergert P, Eisenberg B, Collier MJ, Nash C, Clough J, Newport D (2016) Renaturing cities using a regionally-focused biodiversity-led multifunctional benefits approach to urban green infrastructure. Environ Sci Policy

Cousins JJ, Hill DT (2021) Green infrastructure, stormwater, and the financialization of municipal environmental governance. J Environ Plann Policy Manage 23(5):581–598

Coutts C (2016) Green infrastructure and public health. Routledge, Abingdon, UK

Cosgrove WJ, Loucks DP (2015) Water management: current and future challenges and research directions. Water Resour Res 51(6):4823–4839

David PA (2001) Path dependence, its critics and the quest for "historical economics" path dependence, its critics and the quest for "historical economics" In: Evolution and path dependence in economic ideas: past and present

Davis M, Cartwright L (2019) Financing for society: assessing the suitability of crowdfunding for the public sector. University of Leeds report

Deely J, Hynes S, Barquin J, Burgess D, Finney G, Silio A, Álvarez-Martínez JM, Bailly D, Ballé-Béganton J (2020) Barrier identification framework for the implementation of blue and green infrastructures. Land Use Policy 99:105108

Depietri Y, McPhearson T (2017) Integrating the grey, green, and blue in cities: nature-based solutions for climate change adaptation and risk reduction. In: Kabisch N, Korn H, Stadler J, Bonn A (eds) Nature-based solutions to climate change adaptation in urban areas. Springer, Cham, pp 91–109

Derkzen ML, Van Teeffelen AJ, Verburg PH (2017) Green infrastructure for urban climate adaptation: how do residents' views on climate impacts and green infrastructure shape adaptation preferences? Landsc Urban Plan 157:106–130

Dhakal KP, Chevalier LR (2016) Urban stormwater governance: the need for a paradigm shift. Environ Manage 57(5):1112–1124

Dhanya B, Ramananda S, Dhyani S (2022) Mainstreaming blue-green infrastructure in policy and planning for urban resilience in the global south: promises and pitfalls. In: Blue-green infrastructure across Asian countries: improving urban resilience and sustainability. Springer, Singapore, pp 499–518

Di Marino M, Tiitu M, Lapintie K, Viinikka A, Kopperoinen L (2019) Integrating green infrastructure and ecosystem services in land use planning. Results from two Finnish case studies. Land Use Policy 82:643–656

Du S, Van Rompaey A, Shi P, Wang JA (2015) A dual effect of urban expansion on flood risk in the Pearl River Delta (China) revealed by land-use scenarios and direct runoff simulation. Nat Hazards 77(1):111–128

Dupras J, Drouin C, André P, Gonzalez A (2015) Towards the establishment of a green infrastructure in the region of Montreal (Quebec, Canada). Plan Pract Res 30(4):355–375

Engel S, Pagiola S, Wunder S (2008) Designing payments for environmental services in theory and practice: an overview of the issues. Ecol Econ 65(4):663–674

Evans AJ, Garrod B, Firth LB, Hawkins SJ, Morris-Webb ES, Goudge H, Moore PJ (2017) Stakeholder priorities for multi-functional coastal defence developments and steps to effective implementation. Mar Policy 75:143–155

Faggian R, Sposito V (2009) Systemic regional development—a system thinking approach. In: Proceedings of the 53rd annual meeting of the ISSS.

Ferraro PJ, Pattanayak SK (2006) Money for nothing? a call for empirical evaluation of biodiversity conservation investments. PLoS Biol 4(4), e(105)

Finewood MH, Matsler AM, Zivkovich J (2019) Green infrastructure and the hidden politics of urban stormwater governance in a post industrial City. Ann Am Assoc Geogr 109(3):909–925

Foster J, Lowe A, Winkelman S (2011) The value of green infrastructure for urban climate adaptation. Center Clean Air Policy 750(1):1–52

Furlong C, Phelan K, Dodson J (2018) The role of water utilities in urban greening: a case study of Melbourne, Australia. Util Policy 53:25–31

Fryd O, Backhaus A, Birch H, Fratini CF, Ingvertsen ST, Jeppesen J, Panduro TE, Roldin M, Jensen MB (2013) Water sensitive urban design retrofits in Copenhagen—40% to the sewer, 60% to the city. Water Sci Technol 67(9):1945–1952

Gentle N, Kierce S, Nitz A (2001) Economic costs of natural disasters in Australia. Aust J Emerg Manage 16(2):38–43

Gersonius B, Van Buuren A, Zethof M, Kelder E (2016) Resilient flood risk strategies: institutional preconditions for implementation. Ecol Soc 21(4):28

Ghofrani Z, Sposito V, Faggian R (2017) A comprehensive review of blue-green infrastructure concepts. Int J Environ Sustain 6(1)

Göpfert C, Wamsler C, Lang W (2019) A framework for the joint institutionalization of climate change mitigation and adaptation in city administrations. Mitig Adapt Strat Glob Change 24:1–21

Govindarajulu D (2014) Urban green space planning for climate adaptation in Indian cities. Urban Clim 10:35–41

Gulsrud NM, Hertzog K, Shears I (2018) Innovative urban forestry governance in Melbourne?: Investigating 'green placemaking' as a nature-based solution. Environ Res 161:158–167

Hamel P, Tan L (2021) Blue-green infrastructure for flood and water quality management in southeast Asia: evidence and knowledge gaps. Environ Manage 69(4):699–718

Hamin EM, Abunnasr Y, Dilthey MR, Judge PK, Kenney MA, Kirshen P, Sheahan TC, DeGroot DJ, Ryan RL, McAdoo BG, Nurse L, Buxton JA, Sutton-Grier AE, Albright EA, Marin MA, Fricke R (2018) Pathways to coastal resiliency: the adaptive gradients framework. Sustainability 10(8):2629

Hansen R, Olafsson AS, van der Jagt APN, Rall E, Pauleit S (2019) Planning multifunctional green infrastructure for compact cities: what is the state of practice? Ecol Ind 96:99–110

Hoang L, Fenner RA (2016) System interactions of stormwater management using sustainable urban drainage systems and green infrastructure. Urban Water J 13(7):739–758

Hoyle H, Jorgensen A, Warren P, Dunnett N, Evans K (2017) "Not in their front yard" the opportunities and challenges of introducing perennial urban meadows: a local authority stakeholder perspective. Urban for Urban Green 25:139–149

IPCC (2014) Climate change widespread, rapid, and intensifying

Johns CM (2019) Understanding barriers to green infrastructure policy and stormwater management in the City of Toronto: a shift from grey to green or policy layering and conversion? J Environ Plann Manage 62(8):1377–1401

Kabisch N, Korn H, Stadler J, Bonn A (2017) Nature-based solutions to climate change adaptation in urban areas: linkages between science, policy and practice, 1st edn. Springer Nature

Kaur R, Gupta K (2022) Blue-Green Infrastructure (BGI) network in urban areas for sustainable storm water management: a geospatial approach. City Environ Interact 16:100087

Keeley M, Koburger A, Dolowitz DP, Medearis D, Nickel D, Shuster W (2013) Perspectives on the use of green infrastructure for stormwater management in Cleveland and Milwaukee. Environ Manag 51:1093–1108

Koc CB, Osmond P, Peters A (2017) Towards a comprehensive green infrastructure typology: a systematic review of approaches, methods and typologies. Urban Ecosyst 20(1):15–35

Kundu A (2011) Trends and processes of urbanisation in India. International Institute for Environment and Development (IIED) urbanization and emerging population issues

Kremer P, Hamstead Z, Haase D, McPhearson T, Frantzeskaki N, Andersson E, Kabisch N (2016) Key insights for the future of urban ecosystem services research. Ecol Soc 21(2):29

Lawson E, Thorne C, Ahilan S, Allen D, Arthur S, Everett G, Fenner R, Glenis V, Guan D, Hoang L, Kilsby C, Lamond J, Mant J, Maskrey S, Mount N, Sleigh A, Smith L, Wright N (2014) Delivering and evaluating the multiple flood risk benefits in blue-green cities: an interdisciplinary approach. WIT Trans Ecol Environ 184:113–124

Leary N, Adejuwon J, Barros V, Batima P, Biagini B, Burton I, Chinvanno S, Cruz R, Dabi D, De Comarmond A, Dougherty B (2012) A stitch in time: general lessons from specific cases. In: Climate change and adaptation. Routledge, pp 1–27

Levy Z, Smardon R, Bays J, Meyer D (2014) A point source of a different color: identifying a gap in United States regulatory policy for 'green' CSO treatment using constructed wetlands. Sustainability 6(5):2392–2412

Li H, Ding L, Ren M, Li C, Wang H (2017) Sponge city construction in China: a survey of the challenges and opportunities. Water (Switzerland) 9(9):594

Liao KH, Deng S, Tan PY (2017) Blue-green infrastructure: new frontier for sustainable urban stormwater management. In: Tan P, Jim C (eds) Greening cities: forms and functions. Springer, pp 203–226

Lin L (2022) Venture capital in the rise of sustainable investment. Eur Bus Organ Law Rev 23(1):187–216

Liu Y, Theller LO, Pijanowski BC, Engel BA (2016) Optimal selection and placement of green infrastructure to reduce impacts of land use change and climate change on hydrology and water quality: an application to the trail creek watershed, Indiana. Sci Total Environ 553:149–163

Malhotra A, Mishra AK, Vyas I (2022) Financing urban infrastructure in India through tax increment financing instruments: a case for smart cities mission. J Public Aff 22(3):e2554

Mansur AV, McDonald RI, Güneralp B, Kim H, Puppim de Oliveira JA, Callaghan CT, Hamel P, Kuiper JJ, Wolff M, Liebelt V, Martins IS, Elmqvist T, Pereira HM (2022) Nature futures for the urban century: integrating multiple values into urban management. Environ Sci Policy 131:46–56

Matthews T, Lo AY, Byrne JA (2015) Reconceptualizing green infrastructure for climate change adaptation: barriers to adoption and drivers for uptake by spatial planners. Landsc Urban Plan 138:155–163

McGurk E, Hynes S, Manton R, Thorne F, Clifford E (2019) Greenways, recreational access and landowner willingness to accept: a contingent valuation study of farmers in Ireland. J Environ Plann Manage 62(13):2375–2392

Mell I (2018) Financing the future of green infrastructure planning: alternatives and opportunities in the UK. Landsc Res 43:751–768

Mell I, Scott A (2023) Definitions and context of blue-green infrastructure. In: ICE manual of blue-green infrastructure. ICE Publishing, pp 3–22

Meran G, Siehlow M, von Hirschhausen C (2021) Integrated water resource management: principles and applications. In: The economics of water: rules and institutions. Springer, pp 23–121

Merk O, Saussier S, Staropoli C, Slack E, Kim JH (2012) Financing green urban infrastructure

Mguni P, Herslund L, Jensen MB (2016) Sustainable urban drainage systems: examining the potential for green infrastructure-based stormwater management for Sub-Saharan cities. Nat Hazards 82:241–257

Musgrave RA, Musgrave PB (1973) Public finance in theory and practice. New York, McGraw Hill, First edition

Myong Koh J (2018) Green infrastructure financing. Palgrave Macmillan

National Research Council (2009) Urban stormwater management in the United States. National Academies Press, Washington, DC

O'Brien P, O'Neill P, Pike A (2019) Funding, financing and governing urban infrastructures. Urban Stud 56(7):1291–1303
O'Donnell EC, Lamond JE, Thorne CR (2017) Recognising barriers to implementation of blue-green infrastructure: a Newcastle case study. Urban Water J 14(9):964–971
Obeng EA, Aguilar FX, Mccann LM (2018) Payments for forest ecosystem services: a look at neglected existence values, the free-rider problem and beneficiaries' willingness to pay. Int for Rev 20(2):206–219
Padigala B (2012) Urbanization and changing green spaces in Indian cities (case study—city of Pune). Int J Geol Earth Environ Sci 2(2):148–156
Parr TB, Smucker NJ, Bentsen CN, Neale MW (2016) Potential roles of past, present, and future urbanization characteristics in producing varied stream responses. Freshw Sci 35(1):436–443
Pataki DE, Carreiro MM, Cherrier J, Grulke NE, Jennings V, Pincetl S, Pouyat RV, Whitlow TH, Zipperer WC (2011) Coupling biogeochemical cycles in urban environments: ecosystem services, green solutions, and misconceptions. Front Ecol Environ 9(1):27–36
Peters H, Landström C (2022) The public and its assets: performing appraisal and advocacy for blue and green infrastructure in London, England. Environ Plann E Nat Space 5(4):2193–2215
Pickett ST, Cadenasso ML, Childers DL, McDonnell MJ, Zhou W (2016) Evolution and future of urban ecological science: ecology in, of, and for the city. Ecosyst Health Sustain 2(7):e01229
Plieninger T, Thapa P, Bhaskar D, Nagendra H, Torralba M, Zoderer BM (2022) Disentangling ecosystem services perceptions from blue infrastructure around a rapidly expanding megacity. Landsc Urban Plan 222:104399
Pontee N, Narayan S, Beck MW, Hosking AH (2016) Nature-based solutions: lessons from around the world. Proc Inst Civ Eng Marit Eng 169(1):29–36
Randhawa A, Kumar A (2017) Exploring sustainability of smart development initiatives in India. Int J Sustain Built Environ 6:701–710
Roberts D, Boon R, Diederichs N, Douwes E, Govender N, Mcinnes A, Mclean C, O'Donoghue S, Spires M (2012) Exploring ecosystem-based adaptation in Durban, South Africa: "learning-by-doing" at the local government coal face. Environ Urban 24(1):167–195
Romero F (2016) Challenges of open space preservation: a texas case study. Town Plann Rev (TPR) 87(2)
Ryan M, O'Donoghue C, Phillips H (2016) Modelling financially optimal afforestation and forest management scenarios using a bio-economic model. Open J for 6(1):19
Sanchez FG, Govindarajulu D (2023) Integrating blue-green infrastructure in urban planning for climate adaptation: lessons from Chennai and Kochi, India. Land Use Policy 124:106455
Sartzetakis ES (2021) Green bonds as an instrument to finance low carbon transition. Econ Chang Restruct 54(3):755–779
Schiappacasse P, Müller B (2015) Planning green infrastructure as a source of urban and regional resilience—towards institutional challenges. Urbani Izziv 26:S13–S24
Schomers S, Matzdorf B (2013) Payments for ecosystem services: a review and comparison of developing and industrialized countries. Ecosyst Serv 6:13–30
Sehrawat S, Shekhar S (2023) Global insights, local realities: BGI challenges and opportunities in Indian urbanization. Eco Cities 4(1):2409
Sengupta D, Mitra S, Shaw R (2022) Blue-green infrastructure: a possible connect to Guwahati smart city. In: North-East research conclave. Springer Nature, Singapore, pp 247–262
Shamsuzzoha A, Nieminen J, Piya S, Rutledge K (2021) Smart city for sustainable environment: a comparison of participatory strategies from Helsinki, Singapore and London. Cities 114:103194
Sharma D, Tomar S (2010) Mainstreaming climate change adaptation in Indian cities. Environ Urban 22(2):451–465
Shruti S, Singh PK, Ohri A (2020) Evaluating the environmental sustainability of smart cities in India: the design and application of the Indian smart city environmental sustainability index. Sustainability 13(1):327

Singh D, Mishra P (2023) Smart city mission and urban environmental sustainability in India. In: The impact of environmental emissions and aggregate economic activity on industry: theoretical and empirical perspectives. Emerald Publishing Limited, pp 291–312

Smiley KT, Sharma T, Steinberg A, Hodges-Copple S, Jacobson E, Matveeva L (2016) More inclusive parks planning: park quality and preferences for park access and amenities. Environ Justice 9(1):1–7

Smith N (1984) Uneven development: nature, capital, and the production of space (Third). The University of Georgia Press, Athens

Smith RM, Pathak P (2018) Urban sustainability in India: Green buildings, AMRUT yojana, and smart cities. In: Metropolitan governance in Asia and the pacific rim: borders, challenges, futures, pp 163–190

Steadman S, Smith HM, Jeffrey P, Wheat P (2024) Exploring public perceptions and support for green infrastructure funding mechanisms: a study of the Oxford–Cambridge Arc, England. J Environ Plan Manag 67:2502–2527

Stefanakis AI, Calheiros CSC, Nikolaou I (2021) Nature-based solutions as a tool in the new circular economic model for climate change adaptation. Circ Econ Sustain 1:303–318

Sun Y, Deng L, Pan SY, Chaing PC, Sable SS, Shah KJ (2020) Integration of green and grey infrastructures for sponge city: Water and energy nexus. Water-Energy Nexus 3:29–40

Sussams LW, Sheate WR, Eales RP (2015) Green infrastructure as a climate change adaptation policy intervention: muddying the waters or clearing a path to a more secure future? J Environ Manage 147:184–193

Thorne CR, Lawson EC, Ozawa C, Hamlin SL, Smith LA (2015) Overcoming uncertainty and barriers to adoption of blue-green infrastructure for urban flood risk management. J Flood Risk Manag 11:S960–S972

Toxopeus H, Polzin F (2021) Reviewing financing barriers and strategies for urban nature-based solutions. J Environ Manage 289:112371

Trencher GP, Yarime M, Kharrazi A (2013) Co-creating sustainability: cross-sector university collaborations for driving sustainable urban transformations. J Clean Prod 50:40–55

Tubridy F (2021) Co-financing green resilient infrastructures in Copenhagen: integrated or superficial design? Landsc Res 46(2):261–272

Udas-Mankikar S, Driver B (2021) Blue-green infrastructure: an opportunity for Indian cities. ORF Occasional Paper, 317

UN DESA, Population Division (2014) World urbanization prospects: The 2014 revision. United Nations Department of Economic and Social Affairs, New York, NY

Vásquez A, Devoto C, Giannotti E, Velásquez P (2016) Green infrastructure systems facing fragmented cities in Latin America: case of Santiago, Chile. Procedia Eng 161:1410–1416

Wang N, Ma M (2021) Public-private partnership as a tool for sustainable development–What literature says? Sustain Dev 29(1):243–258

Well F, Ludwig F (2020) Blue–green architecture: a case study analysis considering the synergetic effects of water and vegetation. Front Archit Res 9(1):191–202

Wihlborg M, Sörensen J, Alkan Olsson J (2019) Assessment of barriers and drivers for implementation of blue-green solutions in Swedish municipalities. J Environ Manage 233:706–718

Wilbers GJ, de Bruin K, Seifert-Dähnn I, Lekkerkerk W, Li H, Budding-Polo Ballinas M (2022) Investing in urban blue-green infrastructure—assessing the costs and benefits of stormwater management in a Peri-urban catchment in Oslo, Norway. Sustainability 14(3):1934

Williams JB, Jose R, Moobela C, Hutchinson DJ, Wise R, Gaterell M (2019) Residents' perceptions of sustainable drainage systems as highly functional blue green infrastructure. Landsc Urban Plan 190:103610

Wunder S (2005) Payments for environmental services: some nuts and bolts

Xu C, Tang T, Jia H, Xu M, Xu T, Liu Z, Long Y, Zhang R (2019) Benefits of coupled green and grey infrastructure systems: evidence based on analytic hierarchy process and life cycle costing. Resour Conserv Recycl 151:104478

Zuniga-Teran AA, Staddon C, de Vito L, Gerlak AK, Ward S, Schoeman Y, Hart A, Booth G (2019) Challenges of mainstreaming green infrastructure in built environment professions. J Environ Plann Manage 63(4):710–732

Part IV
Summary

Chapter 15
Urban Blue-Green Infrastructure: Steps to Move Forward for Food Security and Climate Disaster Resilience

Mosa Lajina Begum, Gulsan Ara Parvin, Md. Anwarul Abedin, and Rajib Shaw

Abstract Urban blue-green infrastructure (BGI) is pivotal in enhancing food safety and disaster risk reduction. By integrating natural systems such as green roofs, urban gardens, and water management systems, cities can address multiple challenges simultaneously. BGI not only improves the resilience of urban areas to flooding and extreme weather events but also contributes to sustainable food production and food security. These systems help mitigate risks associated with urban heat islands, reduce runoff and pollution, and support biodiversity, which is essential for robust food systems. Effective BGI can transform urban environments by creating multi-functional spaces that serve ecological, social, and economic purposes. For instance, green roofs and walls reduce stormwater runoff, while urban gardens enhance local food production and reduce reliance on long-distance food transportation. Moreover, the incorporation of blue infrastructure, like rain gardens and permeable pavements, plays a critical role in managing stormwater and reducing flood risks.

Keywords Urban BGI initiatives · Urban management plan · BGI policies for cities

M. L. Begum · Md. A. Abedin (✉)
Department of Soil Science, Bangladesh Agricultural University, Mymensingh, Bangladesh
e-mail: m.a.abedin@bau.edu.bd

G. A. Parvin
College of Policy Science, Ritsumeikan University, Osaka-Ibaraki Campus, Osaka, Japan
e-mail: niruurp@yahoo.com

R. Shaw
Graduate School of Media and Governance, Keio University, Minato, Japan

M. G. Kibria et al. (eds.), *Urban Blue-Green Infrastructure Approach for Food Security and Climate Disaster Resilience*, Disaster Risk Reduction,
https://doi.org/10.1007/978-981-96-8746-6_15

15.1 Introduction

Cities have several difficulties as their populations grow, such as food shortages and increased vulnerability to natural disasters (Zhou et al. 2014). From this viewpoint, the incorporation of blue-green infrastructure (BGI) offers a revolutionary remedy to these urgent issues (de Oliveira et al. 2022). Blue-green infrastructure denotes a system of natural and semi-natural elements, including parks, green roofs, urban forests, wetlands, and water bodies, which enhance urban biodiversity and aesthetics while also contributing significantly to climate adaptation and food supply (Ghofrani et al. 2017). These methods reduce pollution and runoff, boost biodiversity—an essential component of resilient food systems and mitigate hazards associated with urban heat islands (Siehr et al. 2022). Urban environments sometimes encounter spatial constraints for traditional agricultural practices, leading to a disconnection between food systems and urban populations. Nonetheless, BGI suggests innovative approaches to bridge this gap. Cities implementing urban agriculture in green spaces can enhance local food security and mitigate climate change impacts (Eigenbrod and Gruda 2015). These actions enhance soil health, augment carbon sequestration, and fortify urban resilience against severe weather occurrences. Moreover, blue-green infrastructure promotes sustainable water management, mitigating the risks of flooding and drought, both exacerbated by climate change (Hamel and Tan 2022). Natural water retention structures, including bioswales and rain gardens, effectively regulate stormwater runoff, improve water quality, and replenish aquifers, ensuring a reliable water supply for urban agriculture.

To fully utilize urban blue-green infrastructure in the future, a well-thought-out and inclusive approach is needed (Ruan et al. 2024). This involves collaboration with municipal planners, agricultural experts, environmentalists, and community stakeholders to develop multifunctional spaces that address ecological, social, and economic requirements. Public policies must also support these activities by providing financing, incentives, and regulatory frameworks to encourage the development of BGI (Khan 2024). To enhance urban food security and climate resilience using blue-green infrastructure, it is essential to examine best practices, case studies, and innovative solutions that have successfully transformed urban environments. By utilizing nature's potential in urban environments, we can establish sustainable, vibrant spaces that both maintain and safeguard our communities against the inevitable challenges posed by climate change.

15.2 Current Status of Urban Blue-Green Infrastructure

Bangladesh confronts enormous problems due to rising urbanization and climate change, making urban blue-green infrastructure (BGI) essential for food security and disaster resilience. Here's an overview of the current status of BGI:

1. **Urbanization Trends**

- **Population Growth**: As of 2023, around 38% of Bangladesh's population lives in urban areas, projected to reach 60% by 2030 (UN-Habitat 2023). This rapid urbanization intensifies pressures on food systems and urban infrastructure.
- **Food Security**: Approximately **25%** of urban residents experience food insecurity, with urban areas contributing only **8%** of total food production (USAID 2019).

2. **Current BGI Initiatives**

- **Green Space Development**: Initiatives in cities like Dhaka aim to increase urban greenery. The Dhaka North City Corporation plans to develop over 100 parks, targeting 10% green space coverage by 2030.
- **Urban Agriculture**: Urban farming has emerged as a key strategy. For example, urban farmers in Dhaka supply about **5%** of the city's vegetable needs, with potential to increase.

15.3 The Role of Blue-Green Infrastructure in Climate Disaster Resilience

Blue-green infrastructure (BGI) plays a critical role in enhancing urban resilience against climate disasters by integrating natural processes into urban planning and management. This multifaceted approach combines green spaces, such as parks and green roofs, with blue elements like rivers and wetlands, creating a synergistic system that mitigates the impacts of climate change while providing ecological, social, and economic benefits (see Table 15.1). Some important role of BGI has been highlighted below.

Table 15.1 Functions of different urban BGI approaches

Green infrastructure	Functions
Urban greenways	The term "greenways" refers to the representation of a shared highway along a stretch of developed land. They promote urban coherence by connecting green and open spaces. Greenways are often established on underutilized railroads, canal pathways, or abandoned industrial zones
Rooftop and kitchen gardening	An exclusive eco-conscious initiative that offers an opportunity to connect with nature while engaging in the cultivation of food within residential spaces or on rooftops
Urban green belt	The green belt's major goals are to protect land from urbanization, preserve designated areas for agriculture and forestry, and provide wildlife habitat
Urban green wedges	The "green wedge," a key new component in park system planning, tries to keep built-up areas from growing beyond defined towns and settlement bounds

1. **Flood Mitigation and Water Management**: One of the most pressing concerns in cities is urban flooding, caused by increased rainfall and inadequate drainage systems. BGI can mitigate flood risks by incorporating natural water retention and drainage systems into the urban fabric. For example:
   - **Constructed wetlands** act as natural sponges, absorbing excess water during heavy rainfall and releasing it slowly into watercourses, reducing flood peaks.
   - **Green roofs** and **rain gardens** capture rainwater, reducing the burden on urban drainage systems.
   - **Permeable pavements** allow water to infiltrate the ground rather than overwhelming stormwater infrastructure.

   These solutions can alleviate the impact of flash floods, while also recharging groundwater and improving water quality.
2. **Heat Island Mitigation**: Cities are particularly vulnerable to heatwaves due to their dense infrastructure, which traps heat. Green infrastructure, such as parks, urban forests, and green roofs, provides shade and cools the surrounding environment through evapotranspiration. Blue infrastructure, such as ponds and lakes, can also help regulate urban temperatures by providing cool areas in times of extreme heat. Cities that integrate blue-green spaces can experience significant reductions in temperature during heatwaves, which is critical for public health, especially for vulnerable populations such as the elderly.
3. **Enhanced Biodiversity and Ecosystem Services**: Blue-green spaces offer vital habitats for urban wildlife, enhancing biodiversity in cities. This biodiversity, in turn, supports ecosystem services such as pollination, air purification, and soil regeneration. For instance, urban wetlands can attract birdlife and aquatic species, while green corridors provide connectivity for terrestrial species. Biodiversity not only contributes to a city's environmental health but also enhances the psychological and physical well-being of its inhabitants by providing access to nature in densely populated areas.

15.4 Urban Blue-Green Infrastructure for Food Security

In addition to bolstering climate resilience, BGI can play a critical role in enhancing urban food security. As cities grow, the need for local food production becomes more pronounced to reduce dependency on global supply chains and mitigate the risks of food shortages during extreme weather events.

- **Urban Agriculture**: Urban agriculture, which includes rooftop gardens, community farms, and edible landscapes, can be integrated into BGI systems. Green roofs can serve a dual purpose by both managing stormwater and providing spaces for food cultivation. Community gardens in urban parks can supplement local food production and foster social cohesion. Vertical farming in urban centers can help produce large quantities of food within limited space, reducing the transportation costs and environmental footprint associated with importing food.

- **Aquaponics and Hydroponics**: Blue-green infrastructure can support innovative agricultural practices such as aquaponics and hydroponics. Aquaponics combines fish farming (aquaculture) with hydroponics (growing plants in water), creating a symbiotic system that efficiently uses water and nutrients. These systems can be integrated into urban environments, using rooftops, basements, or underutilized spaces to grow food year-round with minimal environmental impact. These techniques reduce water usage and allow for food production even in cities with limited arable land, contributing to more resilient urban food systems.
- **Integration of Food Forests and Green Corridors**: Cities can incorporate food forests and green corridors that provide edible plants, fruits, and nuts while also serving ecological functions. For example, trees that provide shade can also be selected for their ability to produce food, such as nut-bearing or fruit-bearing species. These multifunctional landscapes enhance both the ecological health of the city and its food security.

15.5 Policy and Governance: Steps to Move Forward

The integration of BGI into urban planning is increasingly recognized as a vital strategy for enhancing climate disaster resilience. The strategic integration of Blue-Green Infrastructure into urban planning is imperative for enhancing climate disaster resilience. By leveraging its multifaceted benefits and fostering collaborative governance, cities can not only mitigate the impacts of climate change but also improve the quality of life for their residents. The policies for urban agriculture in Bangladesh has been highlighted in Table 15.2.

The possible strategies to promote and sustain urban BGI has been highlighted below.

1. **Integrate BGI into Urban Planning and Policy**:
   - **Adopt Comprehensive BGI Policies**: Develop and implement policies that mandate the inclusion of blue-green infrastructure in urban development projects. Ensure that zoning laws and land-use regulations support the integration of green and blue elements.
   - **Incorporate BGI in Climate Action Plans**: Include BGI strategies in municipal climate action plans to address both food safety and disaster risk reduction. This ensures a holistic approach to urban resilience and sustainability.
2. **Promote and Support Urban Agriculture**:
   - **Facilitate Access to Urban Land**: Create policies and provide incentives to convert underutilized urban spaces into community gardens, urban farms, and food forests.
   - **Support Local Food Networks**: Strengthen local food networks and supply chains through grants, training programs, and resources for urban farmers and gardeners.

Table 15.2 Policies for the agriculture sector and urban agriculture in Bangladesh (Mandal 2006)

Sl no.	Policies	Key objectives and policy initiatives	Ministry of implementation
1	National Agriculture Policy (NAP), 1999	Improving credit, marketing, and agro-based enterprises; safeguarding the interests of small farmers; ensuring food security; increasing land productivity and income; implementing integrated pest management; maintaining smooth input supplies; setting fair output pricing	Ministry of Agriculture
2	National Land use policy	Minimize agricultural loss, prevent indiscriminate land use, establish regional land use standards, rationalize land acquisition, and align land use with the natural environment	Ministry of Land
3	Environment Policy 1992 and Implementation Programme	This includes protecting the environment, controlling pollution, utilizing natural resources sustainably, and participating in global environmental projects	Ministry of Forests and Environment
4	National Food Policy, 2004	Aim to provide reliable food security, a constant supply of safe and nutritious food at affordable rates and increase access and purchasing power for people	Ministry of Food
5	Detailed Area Plan (DAP)	Create an appealing, functional, and adaptable urban design; start a public sector action plan; professionally prepare and provide a database; lead private sector development; reassure future residents and investors; develop infrastructure guidelines	Rajdhani Unnayan Kartripakkhya

3. **Enhance Community Engagement and Capacity Building**:
    - **Foster Community Participation**: Involve local communities in the planning, design, and maintenance of BGI projects. Community engagement ensures that these initiatives meet local needs and gain public support.
    - **Educational Programs**: Develop educational initiatives that raise awareness about the benefits of BGI and train community members in sustainable practices and maintenance techniques.
4. **Ensure Long-Term Maintenance and Management**:
    - **Establish Maintenance Protocols**: Create detailed maintenance plans for BGI projects, outlining responsibilities, schedules, and budgets. Include provisions for routine inspections and repairs.
    - **Allocate Funding**: Secure sustainable funding sources for the upkeep of BGI systems, including public–private partnerships, grants, and municipal budgets.
5. **Implement Monitoring and Evaluation Frameworks**:

- **Develop Metrics and Indicators**: Define clear metrics and indicators to assess the performance of BGI in terms of food safety and disaster risk reduction. Regularly collect and analyze data to measure effectiveness.
- **Adapt Strategies Based on Feedback**: Use monitoring results to refine and adjust BGI strategies, ensuring they continue to meet evolving urban needs and environmental conditions.

6. **Encourage Multi-Sector Collaboration**:

- **Build Partnerships**: Facilitate collaboration between government agencies, private sector entities, research institutions, and non-governmental organizations to enhance BGI projects. Shared resources and expertise can lead to more effective and innovative solutions.
- **Leverage Funding Opportunities**: Explore various funding sources, including federal and state grants, philanthropic contributions, and private investments, to support BGI initiatives.

7. **Promote Innovation and Adaptation**:

- **Stay Updated with Trends**: Keep abreast of the latest research, technologies, and practices in blue-green infrastructure. Adapt these innovations to local conditions and emerging challenges.
- **Pilot Projects and Best Practices**: Implement pilot projects to test new BGI approaches and gather data on their effectiveness. Share successful case studies and best practices to encourage broader adoption.

15.6 Conclusions

Urban blue-green infrastructure presents a promising path forward for cities seeking to build climate disaster resilience and enhance food security. By rethinking how water and green spaces are integrated into urban planning, cities can mitigate the impacts of extreme weather events while ensuring sustainable food production for their growing populations. Through inclusive policy-making, innovative financing, and strong community engagement, cities can transform into resilient, food-secure environments that thrive in the face of climate challenges. By following the above mentioned steps, cities can effectively harness the benefits of blue-green infrastructure to enhance food safety and reduce disaster risks, contributing to the creation of more sustainable, resilient, and livable urban environments.

References

de Oliveira JAP, Bellezoni RA, Shih WY, Bayulken B (2022) Innovations in urban green and blue infrastructure: tackling local and global challenges in cities. J Clean Prod 362:132355

Eigenbrod C, Gruda N (2015) Urban vegetable for food security in cities: a review. Agron Sustain Dev 35:483–498

Ghofrani Z, Sposito V, Faggian R (2017) A comprehensive review of blue-green infrastructure concepts. Int J Environ Sustain 6(1)

Hamel P, Tan L (2022) Blue–green infrastructure for flood and water quality management in Southeast Asia: evidence and knowledge gaps. Environ Manage 69(4):699–718

Khan S (2024) Reactivation of river corridors as an element of Blue Green Infrastructure (BGI), using Nature-based Solutions (NbS), for urban flood risk resilience in Dresden, Germany

Mandal MAS (2006) A synthesis of agricultural policies in Bangladesh: agriculture sector review. Ministry of Agriculture, Dhaka, Bangladesh

Ruan T, Paavola J, Chan FKS, Xu Y, Baldacchini C, Calfapietra C (2024) A lack of focus on data sharing, stakeholders, and economic benefits in current global green infrastructure planning. J Environ Manage 351:119849

Siehr SA, Sun M, Aranda Nucamendi JL (2022) Blue-green infrastructure for climate resilience and urban multifunctionality in Chinese cities. Wiley Interdisc Rev Energy Environ 11(5):e447

United Nations Human Settlements Programme (UN-Habitat) (2023) Unlocking the potential of cities: financing sustainable urban development

USAID (2019) Agriculture and food security. https://www.usaid.gov/bangladesh/agriculture-and-food-security

Zhou Y, Li N, Wu W, Wu J, Shi P (2014) Local spatial and temporal factors influencing population and societal vulnerability to natural disasters. Risk Anal 34(4):614–639

MIX
Papier aus verantwortungsvollen Quellen
Paper from responsible sources
FSC® C105338

If you have any concerns about our products,
you can contact us on
ProductSafety@springernature.com

In case Publisher is established outside the EU,
the EU authorized representative is:
Springer Nature Customer Service Center GmbH
Europaplatz 3, 69115 Heidelberg, Germany

Printed by Libri Plureos GmbH
in Hamburg, Germany